**Reviews in Modern Astronomy**
**Vol. 23**

*Edited by*
*Regina von Berlepsch*

## *The Series Reviews in Modern Astronomy*

**Vol. 22: Deciphering the Universe through Spectroscopy**

2010

ISBN: 978-3-527-41055-2

**Vol. 21: Formation and Evolution of Cosmic Structures**

2009

ISBN: 978-3-527-40910-5

**Vol. 20: Cosmic Matter**

2008

ISBN: 978-3-527-40820-7

**Vol. 19: The Many Facets of the Universe - Revelations by New Instruments**

2006

ISBN: 978-3-527-40662-3

**Vol. 18: From Cosmological Structures to the Milky Way**

2005

ISBN: 978-3-527-40608-1

**Vol. 17: The Sun and Planetary Systems – Paradigms for the Universe**

2004

ISBN: 978-3-527-40476-6

**Vol. 16: The Cosmic Circuit of Matter**

2003

ISBN: 978-3-527-40451-3

**Vol. 15: Astronomy with Large Telescopes from Ground and Space**

2002

ISBN: 978-3-527-40404-9

# Reviews in Modern Astronomy
# Vol. 23

## Zooming in: The Cosmos at High Resolution

*Edited by*
*Regina von Berlepsch*

WILEY-VCH Verlag GmbH & Co. KGaA

**Edited on behalf of the Astronomische Gesellschaft by**

***Regina von Berlepsch***
Leibniz Institute for Astrophysics Potsdam
Potsdam, Germany
RBerlepsch@aip.de

**Cover**
Artist conception of the Milky Way (R. Hurt: NASA/JPL-Caltech/SSC) showing all sources currently measured (green), including unpublished sources, and all sources observed in the first year of BeSSeL (red), based on their kinematic distances (A. Brunthaler et al.; this book).



**Library of Congress Card No.:**
applied for

**British Library Cataloguing-in-Publication Data**
A catalogue record for this book is available from the British Library.

**Bibliographic information published by the Deutsche Nationalbibliothek**
The Deutsche Nationalbibliothek lists this publication in the Deutsche Nationalbibliografie; detailed bibliographic data are available on the Internet at <http://dnb.d-nb.de>.



**Composition** Uwe Krieg, Berlin

**Printing and Binding** Strauss GmbH, Mörlenbach

**Cover Design** Schulz Grafik-Design, Fußgönheim

Printed in the Federal Republic of Germany
Printed on acid-free paper

**Print ISBN:** 978-3-527-41113-9

# Preface

The annual series *Reviews in Modern Astronomy* of the ASTRONOMISCHE GESELLSCHAFT was established in 1988 in order to bring the scientific events of the meetings of the Society to the attention of the worldwide astronomical community. *Reviews in Modern Astronomy* is devoted to the Karl Schwarzschild Lectures, the Ludwig Biermann Award Lectures, the invited reviews, and to the Highlight Contributions from leading scientists reporting on recent progress and scientific achievements at their respective research institutes.

The Karl Schwarzschild Lectures constitute a special series of invited reviews delivered by outstanding scientists who have been awarded the Karl Schwarzschild Medal of the Astronomische Gesellschaft, whereas excellent young astronomers are honoured by the Ludwig Biermann Prize.

Volume 23 continues the series with fourteen invited reviews and Highlight Contributions which were presented during the International Scientific Conference of the Society on "Zooming in: The Cosmos at High Resolution" held in Bonn, Germany, September 13 to 17, 2010.

The Karl Schwarzschild medal 2010 was awarded to Professor Michel Mayor, Genf. His lecture with the title "Exoplanets: The road to Earth twins" opened the meeting.

The talk presented by the Ludwig Biermann Prize winner 2010, Dr. Maryam Modjaz, Berkeley, dealt with the topic "Stellar Forensics with the Supernova-GRB connection".

In 2010 the Doctoral Thesis Award was established by the Astronomische Gesellschaft to honor the author of the most outstandig Doctoral Thesis of the past year. The first awardee was Hans Moritz Günther. His lecture with the title "Accretion, jets and winds: High-energy emission from young stellar objects" was one of the highlights of the conference.

Other contributions to the meeting published in this volume discuss, among other subjects, the gas history of the universe, the facility for antiproton and ion research, the Bar and Spiral Structure Legacy (BeSSeL) survey and star formation at high resolution.

A report on the Herschel Key Program "Water in star-forming regions with Herschel" completes this volume.

The editor would like to thank the lecturers for their stimulating presentations. Thanks also to the local organizing committee from the Argelander Institute for Astronomy and the Max Planck Institute for Radio Astronomy.

Potsdam, Mai 2011 *Regina v. Berlepsch*

The ASTRONOMISCHE GESELLSCHAFT awards the **Karl Schwarzschild Medal**. Awarding of the medal is accompanied by the Karl Schwarzschild lecture held at the scientific annual meeting and the publication. Recipients of the Karl Schwarzschild Medal are

1959 Martin Schwarzschild:
Die Theorien des inneren Aufbaus der Sterne.
Mitteilungen der AG 12, 15

1963 Charles Fehrenbach:
Die Bestimmung der Radialgeschwindigkeiten
mit dem Objektivprisma.
Mitteilungen der AG 17, 59

1968 Maarten Schmidt:
Quasi-stellar sources.
Mitteilungen der AG 25, 13

1969 Bengt Strömgren:
Quantitative Spektralklassifikation und ihre Anwendung
auf Probleme der Entwicklung der Sterne und der Milchstraße.
Mitteilungen der AG 27, 15

1971 Antony Hewish:
Three years with pulsars.
Mitteilungen der AG 31, 15

1972 Jan H. Oort:
On the problem of the origin of spiral structure.
Mitteilungen der AG 32, 15

1974 Cornelis de Jager:
Dynamik von Sternatmosphären.
Mitteilungen der AG 36, 15

1975 Lyman Spitzer, jr.:
Interstellar matter research with the Copernicus satellite.
Mitteilungen der AG 38, 27

1977 Wilhelm Becker:
Die galaktische Struktur aus optischen Beobachtungen.
Mitteilungen der AG 43, 21

1978 George B. Field:
Intergalactic matter and the evolution of galaxies.
Mitteilungen der AG 47, 7

1980 Ludwig Biermann:
Dreißig Jahre Kometenforschung.
Mitteilungen der AG 51, 37

1981 Bohdan Paczynski:
Thick accretion disks around black holes.
Mitteilungen der AG 57, 27

1982 Jean Delhaye:
Die Bewegungen der Sterne
und ihre Bedeutung in der galaktischen Astronomie.
Mitteilungen der AG 57, 123

1983 Donald Lynden-Bell:
Mysterious mass in local group galaxies.
Mitteilungen der AG 60, 23

1984 Daniel M. Popper:
Some problems in the determination
of fundamental stellar parameters from binary stars.
Mitteilungen der AG 62, 19

1985 Edwin E. Salpeter:
Galactic fountains, planetary nebulae, and warm H I.
Mitteilungen der AG 63, 11

1986 Subrahmanyan Chandrasekhar:
The aesthetic base of the general theory of relativity.
Mitteilungen der AG 67, 19

1987 Lodewijk Woltjer:
The future of European astronomy.
Mitteilungen der AG 70, 21

1989 Sir Martin J. Rees:
Is there a massive black hole in every galaxy.
Reviews in Modern Astronomy 2, 1

1990 Eugene N. Parker:
Convection, spontaneous discontinuities,
and stellar winds and X-ray emission.
Reviews in Modern Astronomy 4, 1

1992 Sir Fred Hoyle:
The synthesis of the light elements.
Reviews in Modern Astronomy 6, 1

1993 Raymond Wilson:
Karl Schwarzschild and telescope optics.
Reviews in Modern Astronomy 7, 1

1994 Joachim Trümper:
X-rays from Neutron stars.
Reviews in Modern Astronomy 8, 1

1995 Henk van de Hulst:
Scaling laws in multiple light scattering under very small angles.
Reviews in Modern Astronomy 9, 1

1996 Kip Thorne:
Gravitational Radiation – A New Window Onto the Universe.
Reviews in Modern Astronomy 10, 1

1997 Joseph H. Taylor:
Binary Pulsars and Relativistic Gravity.
not published

1998 Peter A. Strittmatter:
Steps to the LBT – and Beyond.
Reviews in Modern Astronomy 12, 1

1999 Jeremiah P. Ostriker:
Historical Reflections
on the Role of Numerical Modeling in Astrophysics.
Reviews in Modern Astronomy 13, 1

2000 Sir Roger Penrose:
The Schwarzschild Singularity:
One Clue to Resolving the Quantum Measurement Paradox.
Reviews in Modern Astronomy 14, 1

2001 Keiichi Kodaira:
Macro- and Microscopic Views of Nearby Galaxies.
Reviews in Modern Astronomy 15, 1

2002 Charles H. Townes:
The Behavior of Stars Observed by Infrared Interferometry.
Reviews in Modern Astronomy 16, 1

2003 Erika Boehm-Vitense:
What Hyades F Stars tell us about Heating Mechanisms
in the outer Stellar Atmospheres.
Reviews in Modern Astronomy 17, 1

2004 Riccardo Giacconi:
The Dawn of X-Ray Astronomy
Reviews in Modern Astronomy 18, 1

2005 G. Andreas Tammann:
The Ups and Downs of the Hubble Constant
Reviews in Modern Astronomy 19, 1

2007 Rudolf Kippenhahn:
Als die Computer die Astronomie eroberten
Reviews in Modern Astronomy 20, 1

2008 Rashid Sunyaev:
The Richness and Beauty of the Physics of Cosmological
Recombination
Reviews in Modern Astronomy 21, 1

2009 Rolf-Peter Kudritzki:
Dissecting galaxies with quantitative spectroscopy
of the brightest stars in the Universe
Reviews in Modern Astronomy 22, 1

2010 Michel Mayor:
Exoplanets: The road to Earth twins
Reviews in Modern Astronomy 23, 1

The **Ludwig Biermann Award** was established in 1988 by the ASTRONOMISCHE GESELLSCHAFT to be awarded in recognition of an outstanding young astronomer. The award consists of financing a scientific stay at an institution of the recipient's choice. Recipients of the Ludwig Biermann Award are

1989 Dr. Norbert Langer (Göttingen),
1990 Dr. Reinhard W. Hanuschik (Bochum),
1992 Dr. Joachim Puls (München),
1993 Dr. Andreas Burkert (Garching),
1994 Dr. Christoph W. Keller (Tucson, Arizona, USA),
1995 Dr. Karl Mannheim (Göttingen),
1996 Dr. Eva K. Grebel (Würzburg) and
Dr. Matthias L. Bartelmann (Garching),
1997 Dr. Ralf Napiwotzki (Bamberg),
1998 Dr. Ralph Neuhäuser (Garching),
1999 Dr. Markus Kissler-Patig (Garching),
2000 Dr. Heino Falcke (Bonn),
2001 Dr. Stefanie Komossa (Garching),
2002 Dr. Ralf S. Klessen (Potsdam),
2003 Dr. Luis R. Bellot Rubio (Freiburg im Breisgau),
2004 Dr. Falk Herwig (Los Alamos, USA),
2005 Dr. Philipp Richter (Bonn),
2007 Dr. Henrik Beuther (Heidelberg) and
Dr. Ansgar Reiners (Göttingen),
2008 Dr. Andreas Koch (Los Angeles),
2009 Dr. Anna Frebel (Cambridge, USA) and
Dr. Sonja Schuh (Göttingen),
2010 Dr. Maryam Modjaz (Berkely),

The **The Doctoral Thesis Award** was established in 2010 by the ASTRONOMISCHE GESELLSCHAFT to honor the author of the most outstandig Doctoral Thesis of the past year. Recipient of the first Doctoral Thesis Award is

2010 Dr. Hans M. Günther (Cambridge/MA),

# Contents

*Karl Schwarzschild Lecture*

# The road to Earth twins[1]

Michel Mayor, Christophe Lovis, Francesco Pepe,
Damien Ségransan and Stèphane Udry

Observatoire de l'Université de Genève
51 ch. des Maillettes, CH-1290 Versoix, Switzerland
michel.mayor@unige.ch

**Abstract**

*A rich population of low-mass planets orbiting solar-type stars on tight orbits has been detected by Doppler spectroscopy. These planets have masses in the domain of super-Earths and Neptune-type objects, and periods less than 100 days. In numerous cases these planets are part of very compact multiplanetary systems. Up to seven planets have been discovered orbiting one single star. These low-mass planets have been detected by the HARPS spectrograph around 30 % of solar-type stars. This very high occurrence rate has been recently confirmed by the results of the Kepler planetary transit space mission. The large number of planets of this kind allows us to attempt a first characterization of their statistical properties, which in turn represent constraints to understand the formation process of these systems. The achieved progress in the sensitivity and stability of spectrographs have already led to the discovery of planets with masses as small as* $1.5\,M_\oplus$.

## 1 The discovery of a rich population of low mass planets on tight orbits

Today, more than 500 extrasolar planets have been discovered. Most of the detected exoplanets have been found by using precise measurements of stellar radial velocities. The planetary mass estimate from Doppler measurements is directly proportional to the amplitude of the stellar reflex motion. Our progress to detect very-low-mass planets are directly related to the progress done to improve the sensitivity and stability of spectrographs. In 1989, the detection of HD 114762 b, a companion of 11 Jupiter masses to a metal deficient F star was obtained with spectrographs allowing Doppler measurements with a precision of some 300 m s$^{-1}$ (Latham et al. 1989). Fifteen years ago, the precision achieved by any team searching for exoplanets was of

[1]This article has already appeared in Astron. Nachr./AN 332, no. 5 (2011).

*Reviews in Modern Astronomy 23: Zoomimg in: The Cosmos at High Resolution.* First Edition.
Edited by Regina von Berlepsch.

the order of $15\,\mathrm{m\,s^{-1}}$. Today, the instrumental precision achieved with the HARPS spectrograph at La Silla Observatory is better than $0.5\,\mathrm{m\,s^{-1}}$ (Mayor et al. 2003). At this level of precision we are mostly limited by the intrinsic variability of stellar velocities induced by diverse phenomena (acoustic modes, granulation, magnetic activity). However, by adopting an improved observing strategy, we have already some indications that planetary signals as small as a tiny fraction of a meter per second are detectable.

This progress in instrumentation and observing strategy have made possible the discovery of a rich population of super-Earths and Neptune-mass planets in tight orbits around solar-type stars (Mayor & Udry 2008).

The name "super-Earth" is used to qualify planets more massive than the Earth but with masses smaller than 10 Earth masses, a category of planets absent in the solar system. We mention here a few landmark discoveries of these low-mass planets orbiting solar-type stars. Limiting ourself to planets in the super-Earth range we can mention: $\mu$ Ara c with a mass of $10.5\,\mathrm{M}_\oplus$ and a period of 9.7 days (Santos et al. 2004b, revised in Pepe et al. 2007), HD 69830 b with a mass of $10.2\,\mathrm{M}_\oplus$ and a period of 8.7 days (Lovis et al. 2006), HD 40307 b, c, d, a system with three super-Earths with masses comprised between 4 and $9\,\mathrm{M}_\oplus$ and periods from 4 to 20 days (Mayor et al. 2009b). We also have to mention the exceptional system around HD 10180, with 7 planets of which one with a mass as small as $1.4\,\mathrm{M}_\oplus$ on a tight orbit with a period of 1.17 day (Lovis et al. 2011). In addition to these early detections of super-Earths orbiting solar-type stars, we also have to mention the discoveries of super-Earths hosted by M dwarfs: GJ 876 d, a planet with a mass of $5.9\,\mathrm{M}_\oplus$ and a period of 1.94 day (Rivera et al. 2005, Correia et al. 2010), GJ 581 c, d, e with masses of 5, 7, and $1.9\,\mathrm{M}_\oplus$ (Udry et al. 2007; Mayor et al. 2009a). It is impressive to see that all these super-Earths are part of rich multi-planetary systems with 3 to 7 planets per system. The remarkable progress of instrumentation in the last 15 years is obvious in Fig. 1. The masses of planetary companions are plotted as a function of the epoch of their discovery. The mass of HD 10180 b (Lovis et al. 2011) is a factor 100 smaller than the mass of 51 Peg b (Mayor & Queloz 1995).

## 2 The HARPS program to search for very low mass planets

HARPS is a vacuum-operated high-resolution spectrograph ($R = 115\,000$), fiber-fed, optimized to provide stellar radial-velocity measurements with extreme precision (Mayor et al. 2003). As a reward for its construction, the HARPS consortium has received guaranteed observing time (GTO) to carry out an extrasolar planet search in the southern hemisphere (500 observing nights over 5 years). More than 60 % of the total HARPS GTO observing time has been devoted to two sub-programs having the aim of detecting very low-mass planets. The first of these sub-programs comprises some 400 stars which are non-active, slow rotators, not in spectroscopic binary systems, and were selected from the large volume-limited sample measured for several years with the CORALIE spectrograph on the 1.2 m-Euler telescope at la Silla Observatory. The second sub-program consists of a volume-limited sample of about

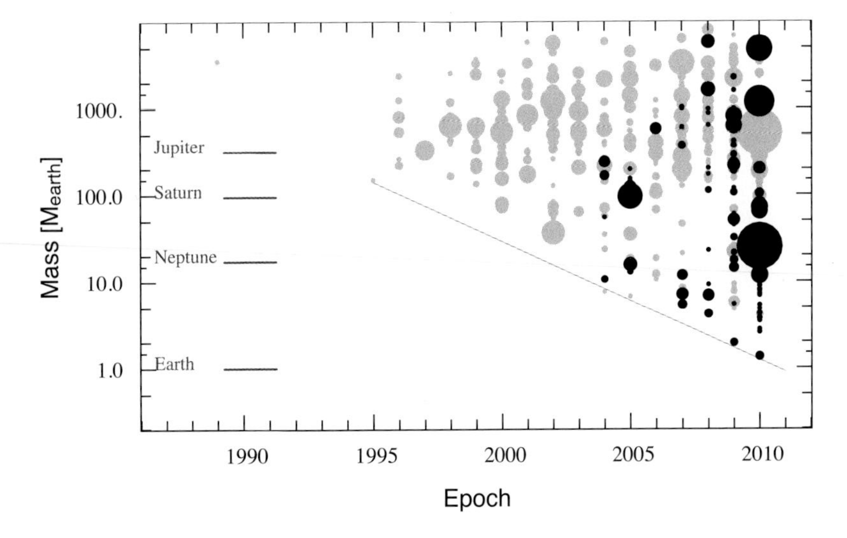


**Figure 1:** (online colour at: www.an-journal.org) Minimum mass of planets detected by Doppler spectroscopy as a function of epoch of discovery. This figure illustrates the impressive progress made in detection sensitivity over the past 15 years. Black symbols indicate HARPS discoveries. Dot size is related to orbital semi-major axis.

120 M dwarfs at the bottom of the main sequence, also selected to be slow rotators and not members of spectroscopic binary systems.

What are the limits presently achieved in terms of radial velocity precision? Several sources of noise can be identified:

- As a result of the efficiency of the cross-correlation technique, a photon noise level of only a fraction of a meter per second is achieved in a few minutes for most of our targets. Sometimes the exposure time is shorter than the typical periods of stellar acoustic modes. In a few minutes, the full amplitude of the stellar velocity variations resulting from acoustic modes could be as large as several meters per second. Long integrations compared to acoustic mode periods are sufficient to have acoustic noise residuals smaller than $0.2\,\mathrm{m\,s^{-1}}$ (rms). For most stars, integrations of 15 minutes are sufficient.

- Dumusque et al. (2011a) have shown that stellar granulation in solar-type stars can induce radial velocity variability comparable to or larger than $1\,\mathrm{m\,s^{-1}}$ on longer timescales compared to acoustic modes. Several measurements spanning several hours are requested to damp the granulation noise.

- Any anisotropies in stellar atmospheres related to magnetic activity will induce radial velocity variations at the stellar rotation period. The amplitude of the radial-velocity jitter is related with stellar chromospheric activity. If we want to search for very low mass planets we need to carefully select "non-active" stars . The reemission in the core of the calcium H and K lines is a good indicator of the chromospheric activity and has been used for the selection of the stellar sample.

- The analysis of the radial velocity variations of several solar-type stars has recently revealed well-defined variations of several $\mathrm{m\,s^{-1}}$ on rather long periods (more than five years). These velocity variations are strongly correlated with the mean shape of absorption lines and chromospheric indicators like Ca II H and K core emission. These variations are related to the stellar analogs of the solar magnetic cycle. This effect has been observed in stars with rather modest chromospheric activity levels (e.g. $\log R'_{\mathrm{HK}}$ around –4.90, see Lovis et al. 2011b). Any long-term drift in stellar radial velocities cannot be a priori attributed to long period planets if a careful check of the long-term behavior of the line bisector and other activity indicators has not been performed.
- Finally, we still have instrumental noise. Lovis & Pepe (2007) have considerably improved the precision of the wavelength of thorium lines as well as the number of lines to be used for the calibration of the spectrograph. Pressure changes in the plasma with the aging of the ThAr calibration lamp induce a very small shift in the wavelengths. As this effect is smaller for thorium lines than argon, we can use this differential effect to correct the aging effect. Long term drifts have thus been reduced below $0.3\,\mathrm{m\,s^{-1}}$ over timescales of several years. The scrambling effect in optical fibers is excellent … but not perfect and some sub-meter per second error could result from imperfect guiding.

The global budget of all these errors is difficult to determine. The best estimation of the lower limit of the quadratic sum of the different components of the noise is provided by the residuals observed around fitted radial velocity curves. Several stars with a very large number of velocity measurements spanning several years have residuals with a dispersion as low as $0.6\,\mathrm{m\,s^{-1}}$ (when binning the data over a few days). For stars with larger chromospheric activity, we can obviously have larger residuals.

This is the precision presently achieved for the HARPS program, for which we have derived preliminary results for the population of low mass planets, as discussed in the next section. If we are searching for low-mass planets on rather long periods, it could be useful to bin the measurements done on $N$ consecutive nights. This procedure could help to damp the noise induced by chromospheric activity, with a time scale comparable to the stellar rotation period. First experiments done on stars with a large number of measurements have shown that the residuals decrease to 0.3–$0.5\,\mathrm{m\,s^{-1}}$ after binning over ten consecutive nights.

## 3 Emerging characteristics of low-mass planets and their host star

We are still far from having a detailed and unbiased view of the population of planets with masses in the range of super-Earths and Neptunes. Nevertheless, we can already notice a few emerging properties. The study of planet hosts themselves also provide additional information to constrain planet formation. In particular the metallicity of the parent stars seems to be of prime importance for models of planetary formation.

## 3.1 The mass distribution

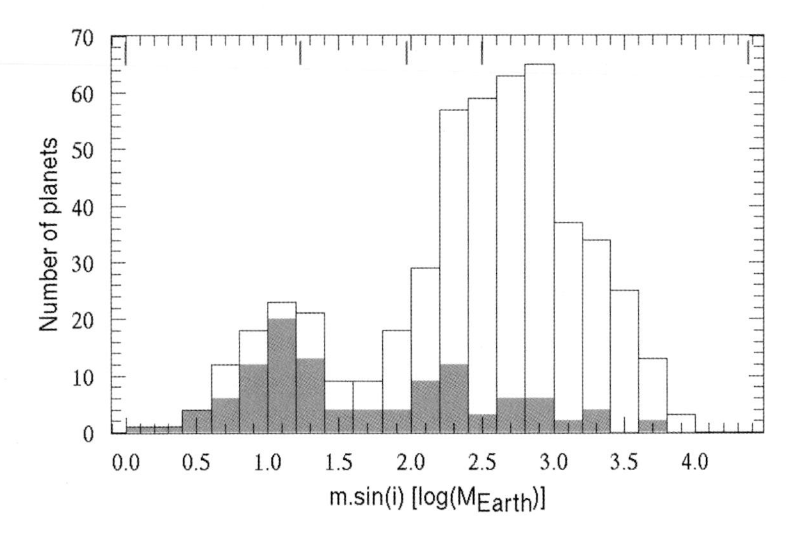


**Figure 2:** (online colour at: www.an-journal.org) Mass distribution of all detected planets. The contribution of the HARPS program (solid histogram) for the detection of very low mass planets is evident.

The mass distribution of all detected planets is illustrated in Fig. 2. In this plot the contribution of the HARPS program for the detection of very low mass planets is evident. Due to the better detection sensitivity of Doppler spectroscopy for massive and/or short period planets, we still have a strong bias against the detection of low-mass planets, especially if they are on long-period orbits.

The bimodal aspect of the mass distribution is a clear indication that the decrease of the distribution for masses less than about one mass of Jupiter is not the result of a detection bias, but is real. The extrapolation by a power-law distribution, as for example $f(m) \sim m^{-1}$, to estimate the number of planets with a mass smaller than the mass of Jupiter is certainly not justified. The observed bimodal shape of the mass distribution from gaseous giant planets to the super-Earth regime provides an interesting constraint for planetary formation scenarios. The planetary formation simulations carried out by Mordasini et al. (2009a,b) also predict a bimodal distribution for that range of planetary mass. In addition these simulations also predict a sharp rise in the mass distribution at a few Earth masses and below. This domain of mass is still at the limit of present instrumental sensitivity. Nevertheless, once again the expected shape of this theoretical mass distribution from 10 down to $1\,M_\oplus$ is clearly not an exponential and any estimate of the frequency of Earth-twins based on an exponential extrapolation is completely unjustified.

## 3.2 The frequency of low-mass multiplanet systems

With the HARPS data presently available from the high-precision sample, we have 48 stars with well-characterized planetary systems. More than 50 % of these systems are multiplanetary. Four of them have 4 planets and the amazing system HD 10180 is the host of 7 planets (Lovis et al. 2011a), one of them having a mass as small as $1.5\,M_\oplus$.

## 3.3 The correlation with the metallicity of host stars

The correlation between the occurrence of gaseous giant planets and the metallicity of host stars is striking. Based on large planetary surveys this correlation is well established by independent teams (Santos, Israelian & Mayor 2001, 2004a; Fischer & Valenti 2005). We have a completely different result if we examine the metallicity of host stars for systems having all planets less massive than $40\,M_\oplus$. We do not have any correlation between the presence of these low-mass planets and the host star metallicity (see Fig. 3), a result already mentioned by Udry et al. (2006) and Sousa et al. (2008), based at that time on a very limited number of stars. With the present study, this lack of correlation with the host star metallicity is robust. The mean metallicity of the 28 planetary systems with planets less massive than $40\,M_\oplus$ is [Fe/H] = –0.12, a metallicity not so different from the mean metallicity of stars in the solar neighborhood.

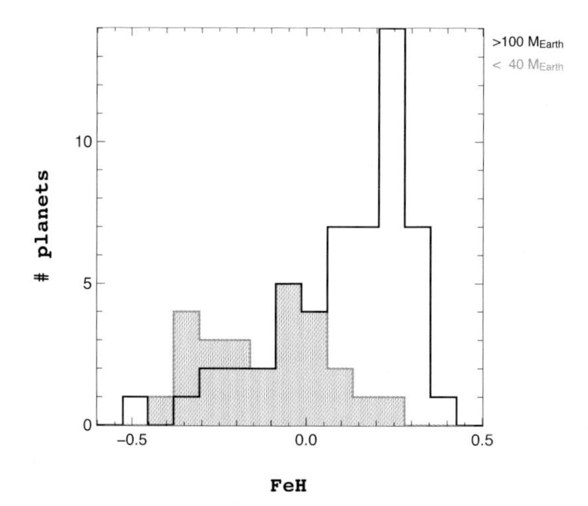


**Figure 3:** (online colour at: www.an-journal.org) Number of planets as a function of host star metallicity. Planets with masses smaller than $40\,M_\oplus$ are hosted by stars of all metallicities, contrary to giant planets whose frequency is strongly dependent on host star metallicity.

## 3.4 The occurrence of low-mass planets orbiting solar-type stars

The occurrence of low-mass planets on tight orbits has been estimated by Lovis et al. (2009). For planets with masses between ~5 and 50 $M_\oplus$ and periods shorter than 100 days, we have detected low-mass planets orbiting about 30 % of the stars in the HARPS sample. A more complete estimate is currently in progress, based on the present, more complete survey.

## 3.5 Searching for Earth-type planets in the habitable zone

The programme devoted to the study of the population of super-Earths and Neptune-type planets is still continuing at la Silla for four additional years after the end of the GTO time. In addition, a new exploratory program has been initiated with the goal of pushing the HARPS precision a little further and try to detect super-Earths in the habitable zone of very nearby G and K dwarfs. An adequate strategy to damp the acoustic and granulation noise sources has been implemented. The sample is limited to only 10 bright non-active stars. Already, low-mass planets have been detected around three stars members of that small sample, see Pepe et al. (2011). The radial velocity signal for one of these planets is as small as $K = 0.56$ m s$^{-1}$. Furthermore, simulations done by Dumusque et al. (2011b) have demonstrated the possibility with the HARPS spectrograph, the present observing strategy and precision, to detect a 2.5 $M_\oplus$ planet orbiting a non-active K dwarf in its habitable zone (see Fig. 4).

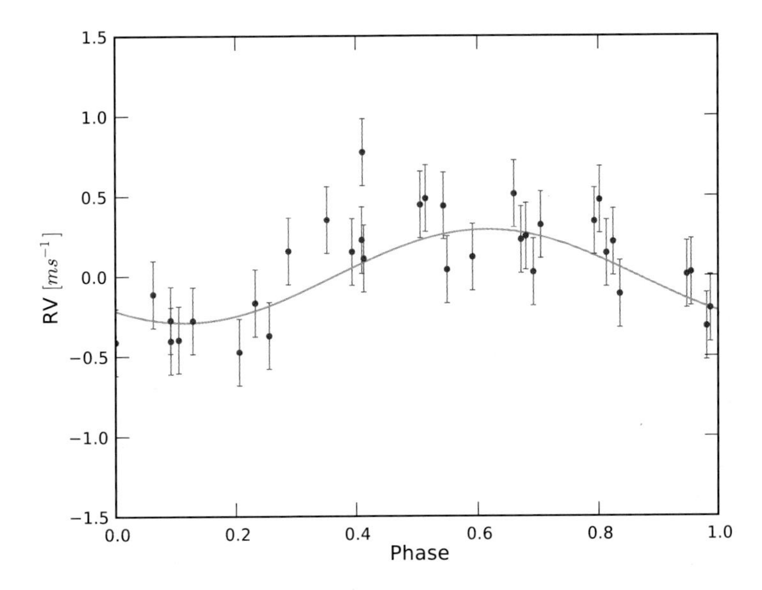


**Figure 4:** (online colour at: www.an-journal.org) Simulation for the detection of a super-Earth of 2.5 $M_\oplus$ in the habitable zone of an inactive K dwarf (from Dumusque et al. 2011b).

Some technical improvements are still feasible to increase the sensitivity and stability of cross-correlation spectrographs like HARPS. A better scrambling of the input beam could be achieved by new optical fibers with octagonal cross sections. These new fibers will strongly diminish the already very small effect of input conditions (guiding errors, variable seeing and focus) on the spectrograph illumination, a mandatory condition to achieve $0.1\,\mathrm{m\,s^{-1}}$ precision. To secure the stability of radial velocity measurements over a span of several years at the level of $0.1\,\mathrm{m\,s^{-1}}$ or better, we must have a calibration device better than the existing ThAr lamps. Developments of laser frequency combs adapted to the resolution and wavelength coverage of HARPS will provide the requested stability (Wilken et al. 2010).

A photon noise on the Doppler signal at the level of $0.1\,\mathrm{m\,s^{-1}}$ requires a rather large telescope size to achieve the needed signal-to-noise ratio in a reasonable exposure time. The ESPRESSO project, presently in development, to be implemented on the 8.2-m VLT telescope at Paranal is designed to achieve the $0.1\,\mathrm{m\,s^{-1}}$ Doppler precision and stability on the long term (Pepe et al. 2010). The ESPRESSO project can also be seen as a precursor for an even more ambitious stable spectrograph, the CODEX project presently at the study phase level for the 42-m E-ELT telescope, to be implemented by ESO at Cerro Armazones (Chile) in the next decade (Pasquini et al. 2010).

We have to keep in mind that for stars with the lowest chromospheric activity, we still do not know the true level of radial velocity jitter. Analysis of the radial velocity scatter of HARPS measurements for non-active stars suggest a minimum jitter of $0.5\,\mathrm{m\,s^{-1}}$ or less. This stellar variability, depending on the changing number and phase of magnetic spots (or other features) will be difficult to model. Preliminary studies show that non-active K dwarfs will be the most suitable targets to search for Earth twins. A large number of Doppler measurements has the potential to overcome the effects of the stellar intrinsic variability and permit detections of planetary signals of $0.2\,\mathrm{m\,s^{-1}}$ or less.

The discovery of radial velocity variations associated with solar cycle analogues with full amplitude as large as $10\,\mathrm{m\,s^{-1}}$ seems a priori to be casting doubts on our ability to detect Earth analogues in the habitable zone. However, using parameters of the cross-correlation function it has been possible to correct the magnetic cycle effects to less than $1\,\mathrm{m\,s^{-1}}$. In addition, for some domain of stellar masses (K dwarfs), we observe that the amplitude of the radial velocity effect is vanishing despite quite noticeable magnetic cycles. Finally, we notice that the periods of magnetic cycles are much longer (about a factor 10) than the expected periods of habitable planets orbiting K dwarfs. We are thus still convinced that Doppler spectroscopy has the potential to detect rocky planets in the habitable zone of K dwarfs.

The medium- or long-term scientific goal to search for chemical signatures of life in the atmospheric spectra of Earth twins via space experiments as the ESA-DARWIN concept will first require identification of targets. It seems that at the moment Doppler spectroscopy is the only method with the potential to detect Earth-type planets in the habitable zone of stars as close as possible to the Sun. The last condition is mandatory, if we want to have a star-planet angular separation large enough for the need of planetary atmosphere spectroscopy, as well as bright enough targets to maximize the signal-to-noise ratio.

From Doppler surveys we know that super-Earths on tight orbits are frequent. We have first hints from microlensing searches that super-Earths could also be frequent at large semi-major axis (Gould et al. 2010). But we do not have any estimate of the frequency of Earth-twins in the habitable zone of solar-type stars and no ideas on their orbital eccentricity distribution. The orbital eccentricity of Earth-twins is also relevant in the frame of life-search experiments. The ESA-PLATO space project is, in that context, the most interesting experiment, complementary to Doppler surveys to explore the domain of Earth-type planets orbiting relatively close stars.

**Acknowledgements**

We would like to thank the Swiss National Science Foundation for its continuous support.

# References

Correia, A., Udry, S., Mayor, M., et al.: 2009, A&A 496, 521

Correia, A.C.M., Couetdic, J., Laskar, J., et al.: 2010, A&A 511, A21

Dumusque, X., Udry, S., Lovis, C., Santos, N.C., Monteiro, M.J.P.F.G.: 2011a, A&A 525, A140

Dumusque, X., Santos, N.C., Udry, S., Lovis, C., Bonfils, X.: 2011b, A&A 527, A82

Fischer, D., Valenti, J.: 2005, ApJ 622, 1102

Gould, A., Dong, S., Gaudi, B.S., et al.: 2010, ApJ 720, 1073

Latham, D.W., Stefanik, R.P., Mazeh, T., Mayor, M., Burki, G.: 1989, Nature 339, 38

Lovis, C., Pepe, F.: 2007, A&A 468, 1115

Lovis, C., Mayor, M., Pepe, F., et al.: 2006, Nature 441, 305

Lovis, C., Mayor, M., Bouchy, F., et al.: 2009, in: F. Pont, D.D. Sasselov, M.J. Holman (eds.), *Transiting Planets*, IAU Symp. 253, p. 502

Lovis, C., Ségransan, D., Mayor, M., et al.: 2011a, A&A 528, A112

Lovis, C., et al.: 2011b, in prep.

Mayor, M., Queloz, D.: 1995, Nature 378, 355

Mayor, M., Udry, S.: 2008, Physica Scripta T 130, 014010

Mayor, M., Pepe, F., Queloz, D., et al.: 2003, The Messenger 114, 20

Mayor, M., Bonfils, X., Forveille, T., et al.: 2009a, A&A 507, 487

Mayor, M., Udry, S., Lovis, C., et al.: 2009b, A&A 493, 639

Mordasini, C., Alibert, Y., Benz, W.: 2009a, A&A 501, 1139

Mordasini, C., Alibert, Y., Benz, W., Naef, D.: 2009b, A&A 501, 1161

Pasquini, L., Cristiani, S., García-López, R., Haehnelt, M., Mayor, M.: 2010, The Messenger 140, 20

Pepe, F., Correia, A.C.M., Mayor, M., et al.: 2007, A&A 462, 769

Pepe, F., Cristiani, S., Rebolo Lopez, R., et al.: 2010, in: I.S. McLean, S.K. Ramsay, H. Takami (eds.), *Ground-based and Airborne Instrumentation for Astronomy III*, SPIE 7735, p. 77350F

Pepe, F., et al.: 2011, A&A, subm.

Rivera, E.J., Lissauer, J.J., Butler, R.P., et al.: 2005, ApJ 634, 625

Santos, N.C., Israelian, G., Mayor, M.: 2001, A&A 373, 1019

Santos, N.C., Israelian, G., Mayor, M.: 2004a, A&A 415, 1153

Santos, N.C., Bouchy, F., Mayor, M., et al.: 2004b, A&A 426, L19

Sousa, S.G., Santos, N.C., Mayor, M., et al.: 2008, A&A 487, 373

Udry, S., Mayor, M., Benz, W., et al.: 2006, A&A 447, 361

Udry, S., Bonfils, X., Delfosse, X., et al.: 2007, A&A 469, L43

Wilken, T., Lovis, C., Manescau, A., et al.: 2010, MNRAS 405, L16

*Ludwig Biermann Award Lecture*

# Stellar forensics with the supernova-GRB connection[1]

Maryam Modjaz

Columbia University, Pupin Physics Laboratories, MC 5247,
550 West 120th Street,
New York, NY 10027, USA
mmodjaz@gmail.com

**Abstract**

*Long-duration gamma-ray bursts (GRBs) and type Ib/c supernovae (SNe Ib/c) are amongst nature's most magnificent explosions. While GRBs launch relativistic jets, SNe Ib/c are core-collapse explosions whose progenitors have been stripped of their hydrogen and helium envelopes. Yet for over a decade, one of the key outstanding questions is what conditions lead to each kind of explosion in massive stars. Determining the fates of massive stars is not only a vibrant topic in itself, but also impacts using GRBs as star formation indicators over distances up to 13 billion light-years and for mapping the chemical enrichment history of the universe. This article reviews a number of comprehensive observational studies that probe the progenitor environments, their metallicities and the explosion geometries of SN with and without GRBs, as well as the emerging field of SN environmental studies. Furthermore, it discusses SN 2008D/XRT 080109 which was discovered serendipitously with the Swift satellite via its X-ray emission from shock breakout and which generated great interest amongst both observers and theorists while illustrating a novel technique for stellar forensics. The article concludes with an outlook on how the most promising venues of research – with the many existing and upcoming large-scale surveys such as PTF and LSST – will shed new light on the diverse deaths of massive stars.*

## 1 Introduction: the importance of stellar forensics

Stripped supernovae (SNe) and long-duration gamma-ray bursts (GRBs) are nature's most powerful explosions from massive stars. They energize and enrich the ISM, and, like beacons, they are visible over large cosmological distances. However, the exact mass and metallicity range of their progenitors is not known, nor the detailed physics of the explosion (see reviews by Woosley & Bloom 2006, and by Smartt 2009). Stripped-envelope SNe (i.e, SN IIb, Ib, Ic, and Ic-bl) are core-collapse events

[1]This article has already appeared in Astron. Nachr./AN 332, no. 5 (2011).

*Reviews in Modern Astronomy 23: Zoomimg in: The Cosmos at High Resolution.* First Edition.
Edited by Regina von Berlepsch.

whose massive progenitors have been stripped of progressively larger amounts of their outermost H and He envelopes (Fig. 1, Clocchiatti et al. 1996; Filippenko 1997). In particular, broad-lined SNe Ic (SNe Ic-bl) are SNe Ic whose line widths approach $30\,000\ \mathrm{km\,s^{-1}}$ around before and around maximum light and whose optical spectra show no trace of H and He.

The exciting connection between long GRBs and SNe Ic-bl and the existence of SNe Ic-bl without observed GRBs, as well as that of GRBs that surprisingly lack SN signatures raises the question of what distinguishes a GRB progenitor from that of an ordinary SN Ic-bl with and without a GRB.

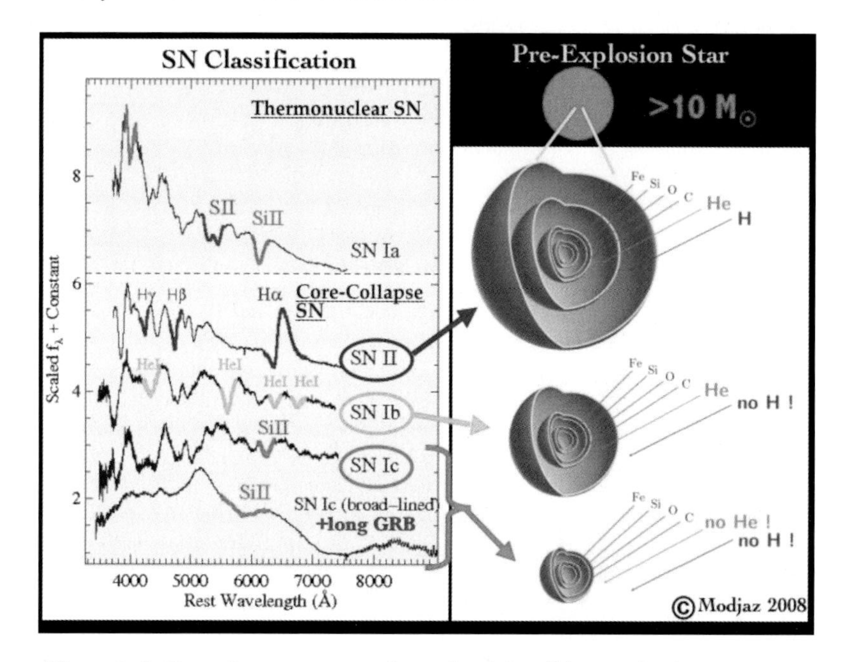


**Figure 1:** (online colour at: www.an-journal.org) Possible mapping between core-collapse SNe types (*left*) and their corresponding progenitor stars (*right*). *Left*: representative observed spectra of different types of SNe. Broad-lined SN Ic are the only type of SNe seen in conjunction with GRBs. Not shown are some of the H-rich members of the SN: SN IIn, and very luminous SN. *Right*: schematic drawing of massive ($\geq$8–10 $M_\odot$) stars before explosion, with different amounts of intact outer layers, showing the "onion-structure" of different layers of elements that result from successive stages of nuclear fusion during the massive stars' lifetimes. The envelope sizes are not drawn to scale; in particular, the outermost hydrogen envelope at the top can be up to 100 times larger than shown. Furthermore, many real massive stars rotate rapidly and are therefore oblate, as well as showing less chemical stratification as drawn here due to convection and overshoot mixing (e.g., see review by Woosley et al. 2002). The bottom star constitute the most stripped (or "naked") star which give rise to SN Ic and sometimes, SN Ic-bl and GRBs, One of the outstanding questions in the field is the exact mechanism with which the outer H and He layers got removed. This figure can be downloaded at http://www.astro.columbia.edu/~mmodjaz/research.html.

Understanding the progenitors of SN Ib/c and of GRB is important on a number of levels:

***Stellar and high-energy astrophysics*:** These stellar explosions leave behind extreme remnants, such as black holes, neutron stars, magnetars, which in themselves are a rich set of phenomena studied over the full wavelength spectrum from gamma-rays to radio. Ideally we would like to construct a map that connects the mass and make-up of a massive star to the kind of death it undergoes and to the kind of remnant it leaves behind. Furthermore, these stellar explosions are sources of gravitational waves and of neutrino emission, and specifically GRBs are leading candidate sites for high-energy cosmic ray acceleration (e.g., Waxman 2004). Thus, it is of broad astrophysical importance to understand the specific progenitor and production conditions for different kinds of cosmic explosions.

***Chemical enrichment history of the universe*:** The universe's first- and second-generation stars were massive. Since GRBs and SN probably contribute differently to the enrichment of heavy elements (e.g., Nomoto et al. 2006; Pruet et al. 2006), determining the fate of massive stars is fundamental to tracing the chemical history of the universe.

***Cosmology*:** GRBs are beacons and can illuminate the early universe. Indeed, until recently, the object with the highest spectroscopic redshift was a GRB, GRB 090423 at $z \sim 8.2$ (Salvaterra et al. 2009; Tanvir et al. 2009), which means that this explosion occurred merely 630 million years after the Big Bang. Thus, a clear understanding of the stellar progenitors of SN and GRBs is an essential foundation for using them as indicators of star formation over cosmological distances.

Various progenitor channels have been proposed for stripped SNe and GRBs: either single massive Wolf-Rayet (WR) stars with main-sequence (MS) masses of $\gtrsim$30 $M_{\odot}$ that have experienced mass loss during the MS and WR stages (e.g., Woosley et al. 1993), or binaries from lower-mass He stars that have been stripped of their outer envelopes through interaction (Fryer et al. 2007; Podsiadlowski et al. 2004, and references therein), possibly given rise to run-away stars as GRB progenitors (e.g., Cantiello et al. 2007; Eldridge et al. 2011). For long GRBs, the main models for a central engine that is powering the GRB include the collapsar model (MacFadyen & Woosley 1999; Woosley 1993) and the magnetar model (e.g., Usov 1992, for a good summary see Metzger et al. 2011), while rapid rotation of the pre-explosion stellar core appears to be a necessary ingredient for both scenarios.

Attempts to directly identify SN Ib/c progenitors in pre-explosion images obtained with the Hubble Space Telescope or ground-based telescopes have not yet been successful (e.g., Gal-Yam et al. 2005; Maund et al. 2005; Smartt 2009), and cannot conclusively distinguish between the two suggested progenitor scenarios. However, the progenitor non-detections of 10 SN Ib/c strongly indicate that the single massive WR progenitor channel (as we observe in the Local Group) cannot be the only progenitor channel for SN Ibc (Smartt 2009). Similar pre-explosion imaging technique is not possible for GRB progenitors given the large distances at which they are observed.

Thus, in order to fully exploit the potential and power of SNe and GRBs, we have to first figure out their stellar progenitors and the explosions conditions that lead to the various forms of stellar death in a massive star, in form of a "stellar forensics" investigation. In the following review, we will be looking at a number of physical properties in order to find those that set apart SN-GRB, which I will discuss in detail in Sect. 2, from SNe without GRBs: geometry of the explosion (Sect. 4), progenitor mass (Sect. 5) and metallicity (Sect. 6), while the role of binaries are discussed through-out, but not that of magnetic fields. In addition, I will discuss the exciting and emerging field of SN metallicity studies as a promising new tool to probe the progenitors of different kinds of SNe and transients and the story of SN 2008D/XRT 080109 (Sect. 7), which generated great interest amongst both observers and theorists while illustrating a novel technique for stellar forensics

Necessarily, this review will not be complete given the page limit, and is driven by the interest and work of the author, so omissions and simplifications will necessarily arise. Furthermore, given the excellent reviews by Woosley & Bloom (2006), and most recently, Hjorth & Bloom (2011), I will concentrate on developments in the field since 2006 and in complimentary areas.

# 2 Solid cases of SN-GRB

While the explanation for GRBs after their initial discovery included a vast array of different theories, intensive follow-up observations of GRBs over the last two decades have established that long-duration soft-spectra GRBs (Kouveliotou et al. 1993), or at least a significant fraction of them, are directly connected with supernovae and result from the cataclysmic death of massive, stripped stars (see review by Woosley & Bloom 2006). The most direct proof of the SN-GRB association comes from spectra taken of the GRB afterglows, where the spectral fingerprint of SN, specifically that of a broad-lined SN Ic, emerges over time in the spectrum of the GRB afterglow. Near maximum light, GRB-SNe appear to show broad absorption lines of O I, Ca II, and Fe II (see Fig. 1), while there is no photospheric spectrum of a confirmed GRB-SN that indicated the presence of H or showed optical lines of He I (see also below).

Below we briefly list the SN-GRB cases, in order of descending quality of data (see also Table 1 in Woosley & Bloom 2006 and detailed discussions in Hjorth & Bloom 2011). The five most solid cases of the SN-GRB connection, with high signal-to-noise and multiple spectra, are usually at low $z$: SN1998bw/GRB980425 at $z = 0.0085$ (Galama et al. 1998), SN2003dh/GRB030329 at $z = 0.1685$ (Hjorth et al. 2003; Matheson et al. 2003; Stanek et al. 2003), SN2003lw/GRB031203 at $z = 0.10058$ (Malesani et al. 2004), SN2006aj/GRB060218 at $z = 0.0335$ (Campana et al. 2006; Cobb et al. 2006; Kocevski et al. 2007; Mirabal et al. 2006; Modjaz et al. 2006; Pian et al. 2006; Sollerman et al. 2006), and most recently, SN2010bh/GRB100316D at $z = 0.0593$ (Chornock et al. 2011; Starling et al. 2011), where the SN spectra lines were visible as early as 2 days after the GRB, (Chornock et al. 2011). Two special SNe, SNe 2008D and 2009bb, and the potential presence of a jet in them will be discussed below.

Again, it is important to note that the spectra of the observed GRB-SNe are not any kind of core-collapse SNe, but specifically those of SN Ic-bl. The fact that there is no longer the large H envelope present when the star explodes as a SN-GRB is a crucial aspect of why and how the jet can punch its way trough the star (Woosley et al. 1993; Zhang et al. 2004). In addition, SN-GRB do not show the optical Helium lines in their spectra. While there is some discussion of He in the spectrum of SN 1998bw/GRB 980425 (Patat et al. 2001), its claim is based on a broad spectral feature at 1 micron, which could be due to lines other than He I $\lambda$10830 Å (Gerardy et al. 2004; Millard et al. 1999; Sauer et al. 2006). The NIR spectrum of the most recent SN2010bh/GRB100316D, did not show the 1 micron He line (Chornock et al. 2011).

What's more, it remains note-worthy and peculiar that almost all of the solid SN-GRB connections are with GRBs that are usually regarded as non-classical: i.e., GRBs that are less beamed (30°–80°), of low gamma-ray luminosity (i.e., $L_\gamma^{\rm iso} \leq 10^{49}$ erg s$^{-1}$), have very soft spectra, and thus, are also called X-ray Flashes (XRFs) or X-ray rich GRBs, being, discussed below, perhaps more common than cosmological GRBs (Cobb et al. 2006; Guetta & Della Valle 2007; Soderberg et al. 2006a). Only GRB 030329 connected with SN 2003dh is the one classical GRB whose kin we see at high $z$. Either those cosmological high-luminosity GRs are rare at low $z$, where we can see the SN signatures spectroscopically, or the SN-GRB connection is confined to only GRBs that are more isotropic and of low luminosity. For reference, a SN with the same large luminosity as SN 1998bw/GRB 980425 will appear at $R \sim 22$ mag at $z = 0.5$, so approaching the limit of obtaining a spectrum with a large-aperture telescope and reasonable exposure times.

The second broad class encompasses cases with only one epoch of low S/N spectra, which are at higher $z$: XRF 020903 at $z = 0.25$ (Bersier et al. 2006; Soderberg et al. 2005), SN 2002lt/GRB 021211 at $z = 1.006$ (Della Valle et al. 2003), SN 2005nc/GRB 050525A at $z = 0.606$ (Della Valle et al. 2006b). The last class of possible SN-GRB connections is based on observing rebrightening in the light curves of GRB afterglows that are consistent with emerging SN light curves, and in some cases with multi-color light curves that constrain the SED. While a a few high-$z$ cases have high-quality data that make them convincing (e.g., most recently Cano et al. 2011; Cobb et al. 2010), there are many more where the data is of lower quality (e.g., Zeh et al. 2004), making the SN interpretation in all cases less secure. The earliest, but more indirect, hints for the existence of a link between GRBs and the death of massive stars was the detection of star-formation features in the host galaxies of GRBs (Djorgovski et al. 1998; Fruchter et al. 1999) and correlation of GRB positions in their host galaxies with starforming regions (Bloom et al. 2002).

While two GRBs, GRBs 060505 and 060614, have been observed without a bright SN (Della Valle et al. 2006a; Fynbo et al. 2006; Gal-Yam et al. 2006), it is debated whether those were indeed bona fide long-GRBs (Bloom et al. 2008; Gehrels et al. 2006; Zhang et al. 2007), posing another challenge to the GRB classification scheme. In any case, it is fair to say, that for any bona fide and un-ambiguous long-GRB with a sufficiently low redshift to enable a spectroscopic SN detection, a broad-lined SN Ic has been detected. This does not apply to X-ray flashes, where multiple searches for SN signatures in low-$z$ XRFs have not yielded clear evidence for associated SNe (Levan et al. 2005; Soderberg et al. 2005).

# 3 Do all SNe Ic-bl have an accompanied GRB?

While the list of SN-GRB connections is short, there is a growing number of SN Ic-bl that are discovered by various SN surveys[2], which are not observed to have an accompanied GRB or to be engine-driven (except SN Ic-bl 2009bb, see below). One plausible explanation may be that all SN Ic-bl have an accompanied GRB but are not observed by us because of viewing-angle effects: our line-of-sight may not intersect the collimated jet of GRB emission for so-called "Off-axis GRBs" and thus we may not detect Gamma-Rays. Various investigations have been trying to address this viewing angle effect and to find Off-axis GRBs.

## 3.1 Search for off-axis GRBs

Even for the most highly-beamed GRB, the GRB jet gets decelerated over time and becomes effectively an isotropic blast wave such that the jet that was initially beamed away from our line of sight produces afterglow emission (so-called "orphan afterglows") which we may see to increase over time scales from months to several years (e.g., Perna & Loeb (1998); van Eerten et al. (2010)). If GRB jets are highly beamed, off-axis GRBs and their subsequent orphan afterglow are a natural prediction. Thus, various wide-field searches have looked for them in the optical and radio wavelengths (e.g., Levinson et al. 2002; Malacrino et al. 2007), but none has been detected at high significance. Along the same vein, specifically Soderberg et al. (2006b) targeted as part of their survey 68 SN Ib/c (including SN Ic-bl) for late-time VLA observations to search for off-axis GRB afterglows, but none of their objects showed evidence for bright, late-time radio emission that could be attributed to off-axis jets coming into our line of sight, until 2009. SN Ic-bl 2009bb was discovered optically and exhibited a large radio luminosity that requires substantial relativistic outflow with $10^3$ more matter coupled to relativistic ejecta than expected from normal core-collapse, and thus, arguing for an engine-driven SN (Pignata et al. 2011; Soderberg et al. 2010b). While no coincident GRB was detected, it is not clear whether it is because there was a weak GRB that went undetected or the GRB was off-axis or there were no gamma-rays produced during the SN. While also SN Ic 2007gr was claimed to indicate an engine-driven explosion without an observed GRB (Paragi et al. 2010), its radio light curves and X-ray data indicate that it may well be an ordinary SN Ib/c explosion (Soderberg et al. 2010a).

## 3.2 Relative rates of SN Ic-bl vs. LGRB

A statistical approach for understanding the SN-GRB connection is to compare the respective rates of broad-lined SN Ic to those of long GRBs and see if they are com-

[2] For a full list of IAUC-announced SNe, see the following link http://www.cfa.harvard.edu/iau/lists/RecentSupernovae.html.

parable. While this line of argument is very reasonable, it is not very conclusive at this point, given the fact that both kinds of rates are uncertain, as we detail now. On the GRB side, the beaming angle is uncertain, given the possible two different populations of GRBs (high-luminosity, highly-beamed vs. low-luminosity and nearly isotropic) – for the SN side, the rates of SN subtypes are not well known, specifically those of SN Ic-bl (Li et al. 2011), as well as how selection effects may enter differently for GRBs and SN searches. Guetta & Della Valle (2007) investigated this question by distinguishing between high- and low-luminosity GRBs and by deriving the SN Ic-bl rate from a heterogenous list of SNe discovered by different surveys. They estimate that the ratio of low-luminosity GRBs to SN Ic-bl is in the range of $\sim$1%–10 %, assuming that SN Ic-bl live in the same environments as SN-GRBs and have the same host galaxy luminosity, $M_B$, which we show below to not hold. Independently, the extensive radio search for off-axis GRBs in 143 optically discovered SN Ib/c (not strictly SN Ic-bl) yields that less than $\sim$1% of SN Ib/c harbor central engines (Soderberg et al. 2010b), thus broadly consistent with the above estimates.

In conclusion, it appears that SN-GRB are intrinsically rare and that certain conditions must be fulfilled for an exploding, massive and stripped star to simultaneously produce a GRB jet and to release a large amount of energy.

# 4 Aspherical explosions: only in SN-GRBs?

One of the fundamental questions in the SN-GRB field is whether aspherical explosions are the exclusive and distinguishing property of GRB-SN, or whether they are generic to the core-collapse process. Besides polarization measurements, late-time spectroscopy is a premier observational tool for studying the geometry of the SN explosion. At late times, (>3–6 months), the whole SN ejecta become optically thin in the continuum and hence affords a deeper view into the core of the explosion than spectra taken during the early photospheric phase. Moreover, the emission line shapes provide information about the velocity distribution of the ejecta (Fransson & Chevalier 1987; Schlegel & Kirshner 1989), and thus its radial extent, since the ejecta are in homologous expansion (where $v_r \propto r$). A radially expanding spherical shell of gas produces a square-topped profile, while a filled uniform sphere produces a parabolic profile. In contrast, a cylindrical ring, or torus, that expands in the equatorial plane gives rise to a "double-peaked" profile as there is very little low-velocity emission in the system, while the bulk of the emitting gas is located at $\pm v_{\rm t}$, where $v_{\rm t}$ is the projected expansion velocity at the torus.

Thus, a number of studies embarked on the difficult undertaking of obtaining such nebular spectra with adequate resolution and signal for a number of stripped SNe, as the objects are usually faint at that stage (19–23 mag) and call for large-aperture telescopes. First, Mazzali et al. (2005) and then Maeda et al. (2007) reported that SN Ic-bl 2003jd and the peculiar SN Ib 2005bf, respectively, displayed a double-peaked profile of [O I] $\lambda\lambda$6300, 6364 Å in nebular spectra. Subsequently, Modjaz et al. (2008b) and Maeda et al. (2008) independently presented a large number of stripped SNe displaying pronounced double-peaked profiles of [O I] $\lambda\lambda$6300, 6364 Å, the strongest line in late-time spectra of SNe without hydrogen, with ve-

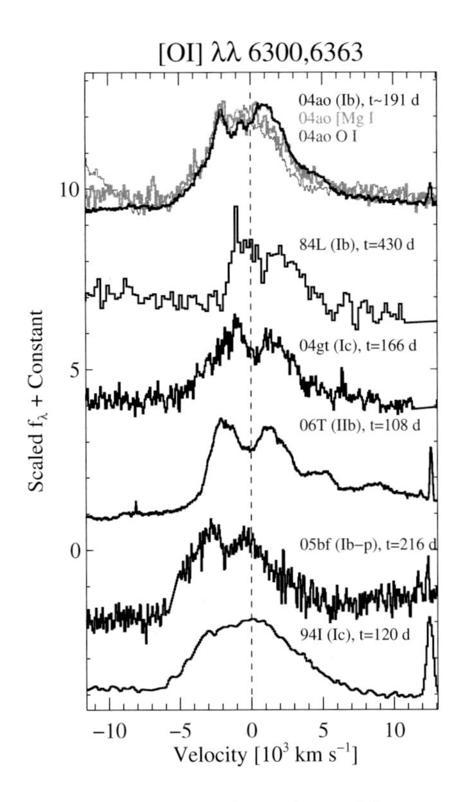


**Figure 2:** (online colour at: www.an-journal.org) Montage of five SNe with double-peaked oxygen profiles in velocity space. SN name, type, and phase of spectrum (with respect to maximum light). SN 1994I (Filippenko et al. 1995), which exhibits a single-peak oxygen line profile, is plotted for comparison at the bottom. The dashed line marks zero velocity with respect to 6300 Å. For SN 2004ao, we plot the scaled profiles of O I $\lambda$7774 Å (in blue) and [Mg I] $\lambda$4571 Å (in red), which are not doublets, but also exhibit the two peaks. As discussed in the text, the two horns are unlikely to be due to the doublet nature of [O I] $\lambda\lambda$6300, 6364 Å. From Modjaz et al. (2008b).

locity separations ranging between 2000 to 4000 $\mathrm{km\,s^{-1}}$ (see Fig. 2). Those profiles were interpreted as indicating an aspherical distribution of oxygen, possibly in a torus or flattened disc seen edge-on, suggesting that strong asphericity is ubiquitous in core-collapse SNe, and not necessarily a signature of an association with a GRB.

For SN spectra with sufficient S/N in some of the additional lines (O I $\lambda$7774, [Mg I] $\lambda$4571) and with multiple epochs (e.g., SNe 2004ao, 2008D, Modjaz et al. 2009; Tanaka et al. 2009b), the double-peaked profiles are unlikely to be caused by known optical depth effects. Furthermore, Taubenberger et al. (2009) presented and analyzed a large set of 98 late-time spectra of a total of stripped SNe (some of which were taken from the literature) and found a rich phenomenology of line structures. The results of their statistical analysis suggest that probably at least half of all stripped SNe are aspherical and that line profiles are indeed determined by the ejecta geometry, with Mg and O similarly distributed within the SN ejecta.

Recently, Milisavljevic et al. (2010) presented high S/N and multi-epoch nebular spectra of a select number of stripped SNe (including published data) and, coupled with detailed spectral line analysis and fitting, raise important questions about the interpretation of the indicated geometry. They suggest that alternative geometries beyond torus or highly flattened disks are possible for some of the SNe, where the double-peaked oxygen profile could be either coming only from preferentially blueshifted emission with internal obscuration in the red, or could consist of two separate emission components (a broad emission source centered around zero velocity and a narrow, blue-shifted source). Future high-S/N, multi-epoch and large-wavelength observations of a large sample of SNe II/IIb/Ib/Ic coupled with radiative transfer models should help to elucidate the observed blue- and redshifts of the line profiles and constrain the exact geometry.

In conclusions, observed double-peaked oxygen lines are not necessarily a proxy of a mis-directed GRB jet and they suggest that asphericities (of whatever exact geometry they may be) are most likely prevalent in normal core-collapse events. This result that asphericities are an ubiquitous feature during core-collapse is in line with similar conclusions based on polarization studies of SN II, Ib and SN Ic (e.g., Leonard & Filippenko 2005; Maund et al. 2009), neutron-star kick velocities (Wang et al. 2006), young SN remnant morphologies (Fesen et al. 2006), and theoretical modeling efforts (Burrows et al. 2006; Dessart et al. 2008; Scheck et al. 2006). Aspherical explosion geometry does not appear to be distinguishing feature of SN-GRBs, though SN-GRBs may have the highest degree of asphericity according to some models (Maeda et al. 2008).

# 5 Progenitor mass as the culprit?

One obvious possibility is that progenitor mass, one of the most fundamental properties of a star, may set apart SN with and without GRBs. Specifically, the SN-GRB progenitors could be of high mass (enough to produce a BH required for the collapsar model) and higher than the progenitors of SN without GRBs. In order to estimate the mass of the SN-GRB progenitors, we have to use a different method than the direct pre-explosion imaging technique (which even with HST's exquisite resolution can only be used for up to $\sim$20 Mpc), since the GRB and SN progenitors are at much larger, cosmological distances and since even in the local universe, detection attempts have failed (Smartt et al. 2009). There are two different techniques for indirectly estimating the main sequence (MS) mass of a SN/GRB progenitor. The first one consists of modeling the spectra and light curves of the individual SN/GRB in order to constrain the ejecta mass and core-mass before explosion and then use stellar evolutionary codes (e.g., see Fig. 1 in Tanaka et al. 2009a) to infer the MS mass, subject to the caveats of uncertain mass loss rates and rotation. The second technique entails studying the stellar population at the SN/GRB position as a proxy for the SN/GRB progenitor.

The first technique has been performed for a small number of SN/GRBs (e.g., Mazzali et al. 2006, for a review see Nomoto et al. 2010; Tanaka et al. 2009a), mostly for the nearby GRB-SN and a few peculiar SN and it suggests that the SN-GRB are from the more massive end of stellar masses ($\sim$20–50 $M_{\odot}$), but not necessarily from the most massive stars. However, so far only data of two SNe Ic-bl without an observed GRB (SNe 1997ef and 2002ap) have been modeled, and thus, it is not clear from this line of research whether the GRB-less SN Ic-bl progenitors are as massive as those of SN-GRB.

On the other hand, the second technique of studying the stellar populations at the explosion sites has been performed on a statistical set of different types of SN, SN-GRB and GRBs (those that are at higher redshifts) by comparing the amount of light at the position of the GRB or SN (after it had faded) to that of the rest of the host galaxy, as a proxy for the amount and mass of star formation. Fruchter et al. (2006) and Svensson et al. (2010) found that GRBs are more concentrated towards the brightest regions of their host galaxies than are SN II (for the same range of high $z$), and took their data to indicate larger progenitor masses for GRBs than for SN II, which is consistent with SN II pre-explosion detections that indicate modest main-sequence progenitor masses of 8–16 $M_{\odot}$ for SN IIP (see Smartt et al. 2009 for a review). Importantly, Kelly et al. (2008) demonstrate, using the same technique as Fruchter et al. (2006), that nearby ($z < 0.06$) SNe Ic are also highly concentrated on the brightest regions within their host galaxies, thus implying similarly high progenitor masses for SNe Ic without GRBs, as for GRBs themselves. Thus, these observations suggest another ingredient for GRB production besides higher mass progenitors. While Anderson & James (2008, 2009) have similar findings, their interpretation differs, as they regard the increased centralization of a SN distribution to imply increased progenitor metallicity, not increased progenitor mass. We will turn to the question of metallicity in the next section.

# 6 Metallicity as the culprit?

Metallicity is expected to influence not only the lives of massive stars but also the outcome of their deaths as supernovae (SNe) and as gamma-ray bursts (GRBs). However, before 2008, there were surprisingly few direct measurements of the local metallicities of SN-GRBs, and virtually none for the various types of core-collapse SNe.

Before delving into the details of the metallicity studies, let us explain what we refer to when using the term "metallicity" Theorists usually refer to the iron mass fraction of the SN progenitor, which is important for setting the mass loss of the pre-explosion massive star, since the bulk of the opacity is provided by iron and its huge number of lines (down to $10^{-3} Z_{\odot}$; Vink & de Koter 2005). Observers, on the other hand, usually measure the oxygen abundance of H II regions of some (usually central) part of the host galaxy or, in the best case scenario, at the SN positions[3]. The

[3]We also note that when we discuss oxygen, we do not refer to oxygen that was released during explosion (Sect. 4), since it usually takes $10^5$–$10^7$ years of settling time for the SN yields to be incorporated into the ISM, and we are observing the environments only months to years after explosion. While there

nebular oxygen abundance is the canonical choice of metallicity indicator for ISM studies, since oxygen is the most abundant metal, only weakly depleted onto dust grains (in contrast to refractory elements such as Mg, Si, Fe, with e.g., Fe being depleted by more than a factor of 10 in Orion; see Simón-Díaz & Stasińska 2011), and exhibits very strong nebular lines in the optical wavelength range (e.g., Kobulnicky & Kewley 2004; Tremonti et al. 2004). Thus, well-established diagnostic techniques have been developed (e.g., Kewley & Ellison 2008; Osterbrock 1989; Pagel et al. 1979).

Due to the short lifetimes of the massive SN/GRB progenitor stars ($t \leq 10^6$ yr for a 20 $M_{\odot}$ star, Woosley et al. 2002), we do not expect them to move far from their birth H II region sites (but see Eldridge et al. 2011; Hammer et al. 2006 and below) and thus take the abundance of the H II region at the SN site to indicate the natal metallicity of the SN/GRB progenitor. In one GRB case, where there is an independent metallicity measurement from absorption-line ratios in the X-ray spectra from the circumburst medium of SN 2006aj/XRF 060218 (Campana et al. 2008) and the common nebular oxygen-abundance measurement (e.g., Modjaz et al. 2006), the two completely independently derived values are in broad agreement. Furthermore, it appears that gas-phase oxygen abundances track the abundances of massive stars, as seen in a number of studies for the Orion nebula (see Simón-Díaz & Stasińska 2011 for a good review) and for blue supergiants in NGC 300 (Bresolin et al. 2009).

When considering oxygen abundance one has to remember the long-standing debate about which diagnostic to use, as there are systematic metallicity offsets between different methods (recombination lines vs. collisionally excited lines vs. "direct" method) and different strong-line diagnostics (see Kewley & Ellison 2008 and Moustakas et al. 2010 for detailed discussions), as well as the debate about the solar oxygen abundance value (Asplund et al. 2009). Nevertheless, the (relative) metallicity trends can be considered robust, if the analysis is performed self-consistently in the same scale, and trends are seen across different scales. We demonstrate the power and potential of this approach in the next subchapters.

## 6.1 Metallicity of SNe with and without GRBs

Many theoretical GRB models favor rapidly rotating massive stars at low metallicity (Hirschi et al. 2005; Langer & Norman 2006; Woosley & Heger 2006; Yoon & Langer 2005) as likely progenitors. Low metallicity seems to be a promising route for some stars to avoid losing angular momentum from mass loss (Crowther & Hadfield 2006; Vink & de Koter 2005) if the mass loss mode is set by line-driven, and therefore, metallicity-driven winds (but see Smith & Owocki 2006 for metallicity-independent mass loss from eruptions in some massive stars). If the stellar core is coupled to the outer envelopes via, e.g., magnetic torques (Spruit 2002), it is able to retain its high angular momentum preferentially at low metallicity. High angular

may be concerns about "self-enrichment", i.e., by evolved stars in H II region before explosion (such that measurements would not reflect the natal metallicity but some self-polluted, higher value), many H II regions do not show clear signs of self-enrichment (Wofford 2009).

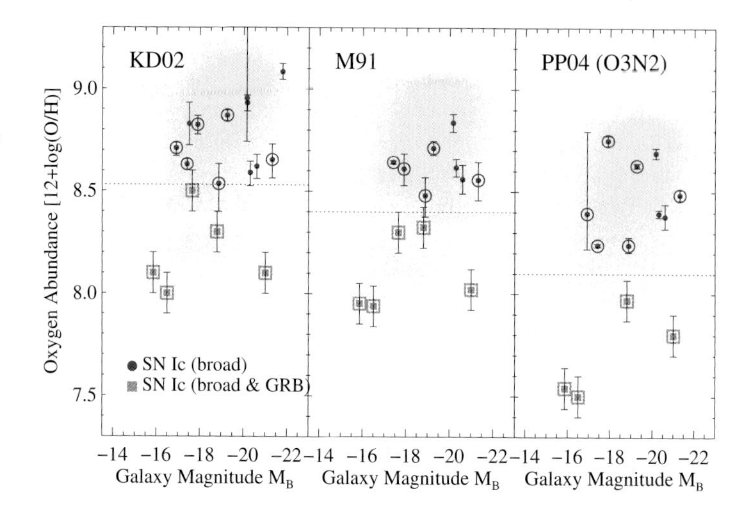


**Figure 3:** (online colour at: www.an-journal.org) Host-galaxy luminosity ($M_B$) and host-galaxy metallicity (in terms of oxygen abundance) at the sites of nearby ordinary broad-lined SNe Ic (blue filled circles) and broad-lined SNe Ic connected with GRBs (red filled squares) in three different, independent metallicity scales. Host environments of SNe-GRBs are more metal poor than host environments of broad-lined SNe Ic where no GRB was observed, for a similar range of host-galaxy luminosities and independent of the abundance scale used. For reference, the yellow points are nuclear values for local star-forming galaxies in SDSS (Tremonti et al 2004), re-calculated in the respective metallicity scales, and illustrate the empirical luminosity-metallicity relationship for galaxies. The host environment of the most recent SN-GRB (Chornock et al. 2011; Starling et al. 2011) is consistent with this trend. From Modjaz et al. (2008a).

momentum in the core appears to be a key ingredient for producing a GRB jet for both the collapsar and the magnetar models.

If the GRB progenitor is supposed to be at low metallicity for minimal mass loss, then how does it remove its outer layers, especially the large Hydrogen envelope, for we do not see any trace of H or He in the spectra of SN-GRB? Either via binaries (e.g., Fryer et al. 2007; Podsiadlowski et al. 2010) or, if the abundances in the star are sufficiently low, and thus, the star rotates rapidly enough, then quasi-chemical homogeneous evolution may set in, where hydrogen gets mix into the burning zones of the star via rotational mixing (Heger & Langer 2000; Langer 1992; Maeder 1987; Maeder & Meynet 2000), such that the star has low hydrogen abundances and a large core mass just before explosion. This mechanism seems plausible for producing a GRB and a broad-lined SN Ic at the same time, though it is debated whether it can explain all observed trends in the VLT FLAMES survey of massive stars at different metallicities (Brott et al. 2011; Frischknecht et al. 2010; Hunter et al. 2009).

Before 2007, a number of studies showed observationally that GRB hosts are of lower luminosity compared to core-collapse SN hosts (Fruchter et al. 2006; Wolf & Podsiadlowski 2007) and, when measurable, of low metallicity (e.g., Fynbo et al.

2003; Modjaz et al. 2006; Prochaska et al. 2004; Sollerman et al. 2005), especially compared to the vast majority of SDSS galaxies (Stanek et al. 2006). The next step was to compare the abundances of SNe Ic-bl with GRBs to SNe Ic-bl intrinsically without GRBs to test whether low metallicity is a necessary condition for GRB production.

In 2007 and 2008 we embarked on directly measuring metallicities of a statistically significant sample of broad-lined SN Ic environments and deriving them in the same fashion, which we presented in our study of Modjaz et al. (2008a), the first of its kind. There, we compared the chemical abundances at the sites of 5 nearby ($z < 0.25$) broad-lined SN Ic that accompany nearby GRBs with those of 12 nearby ($z < 0.14$) broad-lined SN Ic that have no observed GRBs. We showed that the oxygen abundances at the GRB sites are systematically lower than those found at the sites of ordinary broad-lined SN Ic (Fig. 3). Unique features of our analysis included presenting new spectra of the host galaxies and analyzing the measurements of both samples in the same set of ways, via three independent metallicity diagnostics, namely those of Kewley & Dopita (2002, KD02), McGaugh (1991, M91), and Pettini & Pagel (2004, PP04). We demonstrated that neither SN selection effects (SN found via targeted vs. non-targeted surveys, for an extensive discussion see Sect. 8) nor the choice of strong-line metallicity diagnostic can cause the observed trend.

Though our sample size was small, the observations (before 2009) were consistent with the hypothesis that low metal abundance is the cause of some massive stars becoming SN-GRB. While each metallicity diagnostic has its own short-coming, if we use the scale of PP04, which has been suggested by Bresolin et al. (2009) to be the strong-line method in most agreement with abundances from stars, then the "cut-off" value would be $0.3Z_{\odot}$. Furthermore, a comparison between the local metallicity of the GRB-SN site and the global host galaxy value via resolved metallicity maps yields that the GRB-SN local values track the global host value, but are also the most metal-poor site of the galaxy (Christensen et al. 2008; Levesque et al. 2011).

This was in 2008 – *now*, however, the case for a metallicity threshold is much less clear. Over the last three years, two "dark" GRBs (Graham et al. 2009; Levesque et al. 2010b) and one radio-relativistic SN (Levesque et al. 2010c; Soderberg et al. 2010b) have been observed at high, super-solar, metallicity. Nevertheless, even if one includes those higher metallicity explosions regardless of whether they share the same progenitor channels as SN-GRBs, Levesque et al. (2010a) show that the $M$-$Z$ relationship for GRBs lies systematically below that of the bulk of the normal starforming galaxies in the corresponding redshift ranges (see their Fig. 1). While there are suggestions that the low-metallicity preference could be partly produced by the newly-discovered relationship between host galaxy metallicity, mass and star formation rate (so called "fundamental" metallicity relation, Mannucci et al. 2010) such that low metallicity galaxies have high star formation rates (Kocevski & West 2010; Mannucci et al. 2011), it does not explain why there are not more GRBs in intermediate- and high-mass galaxies (Kocevski & West 2010; Kocevski et al. 2009) and the observed evolution in the GRB rate density with increasing redshifts (Butler et al. 2010).

An obvious test will be to construct the $M$-$Z$ relationship for other explosions that track massive star formation, such as normal SN Ib and SN Ic, found in the same fashion as GRBs, namely from non-targeted surveys such as the Palomar Transient Factory (see Sect. 8) with those of GRB.

## 6.2 Metallicity of various types of CCSN

For illuminating the SN-GRB connection and pursuing stellar forensics on the specific SNe Ic-bl with and without GRBs, it is also important to gain an understanding of the progenitors of "normal" stripped SNe. Here too, the two outstanding progenitor channels are either single massive Wolf-Rayet (WR) stars with main-sequence (MS) masses of $\gtrsim$30 $M_{\odot}$ that have experienced mass loss during the MS and WR stages (e.g., Woosley et al. 1993), or binaries from lower-mass He stars that have been stripped of their outer envelopes through interaction (Podsiadlowski et al. 2004, and references therein), or a combination of both. Attempts to directly identify SN Ib/c progenitors in pre-explosion images have not yet been successful (e.g., Gal-Yam et al. 2005; Maund et al. 2005; Smartt 2009).

A more indirect but very powerful approach is to study the environments of a large sample of CCSNe in order to discern systematic trends that characterize their stellar populations. Discussed already was the study of the amount of blue light at the SN position (Sect. 5), which indicates that the SN Ibc are more concentrated towards the brightest regions of their host galaxies than SN II (Anderson & James 2008; Kelly et al. 2008), possibly suggesting the progenitors of SNe Ib/c may thus be more massive than those of SNe II, which are $\sim$8–16 $M_{\odot}$ (see Smartt 2009 for a review).

There exist a few metallicity studies of CCSN host environment that either *indirectly* probe the metallicities of a *large* set of SNe II, Ib, and Ic, or that *directly* probe the local metallicity of a *small* and select set of interesting/peculiar stripped SNe. Nevertheless, interesting trends have emerged: Studies to measure the metallicity by using the SN host-galaxy luminosity as a proxy (Arcavi et al. 2010; Prantzos & Boissier 2003), or by using the metallicity of the galaxy center measured from Sloan Digital Sky Survey (SDSS) spectra (Prieto et al. 2008) to extrapolate to that at the SN position (Boissier & Prantzos 2009) find a) that host galaxies of SNe Ib/c found in targeted surveys seem to be in more luminous and more metal-rich galaxies than those of SNe II (Boissier & Prantzos 2009; Prieto et al. 2008) and b) that SNe Ic are missing in low-luminosity and presumably, low-metallicity galaxies, while SN II, Ib, and Ic-bl are abundant there (Arcavi et al. 2010). Those prior metallicity studies do not directly probe the local environment of each SN (which is different from the galaxy center due to metallicity gradients) nor do some differentiate between the different SN subtypes.

In Modjaz et al. (2011), we presented the largest existing set of host-galaxy spectra with H II region emission lines at the sites of 35 stripped-envelope core-collapse SNe and including those from the literature and from Modjaz et al. (2008a), we analyzed the metallicity environments of a total of 47 stripped SNe. We derived local oxygen abundances in a robust manner in order to constrain the SN Ib/c progenitor population. We obtained spectra at the SN sites, included SNe from targeted and untargeted surveys, and performed the abundance determinations using the same three different oxygen-abundance calibrations as in Modjaz et al. (2008a).

We found that the sites of SNe Ic (the demise of the most heavily stripped stars having lost both H and He layers) are systematically more metal-rich than those of SNe Ib (arising from stars that retained their He layer) in all calibrations. A Kolmogorov-Smirnov test yields the low probability of 1% that SN Ib and SN Ic environment abundances, which are different on average by $\sim$0.2 dex (in the Pettini & Pagel scale), are drawn from the same parent population. Broad-lined SNe Ic (without GRBs) occur at metallicities between those of SNe Ib and SNe Ic. Lastly, we found that the host-galaxy central oxygen abundance is not a good indicator of the local SN metallicity (introducing differences up to 0.24 dex), and concluded that large-scale SN surveys need to obtain local abundance measurements in order to quantify the impact of metallicity on stellar death.

A reasonable suggestion for why the environments of SNe Ic are more metal rich than those of SNe Ib is that metallicity-driven winds (Crowther & Hadfield 2006; Vink & de Koter 2005) in the progenitor stars prior to explosion are responsible for removing most, if not all, of the He layer whose spectroscopic nondetection distinguishes SNe Ic from SNe Ib. This explanation may favor the single massive WR progenitor scenario as the dominant mechanism for producing SNe Ib/c (Woosley et al. 1993), at least for those in large star-forming regions. While the binary scenario has been suggested as the dominant channel for numerous reasons (see Smartt 2009 for a review; Smith et al. 2011), we cannot assess it in detail, since none of the theoretical studies (e.g., Eldridge et al. 2008, and references therein) predict the metallicity dependence of the subtype of stripped SN. However, our results are consistent with the suggestion of Smith et al. (2011) that SNe Ic may come from stars with higher metallicities (and masses) than SNe Ib, even if they are in binaries. Furthermore, the finding that the metallicity environments for SN Ic-bl are different from those of SN Ic indicates that their progenitors may be physically different (perhaps because of magnetic fields or other factors) and the observed high velocities in SN Ic-bl are probably not only due to viewing-angle effects.

Most recently, similar studies for SN Ib and SN Ic were conducted by Anderson et al. (2010) and Leloudas et al. (2011). While they do find small differences between SN Ib and SN Ic, with SN Ic in slightly more metal-rich environments, they conclude that their findings are not statically significant. While the reasons for these different metallicity findings in different studies are not yet clear, some of the aspects of their studies may complicate direct SN Ib vs. SN Ic metallicity comparisons with statistical power. For example, historical SNe Ib/c without firm subtype classifications (e.g., SNe 1962L, 1964L) from only targeted surveys, some with incorrect SN offsets as announced in the IAUC (e.g., for SNe 1987M and 2002ji; S. Van Dyk 2010, private communication) are included (Anderson et al. 2010) and an unequal number of SN Ib (14) and SN Ic (5) in Leloudas et al. (2011). In any case, definite answers should be provided by future environmental metallicities studies using a very large SN crops from the same, homogeneous and galaxy-unbiased survey, such as the one we are undertaking (see Sect. 8), which should be ideally suited to determine the environmental conditions that influence the various kinds of massive stellar deaths in an unbiased fashion.

### 6.3 SN and GRB host metallicity measurements as a rapidly expanding field

Not only for stripped SNe but also for other kinds of SNe and transients have metallicity studies emerged as a promising tool to probe their progenitor and explosion conditions. Another class of CCSN that has piqued a lot of interest in the past few years is the emerging field of over-luminous SNe, i.e., SNe defined as more luminous in absolute magnitudes than $M_V \sim -21$ mag (Ofek et al. 2007; Quimby et al. 2011; Smith et al. 2007), that are being discovered in wide-field surveys. Is is hotly debated what powers their optical brilliance, whether it's due to circumstellar interaction, large amount of synthesized $^{56}$Ni during the explosion of a pair-instability SN or the birth of a magnetar (Kasen & Bildsten 2010).

Host galaxy studies show that their host galaxies are of low luminosity, highly starforming and blue (Neill et al. 2011), similar to GRB-host galaxies, and similarly of low-metallicity, when measured (Stoll et al. 2011), except the host of SN 2006gy, first to be claimed as a pair-instability SN (Ofek et al. 2007; Smith et al. 2007). Furthermore, the best candidate for a pair-instability SN, SN 2007bi (Gal-Yam et al. 2009) has a host galaxy with a metallicity of $12 + \log(\mathrm{O/H})_{\mathrm{M91}} = 8.15 \pm 0.15 \sim 0.3 Z_\odot$ (Young et al. 2010), so it is a subsolar galaxy, but not of extreme subsolar metallicity, as one might expect from Pop. III stars in the high-$z$ universe. Thus, if SN 2007bi is representative of pair-instability SNe, then they should be found frequently during current and next generation of wide-area surveys, which have enough volume to discover rare transients.

Furthermore, even for SN Ia, which arise from the thermonuclear explosion of a white dwarf at or near the Chandraskhar mass limit, host galaxy studies have uncovered trends for SN Ia luminosity with host galaxy morphology (e.g., Hamuy et al. 1996) and mass (e.g., Kelly et al. 2010; Sullivan et al. 2010), where more luminous SN Ia tend to be in more luminous and (assuming the luminosity-metallicity relationship for galaxies) metal-rich galaxies, which is consistent with measured metallicity studies (Gallagher et al. 2008). However, for SN Ia with their long delay times (200 Million yrs to a few Gigayears) and the associated large offsets between birth and explosion sites, it is not clear whether measuring the gas-phase metallicity (which reflects that of the currently starforming gas) at the SN position really reflects the natal metallicity of the old progenitor (Bravo & Badenes 2011). Nevertheless, integrated metallicities from stars in the host galaxy (Gallagher et al. 2008) or those of dwarf galaxies (Childress et al. 2011), which usually have a small spread in metallicities, may still be revealing.

## 7 SN 2008D/XRT080109: stellar forensics by witnessing the death throes of a stripped star

A complimentary stellar forensics tool is to catch the massive star during its death throes. This happened for SN 2008D/XRT080109 (where XRT stands for X-ray transient), which was discovered by Soderberg et al. (2008). From the early light of the explosion, one can reconstruct a massive star's pre-explosion composition and ra-

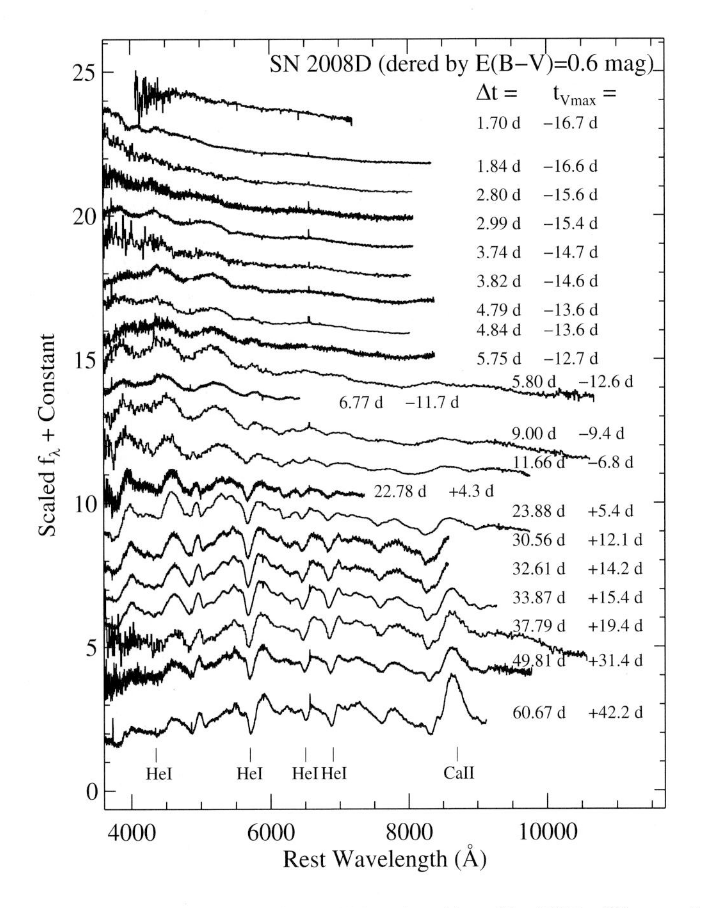


**Figure 4:** Spectral evolution of SN 2008D, dereddened by $E(B-V)_{\rm host} = 0.6$ mag and labeled with respect to date of shock breakout (indicated by $\Delta t$), and to date of $V$-band maximum (indicated by $t_{V\rm max}$). Note the fleeting double-absorption feature around 4000 Å in our early spectrum at $\Delta t$ = 1.84 d. The characteristic optical He lines (due to blueshifted He I $\lambda\lambda$ 4471, 5876, 6678, 7061 Å) become visible starting $t \sim$ 12 d or $t_{V\rm max} \sim$ 6 d. From Modjaz et al. (2009).

dius, which provides a powerful tool to closely investigate a single star out to cosmological distances. Besides its utility, the story of SN 2008D/XRT 080109 reminds us of the importance of serendipity to science. SN 2008D/XRT 080109 was discovered by Soderberg et al. (2008) in X-rays via the Swift satellite, because they were monitoring another SN, SN 2007uy, in the *same* galaxy, when suddenly XRT080109 erupted and lasted ~600 s in X-rays. Furthermore, our program was also monitoring SN 2007uy in the optical and NIR from the ground, providing us with stringent limits on the optical emission hours before the onset of X-ray transient.

In Modjaz et al. (2009), we gathered extensive panchromatic observations (X-ray, UV, Optical, NIR) from 13 different telescopes to determine the nature of SN 2008D, its accompanying Swift X-ray transient 080109, and its progenitor (see also

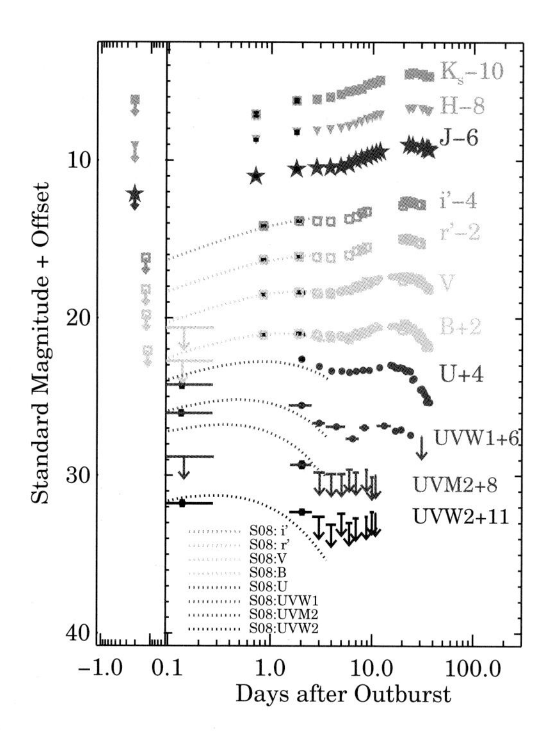


**Figure 5:** (online colour at: www.an-journal.org) Observed optical and NIR light curves of SN 2008D after the onset (*right part*) of XRT 080109, which we adopt as the time of shock breakout. The filled circles show $UVW1$, $UVM2$, $UVW1$, $U$, $B$, $V$ data from Swift/UVOT, the empty circles are $BV$ data from KAIT, while the empty squares are data from the FLWO 1.2-m telescope. $JHKs$ data (filled stars, triangles, and squares) are from PAIRITEL. Note the very early optical data points ($\Delta t = 0.84$ d after shock breakout) from the FLWO 1.2 m telescope, as well as NIR data (at $\Delta t = 0.71$ d) from PAIRITEL, amongst the earliest data of a SN Ib to date. Swift/UVOT upper limits are indicated by the arrows. We also plot the pre-explosion upper limits derived from the 1.2-m CfA and the PAIRITEL data (*left part*). The data have not been corrected for extinction. The two-component light curve is pronounced in the blue, where the 1st peak is due to thermal emission from the expanding and cooling stellar layers and the 2nd peak due to powering by $^{56}$Ni. We note that our earliest ground-based data points are consistent with the light-curve fits by Soderberg et al. (2008, S08; dotted lines), who use the envelope BB emission model from Waxman et al. (2007). From Modjaz et al. (2009).

Soderberg et al. 2008). We first established that SN 2008D is a spectroscopically normal SN Ib (i.e., showing conspicuous He lines, see Fig. 4), which implies the progenitor star had an intact He layer, but had not retained its outermost H envelope. For the first time, the very early-time peak (at $\sim$1 day, see Fig. 5) could be observed for this kind of SN, from which one can deduce the progenitor radius, since that peak is due to black-body emission from the cooling and expanding stellar envelope.

Using our reliable and early-time measurements of the bolometric output of this SN in conjunction with models by Waxman et al. (2007) and Chevalier & Fransson (2008), as well as published values of kinetic energy and ejecta mass, we derived a progenitor radius of $1.2 \pm 0.7\ R_{\odot}$ and $12 \pm 7\ R_{\odot}$, respectively, the latter being more in line with typical WN stars. We furthermore showed that the observed X-ray emission by which it was discovered (Soderberg et al. 2008) is different from those of X-ray flashes, the weaker cousins of GRBs, which demonstrates that even normal SN Ib, surprisingly, can give rise to high-energy phenomena (but see Mazzali et al. 2008).

Lastly, our spectra obtained at three and four months after maximum light show double-peaked oxygen lines that we associate with departures from spherical symmetry, as has been suggested for the inner ejecta of a number of SN Ib cores. Our detailed observations and their analysis, as well as those of others (Malesani et al. 2009; Maund et al. 2009; Mazzali et al. 2008; Soderberg et al. 2008; Tanaka et al. 2009b) have inspired a number of theorists to develop sophisticated models of SN shock breakout including relativistic-mediated shocks (Katz et al. 2010), asphericity (Couch et al. 2011) and the impact of a wind (Balberg & Loeb 2011) to explain its X-ray shock breakout, as well as refine models of the subsequently cooling envelope (Nakar & Sari 2010; Rabinak & Waxman 2011) aimed at reproducing the optical observations.

# 8 The future is now: the golden age of transient surveys and corresponding host galaxy studies

We are embarking on the golden age of transients, i.e., diverse explosive phenomena from both massive stars and compact objects, which a number of large-scale surveys are starting to harvest. These innovative surveys include the Palomar Transient Factory (PTF[4], Rau et al. 2009), Catalina Real-Time Transient Survey (CRTS[5], Drake et al. 2009), PanSTARRS[6] (Tonry & Pan-STARRS Team 2005) and The Chilean Automatic Supernova Search (CHASE[7], Pignata et al. 2009, (for a comparison for 2010, see Gal-Yam & Mazzali 2011), and in the future, LSST, which is a major US undertaking and number one ranked ground-based project during the 2010 Decadal Survey. Because of different survey modes and detection techniques, these innovative surveys are finding known types of SNe in large numbers and with relatively little bias, as well as rare, yet astrophysically interesting events in statistically large numbers. The main innovations in survey mode are: very large field-of-view (e.g., 7.8 square-degrees for PTF, up to 4000 times larger than traditional surveys), galaxy-untargeted, with different cadences, less bias towards bright cores of galaxies.

In contrast, traditional SN surveys usually have small fields of view and thus, specifically target luminous galaxies that contain many stars in order to increase their odds of finding those that explode as SNe. For example, the prolific Lick Observatory SN Search (LOSS, Filippenko et al. 2001; Li et al. 2011) monitors a list of

[4] http://www.astro.caltech.edu/ptf/

[5] http://crts.caltech.edu/

[6] http://ps1sc.org/transients/

[7] http://www.das.uchile.cl/proyectoCHASE/

galaxies that have a mean (median) value at the galaxy magnitude of $M_B = -19.9$ ($M_B = -20.1$) mag. But because more luminous galaxies are more metal-rich (Tremonti et al. 2004, see also Fig. 3) such targeted SN surveys are probably biased towards finding SN in high-metallicity galaxies. However, the new surveys alleviate this galaxy- and metallicity-bias and PTF has already found over 1090 spectroscopically ID'ed SNe (as of April 2011) in this galaxy-untargeted fashion, probing SNe and transients in all kinds of galactic environments (except in highly obscured ones, of course).

The difference in survey mode appears to be important: using core-collapse SN discovered with PTF, Arcavi et al. (2010) showed that different populations of galaxies may be hosting different types of stripped SNe. The least stripped SN (SN IIb) were found in the low-luminosity galaxies (dwarfs), which Arcavi et al. took to indicate a metallicity effect, where the massive progenitor at low-metallicity did not have sufficiently winds to remove its He layer in order to explode as a SN Ic. While these findings are in line with Modjaz et al. (2011), in order to verify them with direct metallicity measurements, as well as for resolving some of the issues discussed in Sect. 6.2, it is necessary to conduct a thorough and extensive host galaxy study with a large single-survey, untargeted, spectroscopically classified, and homogeneous collection of stripped SNs, something we are currently undertaking.

For core-collapse SNe and transients, the next big frontier is to hunt for the lowest metallicity host galaxies, in order to fully confront theoretical predictions of the impact of metallicity on stellar death (Heger et al. 2003; O'Connor & Ott 2011) with observations, via the untargeted surveys or specifically low-luminosity-targeted surveys.

**Acknowledgements**

I thank the Astronomische Gesellschaft and its selection committee for awarding me the 2010 Ludwig-Biermann Award and for giving me the opportunity to participate in the 2010 annual meeting in Bonn. Furthermore, I thank my many collaborators over the years and at various institutions for their fruitful collaboration. M.M. acknowledges support by the Hubble Fellowship grant HST-HF-51277.01-A, awarded by STScI, which is operated by AURA under NASA contract NAS5-26555. This research has made use of NASA's Astrophysics Data System.

## References

Anderson, J.P., James, P.A.: 2008, MNRAS 390, 1527

Anderson, J.P., James, P.A.: 2009, MNRAS 399, 559

Anderson, J.P., Covarrubias, R.A., James, P.A., Hamuy, M., Habergham, S.M.: 2010, MNRAS 407, 2660

Arcavi, I., Gal-Yam, A., Kasliwal, M.M., et al.: 2010, ApJ 721, 777

Asplund, M., Grevesse, N., Sauval, A.J., Scott, P.: 2009, ARA&A 47, 481

Balberg, S., Loeb, A.: 2011, MNRAS, in press

Bersier, D., Fruchter, A.S., Strolger, L.-G., et al.: 2006, ApJ 643, 284

Bloom, J.S., Kulkarni, S.R., Djorgovski, S.G.: 2002, AJ 123, 1111

Bloom, J.S., Butler, N.R., Perley, D.A.: 2008, in: M. Galassi, D. Palmer, E. Fenimore (eds.), *Gamma-Ray Bursts 2007*, AIPC 1000, p. 11

Boissier, S., Prantzos, N.: 2009, A&A 503, 137

Bravo, E., Badenes, C.: 2011, MNRAS, in press

Bresolin, F., Gieren, W., Kudritzki, R., Pietrzyński, G., Urbaneja, M.A., Carraro, G.: 2009, ApJ 700, 309

Brott, I., Evans, C.J., Hunter, I., et al.: 2011, A&A, in press, astro-ph/1102.0766

Burrows, A., Livne, E., Dessart, L., Ott, C.D., Murphy, J.: 2006, ApJ 640, 878

Butler, N.R., Bloom, J.S., Poznanski, D.: 2010, ApJ 711, 495

Campana, S., et al.: 2006, Nature 442, 1008

Campana, S., Panagia, N., Lazzati, D., et al.: 2008, ApJ 683, L9

Cano, Z., et al.: 2011, MNRAS, in press, astro-ph/1012.1466

Cantiello, M., Yoon, S., Langer, N., Livio, M.: 2007, A&A 465, L29

Chevalier, R.A., Fransson, C.: 2008, ApJ 683, L135

Childress, M., Aldering, G., Aragon, C., et al.: 2011, ApJ 733, 3

Chornock, R., Berger, E., Levesque, E.M., et al.: 2011, ApJ, submitted, astro-ph/1004.2262

Christensen, L., Vreeswijk, P.M., Sollerman, J., Thöne, C.C., Le Floc'h, E., Wiersema, K.: 2008, A&A 490, 45

Clocchiatti, A., Wheeler, J.C., Brotherton, M.S., Cochran, A.L., Wills, D., Barker, E.S., Turatto, M.: 1996, ApJ 462, 462

Cobb, B.E., Bailyn, C.D., van Dokkum, P.G., Natarajan, P.: 2006, ApJ 645, L113

Cobb, B.E., Bloom, J.S., Perley, D.A., Morgan, A.N., Cenko, S.B., Filippenko, A.V.: 2010, ApJ 718, L150

Couch, S.M., Pooley, D., Wheeler, J.C., Milosavljević, M.: 2011, ApJ 727, 104

Crowther, P.A., Hadfield, L.J.: 2006, A&A 449, 711

Della Valle, M., et al.: 2003, A&A 406, L33

Della Valle, M., Chincarini, G., Panagia, N., et al.: 2006a, Nature 444, 1050

Della Valle, M., Malesani, D., Bloom, J.S., et al.: 2006b, ApJ 642, L103

Dessart, L., Burrows, A., Livne, E., Ott, C.D.: 2008, ApJ 673, L43

Djorgovski, S.G., Kulkarni, S.R., Bloom, J.S., Goodrich, R., Frail, D.A., Piro, L., Palazzi, E.: 1998, ApJ 508, L17

Drake, A.J., Djorgovski, S.G., Mahabal, A., et al.: 2009, ApJ 696, 870

Eldridge, J.J., Izzard, R.G., Tout, C.A. 2008, MNRAS 384, 1109

Eldridge, J.J., Langer, N., Tout, C.A.: 2011, MNRAS, in press, astro-ph/1103.1877

Fesen, R.A., Hammell, M.C., Morse, J., et al.: 2006, ApJ 645, 283

Filippenko, A.V.: 1997, ARA&A 35, 309

Filippenko, A.V., et al.: 1995, ApJ 450, L11

Filippenko, A.V., Li, W.D., Treffers, R.R., Modjaz, M.: 2001, in: B. Paczynski, W.-P. Chen, C. Lemme (eds.), *Small Telescope Astronomy on Global Scales*, ASPC 246, IAU Coll. 183, p. 121

Fransson, C., Chevalier, R.A.: 1987, ApJ 322, L15

Frischknecht, U., Hirschi, R., Meynet, G., Ekström, S., et al.: 2010, A&A 522, A39

Fruchter, A.S., et al.: 1999, ApJ 519, L13

Fruchter, A.S., Levan, A.J., Strolger, L., et al.: 2006, Nature 441, 463

Fryer, C.L., Mazzali, P.A., Prochaska, J., et al.: 2007, PASP 119, 1211

Fynbo, J.P.U., Jakobsson, P., Möller, P., et al.: 2003, A&A 406, L63

Fynbo, J.P.U., Watson, D., Thoene, C.C., et al.: 2006, Nature 444, 1047

Galama, T.J., et al.: 1998, Nature 395, 670

Gallagher, J.S., Garnavich, P.M., Caldwell, N., et al.: 2008, ApJ 685, 752

Gal-Yam, A., Mazzali, P.: 2011, astro-ph/1103.5165

Gal-Yam, A., Fox, D.B., Kulkarni, S.R., et al.: 2005, ApJ 630, L29

Gal-Yam, A., Fox, D., Price, P., et al.: 2006, Nature 444, 1053

Gal-Yam, A., Mazzali, P., Ofek, E.O., Nugent, P.E., et al.: 2009, Nature 462, 624

Gehrels, N., Norris, J.P., Barthelmy, S.D., et al.: 2006, Nature 444, 1044

Gerardy, C.L., Fesen, R.A., Marion, G.H., et al.: 2004, in: P. Höflich, P. Kumar, J.C. Wheeler (eds.), *Cosmic Explosions in Three Dimensions*, p. 57

Graham, J.F., Fruchter, A.S., Kewley, L.J., et al.: 2009, in: C. Meegan, C. Kouveliotou, N. Gehrels (eds.), *Gamma-Ray Burst*, AIPC 1133, p. 269

Guetta, D., Della Valle, M.: 2007, ApJ 657, L73

Hammer, F., Flores, H., Schaerer, D., Dessauges-Zavadsky, M., Le Floc'h, E., Puech, M.: 2006, A&A 454, 103

Hamuy, M., Phillips, M.M., Suntzeff, N.B., et al.: Krzeminski, W.: 1996, AJ 112, 2408

Heger, A., Langer, N.: 2000, ApJ 544, 1016

Heger, A., Fryer, C.L., Woosley, S.E., Langer, N., Hartmann, D.H.: 2003, ApJ 591, 288

Hirschi, R., Meynet, G., Maeder, A.: 2005, A&A 443, 581

Hjorth, J., Bloom, J.S.: 2011, in: C. Kouveliotou, R.A.M.J. Wijers, S.E. Woosley (eds.), in: *Gamma-Ray Bursts*, 2011, Chapter 9, astro-ph/1104.2274

Hjorth, J., et al.: 2003, Nature 423, 847

Hunter, I., Brott, I., Langer, N., et al.: 2009, A&A 496, 841

Kasen, D., Bildsten, L.: 2010, ApJ 717, 245

Katz, B., Budnik, R., Waxman, E.: 2010, ApJ 716, 781

Kelly, P.L., Kirshner, R.P., Pahre, M.: 2008, ApJ 687, 1201

Kelly, P.L., Hicken, M., Burke, D.L., Mandel, K.S., Kirshner, R.P.: 2010, ApJ 715, 743

Kewley, L.J., Dopita, M.A.: 2002, ApJS 142, 35

Kewley, L.J., Ellison, S.L.: 2008, ApJ 681, 1183

Kobulnicky, H.A., Kewley, L.J.: 2004, ApJ 617, 240

Kocevski, D., West, A.A.: 2010, astro-ph/1011.4060

Kocevski, D., Modjaz, M., Bloom, J.S., et al.: 2007, ApJ 663, 1180

Kocevski, D., West, A.A., Modjaz, M.: 2009, ApJ 702, 377

Kouveliotou, C., Meegan, C.A., Fishman, G.J., et al.: 1993, ApJ 413, L101

Langer, N.: 1992, A&A 265, L17

Langer, N., Norman, C.A.: 2006, ApJ 638, L63

Leloudas, G., Gallazzi, A., Sollerman, J., et al.: 2011, A&A, in press, astro-ph/1102.2249

Leonard, D.C., Filippenko, A.V.: 2005, in: M. Turatto, S. Benetti, L. Zampieri, W. Shea (eds.), *1604-2004: Supernovae as Cosmological Lighthouses*, ASPC 342, p. 330

Levan, A., Nugent, P., Fruchter, A., et al.: 2005, ApJ 624, 880

Levesque, E.M., Berger, E., Kewley, L.J., Bagley, M.M.: 2010a, AJ 139, 694

Levesque, E.M., Kewley, L.J., Graham, J.F., Fruchter, A.S. 2010b, ApJ 712, L26

Levesque, E.M., Soderberg, A.M., Foley, R.J., et al.: 2010c, ApJ 709, L26

Levesque, E.M., Berger, E., Soderberg, A.M., Chornock, R.: 2011, ApJL, submitted, astro-ph/1104.2865

Levinson, A., Ofek, E.O., Waxman, E., Gal-Yam, A.: 2002, ApJ 576, 923

Li, W., Leaman, J., Chornock, R., et al: 2011, MNRAS 412, 1441

MacFadyen, A.I., Woosley, S.E.: 1999, ApJ 524, 262

Maeda, K., Tanaka, M., Nomoto, K., et al.: 2007, ApJ 666, 1069

Maeda, K., Kawabata, K., Mazzali, P.A., et al.: 2008, Science 319, 1220

Maeder, A.: 1987, A&A 173, 247

Maeder, A., Meynet, G.: 2000, ARA&A 38, 143

Malacrino, F., Atteia, J., Boër, M., et al.: 2007, A&A 464, L29

Malesani, D., et al.: 2004, ApJ 609, L5

Malesani, D., Fynbo, J.P.U., Hjorth, J., et al.: 2009, ApJ 692, L84

Mannucci, F., Cresci, G., Maiolino, R., Marconi, A., Gnerucci, A.: 2010, MNRAS 408, 2115

Mannucci, F., Salvaterra, R., Campisi, M.A.: 2011, MNRAS, in press

Matheson, T., et al.: 2003, ApJ 599, 394

Maund, J.R., Smartt, S.J., Schweizer, F.: 2005, ApJ 630, L33

Maund, J.R., Wheeler, J.C., Baade, D., Patat, F., Höflich, P., Wang, L., Clocchiatti, A.: 2009, ApJ 705, 1139

Mazzali, P.A., Kawabata, K.S., Maeda, K., et al.: 2005, Sci 308, 1284

Mazzali, P.A., Deng, J., Pian, E., et al.: 2006, ApJ 645, 1323

Mazzali, P.A., Valenti, S., Della Valle, M., et al.: 2008, Sci 321, 1185

McGaugh, S.S.: 1991, ApJ 380, 140

Metzger, B.D., Giannios, D., Thompson, T.A., Bucciantini, N., Quataert, E.: 2011, MNRAS, in press, astro-ph/1012.0001

Milisavljevic, D., Fesen, R.A., Gerardy, C.L., Kirshner, R.P., Challis, P.: 2010, ApJ 709, 1343

Millard, J., Branch, D., Baron, E., et al.: 1999, ApJ 527, 746

Mirabal, N., Halpern, J.P., An, D., Thorstensen, J.R., Terndrup, D.M.: 2006, ApJ 643, L99

Modjaz, M., Stanek, K.Z., Garnavich, P.M., et al.: 2006, ApJ 645, L21

Modjaz, M., Kewley, L., Kirshner, R.P., et al.: 2008a, AJ 135, 1136

Modjaz, M., Kirshner, R.P., Blondin, S., Challis, P., Matheson, T.: 2008b, ApJ 687, L9

Modjaz, M., Li, W., Butler, N., et al.: 2009, ApJ 702, 226

Modjaz, M., Kewley, L., Bloom, J.S., Filippenko, A.V., Perley, D., Silverman, J.M.: 2011, ApJ 731, L4

Moustakas, J., Kennicutt, Jr., R.C., Tremonti, C.A., Dale, D.A., Smith, J., Calzetti, D.: 2010, ApJS 190, 233

Nakar, E., Sari, R.: 2010, ApJ 725, 904

Neill, J.D., Sullivan, M., Gal-Yam, A., et al.: 2011, ApJ 727, 15

Nomoto, K., Tominaga, N., Umeda, H., Maeda, K., Ohkubo, T., Deng, J.: 2006, Nucl. Phys. A 777, 424

Nomoto, K., Tanaka, M., Tominaga, N., Maeda, K.: 2010, New A Rev. 54, 191

O'Connor, E., Ott, C.D.: 2011, ApJ 730, 70

Ofek, E.O., Cameron, P.B., Kasliwal, M.M., et al.: 2007, ApJ 659, L13

Osterbrock, D.E.: 1989, *Astrophysics of Gaseous Nebulae and Active Galaxies*, University Science Books, Mill Vallery, CA

Pagel, B.E.J., Edmunds, M.G., Blackwell, D.E., Chun, M.S., Smith, G.: 1979, MNRAS 189, 95

Paragi, Z., Taylor, G.B., Kouveliotou, C., et al.: 2010, Nature 463, 516

Patat, F., et al.: 2001, ApJ 555, 900

Perna, R., Loeb, A.: 1998, ApJ 509, L85

Pettini, M., Pagel, B.E.J.: 2004, MNRAS 348, L59

Pian, E., et al.: 2006, Nature 442, 1011

Pignata, G., Maza, J., Antezana, R., et al.: 2009, in: G. Giobbi et al. (eds.), *Probing Stellar Populations out to the Distant Universe: CEFALU 2008*, AIPC 1111, p. 551

Pignata, G., Stritzinger, M., Soderberg, A., et al.: 2011, ApJ 728, 14

Podsiadlowski, P., Ivanova, N., Justham, S., Rappaport, S.: 2010, MNRAS 406, 840

Podsiadlowski, P., Langer, N., Poelarends, A.J.T., Rappaport, S., Heger, A., Pfahl, E.: 2004, ApJ 612, 1044

Prantzos, N., Boissier, S.: 2003, A&A 406, 259

Prieto, J.L., Stanek, K.Z., Beacom, J.F.: 2008, ApJ 673, 999

Prochaska, J.X., Bloom, J.S., Chen, H.-W., et al.: 2004, ApJ 611, 200

Pruet, J., Hoffman, R.D., Woosley, S.E., Janka, H., Buras, R.: 2006, ApJ 644, 1028

Quimby, R.M., Kulkarni, S.R., Kasliwal, M.M., et al.: 2011, Nature, accepted, astro-ph/0910.0059

Rabinak, I., Waxman, E.: 2011, ApJ 728, 63

Rau, A., Kulkarni, S.R., Law, N.M., et al.: 2009, PASP 121, 1334

Salvaterra, R., Della Valle, M., Campana, S., et al.: 2009, Nature 461, 1258

Sauer, D.N., Mazzali, P.A., Deng, J., Valenti, S., Nomoto, K., Filippenko, A.V.: 2006, MNRAS 369, 1939

Scheck, L., Kifonidis, K., Janka, H.-T., Müller, E.: 2006, ApJ 457, 963

Schlegel, E.M., Kirshner, R.P.: 1989, AJ 98, 577

Simón-Díaz, S., Stasińska, G.: 2011, A&A 526, A48

Smartt, S.J.: 2009, ARA&A 47, 63

Smartt, S.J., Eldridge, J.J., Crockett, R.M., Maund, J.R.: 2009, MNRAS 395, 1409

Smith, N., Owocki, S.P.: 2006, ApJ 645, L45

Smith, N., Li, W., Foley, R.J., et al.: 2007, ApJ 666, 1116

Smith, N., Li, W., Filippenko, A.V., Chornock, R.: 2011, MNRAS 412, 1522

Soderberg, A.M., Berger, E., Page, K.L., et al.: 2008, Nature 453, 469

Soderberg, A.M., Brunthaler, A., Nakar, E., Chevalier, R.A., Bietenholz, M.F.: 2010a, ApJ 725, 922

Soderberg, A.M., Chakraborti, S., Pignata, G., et al.: 2010b, Nature 463, 513

Soderberg, A.M., Kulkarni, S.R., Fox, D.B., et al.: 2005, ApJ 627, 877

Soderberg, A.M., Kulkarni, S.R., Nakar, E., et al.: 2006a, Nature 442, 1014

Soderberg, A.M., Nakar, E., Berger, E., Kulkarni, S.R.: 2006b, ApJ 638, 930

Sollerman, J., Östlin, G., Fynbo, J.P.U., Hjorth, J., Fruchter, A., Pedersen, K.: 2005, New A 11, 103

Sollerman, J., et al.: 2006, A&A 454, 503

Spruit, H.C.: 2002, A&A 381, 923

Stanek, K.Z., et al.: 2003, ApJ 591, L17

Stanek, K.Z., et al.: 2006, Acta Astron. 56, 333

Starling, R.L.C., Wiersema, K., Levan, A.J., et al.: 2011, MNRAS 411, 2792

Stoll, R., Prieto, J.L., Stanek, K.Z., et al.: 2011, ApJ 730, 34

Sullivan, M., Conley, A., Howell, D.A., et al.: 2010, MNRAS 406, 782

Svensson, K.M., Levan, A.J., Tanvir, N.R., Fruchter, A.S., Strolger, L.: 2010, MNRAS 405, 57

Tanaka, M., Tominaga, N., Nomoto, K., et al.: 2009a, ApJ 692, 1131
Tanaka, M., Yamanaka, M., Maeda, K., et al.: 2009b, ApJ 700, 1680
Tanvir, N.R., Fox, D.B., Levan, A.J., et al.: 2009, Nature 461, 1254
Taubenberger, S., Valenti, S., Benetti, S., et al.: 2009, MNRAS 397, 677
Tonry, J., Pan-STARRS team: 2005, BAAS 37, 1363
Tremonti, C.A., Heckman, T.M., Kauffmann, G., et al.: 2004, ApJ 613, 898
Usov, V.V.: 1992, Nature 357, 472
van Eerten, H., Zhang, W., MacFadyen, A.: 2010, ApJ 722, 235
Vink, J.S., de Koter, A.: 2005, A&A 442, 587
Wang, C., Lai, D., Han, J.L.: 2006, ApJ 639, 1007
Waxman, E.: 2004, ApJ 606, 988
Waxman, E., Mészáros, P., Campana, S.: 2007, ApJ 667, 351
Wofford, A.: 2009, MNRAS 395, 1043
Wolf, C., Podsiadlowski, P.: 2007, MNRAS 375, 1049
Woosley, S.E.: 1993, ApJ 405, 273
Woosley, S.E., Bloom, J.S.: 2006, ARA&A 44, 507
Woosley, S.E., Heger, A.: 2006, ApJ 637, 914
Woosley, S.E., Langer, N., Weaver, T.A.: 1993, ApJ 411, 823
Woosley, S.E., Heger, A., Weaver, T.A.: 2002, Reviews of Modern Physics 74, 1015
Yoon, S.-C., Langer, N.: 2005, A&A 443, 643
Young, D.R., Smartt, S.J., Valenti, S., et al.: 2010, A&A 512, A70
Zeh, A., Klose, S., Hartmann, D.H.: 2004, ApJ 609, 952
Zhang, B., Zhang, B., Liang, E., Gehrels, N., Burrows, D.N., Mészáros, P.: 2007, ApJ 655, L25
Zhang, W., Woosley, S.E., Heger, A.: 2004, ApJ 608, 365

*Doctoral Thesis Award Lecture 2010*

# Accretion, jets and winds: High-energy emission from young stellar objects[1]

Hans Moritz Günther

Harvard-Smithsonian Center for Astrophysics
60 Garden Street
Cambridge, MA 02138, USA
hguenther@cfa.harvard.edu

**Abstract**

*This article summarizes the processes of high-energy emission in young stellar objects. Stars of spectral type A and B are called Herbig Ae/Be (HAeBe) stars in this stage, all later spectral types are termed classical T Tauri stars (CTTS). Both types are studied by high-resolution X-ray and UV spectroscopy and modeling. Three mechanisms contribute to the high-energy emission from CTTS: 1) CTTS have active coronae similar to main-sequence stars, 2) the accreted material passes through an accretion shock at the stellar surface, which heats it to a few MK, and 3) some CTTS drive powerful outflows. Shocks within these jets can heat the plasma to X-ray emitting temperatures. Coronae are already well characterized in the literature; for the latter two scenarios models are shown. The magnetic field suppresses motion perpendicular to the field lines in the accretion shock, thus justifying a 1D geometry. The radiative loss is calculated as optically thin emission. A mixture of shocked and coronal gas is fitted to X-ray observations of accreting CTTS. Specifically, the model explains the peculiar line-ratios in the He-like triplets of* Ne IX *and* O VII. *All stars require only small mass accretion rates to power the X-ray emission. In contrast, the HAeBe HD 163296 has line ratios similar to coronal sources, indicating that neither a high density nor a strong UV-field is present in the region of the X-ray emission. This could be caused by a shock in its jet. Similar emission is found in the deeply absorbed CTTS DG Tau. Shock velocities between 400 and 500 km $^{-1}$ are required to explain the observed spectrum.*

## 1 Introduction

Stars and planetary systems form by gravitational collapse of large molecular clouds. Mass infall leads to the formation of a proto-star, which is deeply embedded in an

[1] This article has already appeared in Astron. Nachr./AN 332, no. 5 (2011).

*Reviews in Modern Astronomy 23: Zoomimg in: The Cosmos at High Resolution.* First Edition.
Edited by Regina von Berlepsch.

envelope of gas and dust. Due to the conservation of angular momentum matter from this envelope does not accrete directly onto the central star, but forms a proto-stellar disk around it. The envelope depletes and the central star becomes visible as stellar evolution proceeds. In this stage the stars are called classical T Tauri stars (CTTS), if they are of low-mass ($< 3M_{\odot}$, spectral type F or later) and Herbig Ae/Be stars (HAeBe), if they are of spectral type A or B. CTTS are cool stars with a convective photosphere, thus they generate magnetic fields in solar-like $\alpha - \Omega$ dynamos. Their field lines can reach out a few stellar radii and couple to the disk at the co-rotation radius, because the energetic radiation from the central star ionizes the upper layers of the disk. Thus, the accreting matter is forced to follow the magnetic field lines (Bertout et al. 1988; Koenigl 1991; Shu et al. 1994; Uchida & Shibata 1984). It is accelerated along the accretion funnel and hits the stellar surface at free fall velocity, so a strong shock develops. This magnetically-funneled accretion scenario explains the wide and unusual emission line profiles observed in CTTS for H$\alpha$ and other Balmer lines (Muzerolle et al. 1998a,b). In fact, CTTS are often defined as young, low-mass stars with an H$\alpha$ equivalent width (EW) $> 10$ Å. Once the disk mass drops and the accretion ceases, the width of the H$\alpha$ line decreases and the stars are classified as weak-line T Tauri stars (WTTS). Eventually, the disk mass is completely absorbed in planets, accreted by the star or driven out of the system by stellar winds and photoevaporation, although transitional disks can exist in WTTS for some time (Padgett et al. 2006).

A wide range of observational evidence supports the magnetically-funneled accretion scenario for CTTS. Eisner et al. (2006) resolved the inner hole in the disk of TW Hya, the closest CTTS, with radio-interferometry. The accretion funnels are not resolved, but there is a very good agreement between observed and modeled line profiles for H$\alpha$ and other hydrogen emission lines, where emission from infalling material causes the blue-shifted wings of those lines (Fang et al. 2009; Muzerolle et al. 1998a,b). The energy from the accretion shock heats an area of the surrounding photosphere of the star to temperatures of the order of 20 000 K. In turn, this in turn emits a hot black-body continuum, which veils the photospheric emission lines (Calvet & Gullbring 1998; Gullbring et al. 2000). The strength of the veiling in the UV and the optical wavelength range is one measure of the accretion rate. Some CTTS with fast rotation have been Doppler-imaged and hot spots on the surface can be seen (Strassmeier et al. 2005; Unruh et al. 1998). Zeeman-Doppler imaging reveals that the Ca infrared triplet originates in region of strong magnetic fields, which, in the magnetically funneled accretion model, are the footpoints of the accretion funnels (Donati et al. 2007, 2008).

CTTS and WTTS are also copious emitters of X-rays and UV radiation; Feigelson & Montmerle (1999) review the knowledge before X-ray grating spectroscopy with *Chandra* and *XMM-Newton*. In this article I will discuss coronal activity, which is common to both CTTS and WTTS (section 2.1) and then summarize some observational characteristics in X-ray and UV spectroscopy, which set CTTS apart from main-sequence (MS) stars (section 2.2) to discuss accretion (section 2.3), winds (section 2.4) and stellar jets (section 2.5) in the following. Section 3 compares the properties of CTTS with HAeBes. I end with a short summary in section 4.

# 2 Classical T Tauri Stars

X-ray emission from T Tauri stars (TTS) was discovered with the *Einstein* satellite (Feigelson & Decampli 1981). Starting with *ROSAT*, X-ray surveys were used as a tool to identify young stars in star forming regions, e.g. in the Taurus molecular cloud (Neuhäuser et al. 1995), the Chameleon star forming region (Alcala et al. 1995) and the Orion star forming cluster (Alcala et al. 1996), because young stars have a high level of X-ray activity. This continues today, mostly with *Chandra* because of its high imaging quality. One especially successful example is the Chandra Orion Ultradeep Project (COUP) with an exposure time of nearly 800 ks (Getman et al. 2005). However, the spectral information from these surveys remains poor by today's standards. Only the closest and brightest CTTS can be observed with high-resolution X-ray grating spectroscopy in a reasonable exposure time.

## 2.1 Coronal activity

For a long time the X-ray emission of CTTS was thought of as a scaled-up version of solar activity, because surveys of star forming regions show variability with a fast rise phase and a longer exponential decay in all types of TTS. CTTS and WTTS essentially share the same variability characteristics such as flare duration or the distribution of the flare luminosity, indicating that the same mechanism is responsible for the X-ray emission (e.g. Getman et al. 2008a,b; Stelzer et al. 2007). The majority of flares is small, and as their energy increases, occurrence becomes rarer. In general, the number of flares $\mathrm{d}N$ in the energy interval $\mathrm{d}E$ follows a power-law with $\mathrm{d}N/\mathrm{d}E \propto E^{-\alpha}$. The value of the exponent $\alpha$ in, e.g., the Taurus molecular cloud is $2.5 \pm 0.5$ (Stelzer et al. 2007), compatible with the solar value within the errors. This supports the idea that the majority of the X-ray emission on CTTS is coronal, just as the emission in WTTS or stars on the MS (see Güdel & Nazé 2009, for a review of stellar X-ray emission).

However, WTTS are on average twice as bright as CTTS (Preibisch et al. 2005; Stelzer & Neuhäuser 2001; Telleschi et al. 2007a), so there must be some difference in the X-ray generation.

## 2.2 Observational peculiarities

The following two subsections describe spectral properties, discovered in high-resolution X-ray spectra, which set CTTS apart from WTTS or MS stars.

### 2.2.1 He-like triplets

X-ray grating spectra show the He-like ions of C, O, Ne, Mg and Si. These ions emit a triplet of lines, which consists of a resonance ($r$), an intercombination ($i$), and a forbidden line ($f$) (Gabriel & Jordan 1969; Porquet et al. 2001). The flux ratios of those lines are temperature and density-sensitive. The $R$- and $G$-ratios ($R = f/i$ and $G = (f+i)/r$) are commonly used to describe the triplet; for high electron densities $n_\mathrm{e}$ or strong UV photon fields the $R$-ratio falls below its low-density limit, because electrons are collisionally or radiatively excited from the upper level of the $f$ to the $i$ line, but the UV field of late-type CTTS is too weak to influence the $R$-ratio. The $G$-ratio is a temperature diagnostic of the emitting plasma.

TW Hya was the first CTTS to be observed by *Chandra*/HETGS (for 50 ks). It showed an exceptional line ratio in the He-like triplets of O VII and Ne IX, which Kastner et al. (2002) interpret as a signature of high density. TW Hya has since been observed for 30 ks with *XMM-Newton* (Stelzer & Schmitt 2004), 120 ks with *Chandra*/LETGS (Raassen 2009) and for 500 ks again with *Chandra*/HETGS (Brickhouse et al. 2010). Together, this makes TW Hya one of the targets with the longest exposure times in the history of X-ray spectroscopy. Other CTTS were observed with grating spectroscopy, too: BP Tau (Schmitt et al. 2005), V4046 Sgr (Günther et al. 2006), RU Lup (Robrade & Schmitt 2007), MP Mus (Argiroffi et al. 2007) and Hen 3-600 (Huenemoerder et al. 2007) and they all show the same indication for high-densities that were first seen in TW Hya. Figures 1 and 5 (middle and right column) show examples for the He-like triplets in CTTS. TW Hya has the most extreme $f/i$ ratio observed so far, but also more typical CTTS like V4046 Sgr differ markedly from the line ratio found in typical WTTS such as TWA 4 (Kastner et al. 2004) and TWA 5 (Argiroffi et al. 2005) and MS stars (Ness et al. 2004), which are compatible with the low-density limit of the $f/i$ line ratio according to the CHIANTI 5.1 database (Dere et al. 1998; Landi et al. 2006).

There are exceptions to the high-densities in CTTS – in the more massive, eponymous T Tau itself the O VII triplet is consistent with the coronal limit (Güdel et al. 2007).

The O VII triplet is free of blends and its $R$-ratio is sensitive to densities of $10^{11} - 10^{12}$ cm$^{-3}$, but *Chandra*/ACIS data has a very low effective area at these wavelength. The Ne IX triplet is often blended with iron lines, particularly Fe XIX and Fe XX. Strong sources provide sufficient signal in *Chandra*/HEG, where most of these blends are resolved, but for lower fluxes or observations with *XMM-Newton* these blends are difficult to remove, if the temperature structure is not very well known. Fortunately, most CTTS show an enhanced Ne abundance and a reduced Fe abundance, alleviating this problem to some extent (section 2.3).

The third interesting triplet is the Mg XI He-like triplet with lines at 9.17 Å, 9.23 Å and 9.31 Å. It can be blended by the higher members of the Ne X Lyman series. This problem is somewhat more serious in CTTS than in other objects, because of the enhanced Ne abundance often found in CTTS. For a few active MS stars Testa et al. (2004) extrapolated the strength of the higher order Ne X lines from the lower ones, which can be easily measured. They then fit the Mg XI triplet, taking the line blends into account. Brickhouse et al. (2010) analyzed their *Chandra*/HETGS observations and found that the blending is not important in this case.

For densities $< 10^{15}$ cm$^{-3}$ the collision time is much longer than the radiative decay time in the Lyman series, thus all excited states decay radiatively and the relative strength of the unabsorbed Ne X lines depends only on the temperature and the collision strength. I calculated the Lyman series up to Ly$\epsilon$ for a grid of temperatures with the CHIANTI 5.1 code (Dere et al. 1998; Landi et al. 2006) and extrapolated that series with a geometric function. In the observed spectra I fit as many members of the Lyman series as possible with the CORA tool (Ness & Wichmann 2002), which employs a maximum likelihood method, using a modified Lorentzian line profile with $\beta = 2.5$ and keeping the line width fixed at the instrumental width of 0.02 Å for the *Chandra*/MEG. The fits to the Mg XI triplet again use CORA. Here, the maximum likelihood is calculated for the sum of the extrapolated Ne X lines, Mg XI triplet and a constant, which represents a combination of background, continuum emission and unresolved lines. The likelihood is minimized with Powell's method by adjusting the individual flux of the triplet members, the constant and a wavelength offset. Both CTTS with archival *Chandra*/HETGS observations, TW Hya and V4046 Sgr, have very low fitted absorbing column densities for the accretion shock in the global fit (see section 2.3.2 and 2.3.3), thus the differential transmission between 9.1 Å and 12.13 Å is negligible. Table 1 gives the line fluxes found and figure 1 shows the fit. The constant is the strongest contributor to the total flux. Some of the Ne lines

**Table 1:** Ne IX Ly series and Mg XI triplet fluxes (errors are $1\sigma$ confidence intervals)

| line | $\lambda$ | TW Hya – Chandra | | V4046 Sgr – Chandra | |
|---|---|---|---|---|---|
| | Å | cts | cts s$^{-1}$ cm$^{-2}$ | cts | cts s$^{-1}$ cm$^{-2}$ |
| Ne X Ly$\alpha$ | 12.13 | $1345 \pm 40$ | $(76. \pm 2.0) \times 10^{-6}$ | $274 \pm 17$ | $(51. \pm 3.0) \times 10^{-6}$ |
| Ne X Ly$\beta$ | 10.24 | $247 \pm 17$ | $(8.3 \pm 0.6) \times 10^{-6}$ | $57 \pm 8$ | $(5.6 \pm 0.6) \times 10^{-6}$ |
| Ne X Ly$\gamma$ | 9.71 | $73 \pm 10$ | $(2.4 \pm 0.3) \times 10^{-6}$ | $25 \pm 6$ | $(2.4 \pm 0.6) \times 10^{-6}$ |
| Ne X Ly$\delta$ | 9.48 | $62 \pm 11$ | $(1.5 \pm 0.3) \times 10^{-6}$ | $25 \pm 8$ | $(2.0 \pm 0.7) \times 10^{-6}$ |
| Mg XI r | 9.17 | $43. \pm 8.6$ | $(8.7 \pm 1.7) \times 10^{-7}$ | $10. \pm 3.8$ | $(7.2 \pm 2.7) \times 10^{-7}$ |
| Mg XI i | 9.23 | $16. \pm 6.4$ | $(3.3 \pm 1.3) \times 10^{-7}$ | $0.1 \pm 1.7$ | $(0.07 \pm 1.2) \times 10^{-7}$ |
| Mg XI f | 9.31 | $16. \pm 6.3$ | $(3.2 \pm 1.3) \times 10^{-7}$ | $5.5 \pm 2.9$ | $(3.9 \pm 2.1) \times 10^{-7}$ |

are clearly overpredicted, e.g. at 9.36 Å for TW Hya, others match the observations within the (large) Poisson errors. In any case, the Mg XI lines are not significantly blended by the Ne Lyman series. For TW Hya, this confirms the finding of Brickhouse et al. (2010).

The low-density limit of the $f/i$ ratio for Mg XI is 2.5 for densities $< 10^{12}$ cm$^{-3}$; the ratio drops below 0.2 for densities $> 10^{14}$ cm$^{-3}$. Given the large statistical error on the line fluxes, neither the high-density nor the low-density limit can be excluded for TW Hya or V4046 Sgr, but at least for V4046 Sgr a low-density scenario is more likely as the $i$ line is absent.

### 2.2.2 Ionization balance

To constrain the temperature and the origin of the plasma at low temperatures, which produces the anomalous $R$-ratios in the He-like triplets a line based method is preferable. Ideally, one would use the $G$-ratio of He-like triplets. Unfortunately, the temperature dependence of the $G$-ratio is weak and thus requires a high signal-to-noise ratio, which is only available for TW Hya. Brickhouse et al. (2010) use the higher order resonance lines of O VII for this purpose.

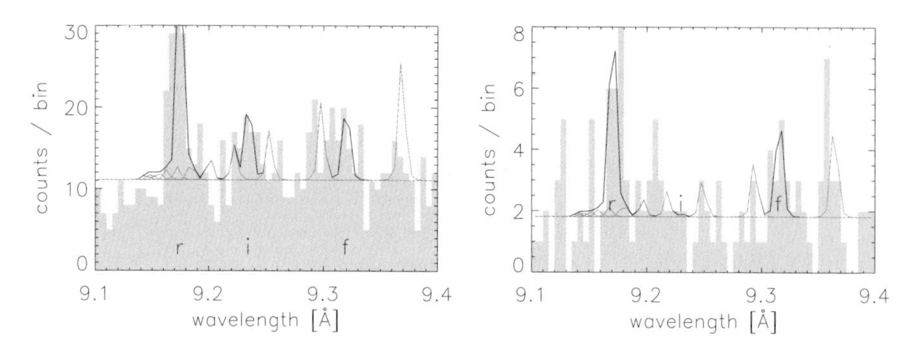


**Figure 1:** Mg XI triplet of TW Hya (top) and V4046 Sgr (bottom). The gray histogram shows the observed counts, the lines shows the best-fit model. Extrapolated lines of the Lyman series a plotted in red/gray (color in electronic version only).

Alternatively, the balance of different ionization stages can be used as a temperature diagnostic. Robrade & Schmitt (2007) and Güdel & Telleschi (2007) constructed a diagram, which shows the ratio of the O VIII to the O VII lines and Günther & Schmitt (2009) added more stars (figure 2). The flux ratios in this diagram are based on unabsorbed luminosities, where the flux has been corrected for the extinction, which was found in a global fit, using the absorption cross-sections from Balucinska-Church & McCammon (1992). This step is necessary, because the extinction varies with wavelength, so different correction factors apply for O VIII at 18.97 Å and the O VII triplet around 21.8 Å. All CTTS are found at the bottom right of the diagram at comparatively high total luminosities. This is a selection bias, because X-ray grating spectroscopy can only be performed for the brightest CTTS in reasonable exposure times. The distance to the TW Hya association is about 57 pc, the next closest regions such as the Taurus-Auriga cloud or the $\rho$ Oph cloud are located at a distance of 130 pc. In contrast, the MS sample from Ness et al. (2004) contains many closer stars and thus reaches down to lower luminosities. Still, the diagram shows an excess of O VII emission in CTTS compared to MS stars of similar luminosity. The peak emission temperature for the O VII lines is around 1-2 MK.

Following the same line of thought, the ratio of O VII and O VI emission will probe the cooler end of the plasma distribution. As no O VI emission lines are found in the X-ray range, they have to be taken from non-simultaneous observations with the *Far Ultraviolet Spectroscopic Explorer (FUSE)*, which covers the wavelength range of the O VI doublet at 1032 and 1038 Å. The O VI fluxes for MS stars are taken from Redfield et al. (2002) and Dupree et al. (2005b), those for CTTS from Günther & Schmitt (2008). The observed *FUSE* fluxes are dereddened with the reddening law of Cardelli et al. (1989). The X-ray and UV observations are taken up to a few years apart, so the line ratios could be influenced by variability. To asses this figure 3 shows two flux ratios for those MS stars which have been observed multiple times in X-rays (Ness et al. 2004). Variability dominates over the formal measurement uncertainty, which is around 10% for most stars. Again, the CTTS are separated from MS stars in this diagram, although less significant than in figure 2. This time, they they lie above

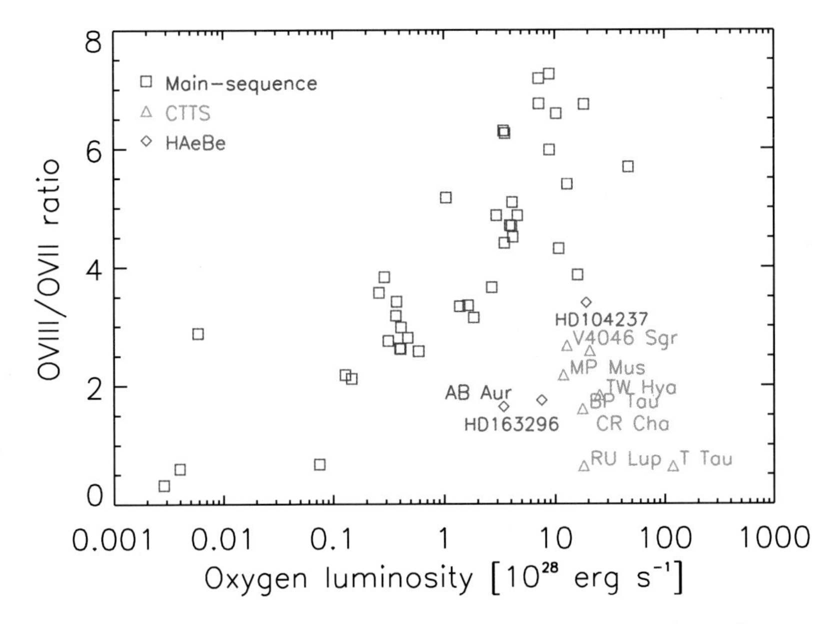


**Figure 2:** Flux ratio of O VIII Ly$\alpha$ and O VII He-like triplet, plotted over the sum of those emission lines. For comparison purposes MS stars from the Ness et al. (2004) sample are shown. Color in electronic version only.

the MS stars, indicating that they have extra O VII emission compared to MS stars with similar luminosity in the oxygen lines. So, CTTS are more luminous in O VII compared to both hotter (O VIII) and cooler (O VI) plasma than MS stars. Given that CTTS and WTTS seems to have very similar coronal emission, CTTS even appear to be hotter, it is unlikely, that this is caused by CTTS being *under*luminous in O VIII and O VI. A much better explanation for this excess of soft X-ray flux is an extra emission component at 1-2 MK (the formation temperature of the O VII He-like triplet) that is present in CTTS, but not in MS stars.

## 2.3 Accretion

One promising candidate for this extra emission is the accretion shock on the star. Here, an overview of accretion shock models is presented in section 2.3.1, then a sample of CTTS it described, where high resolution X-ray spectra indicate the presence of an accretion component (section 2.3.2). Section 2.3.3 shows model fits to this dataset. A comparison between our accretion shock model (Günther et al. 2007) and similar models in the literature is given in section 2.3.4.

### 2.3.1 Accretion models

The inner hole in the accretion disk is at least a few stellar radii wide, and thus the accretion impacts on the star close to free-fall velocity $v_0$. For a star with mass $M_*$

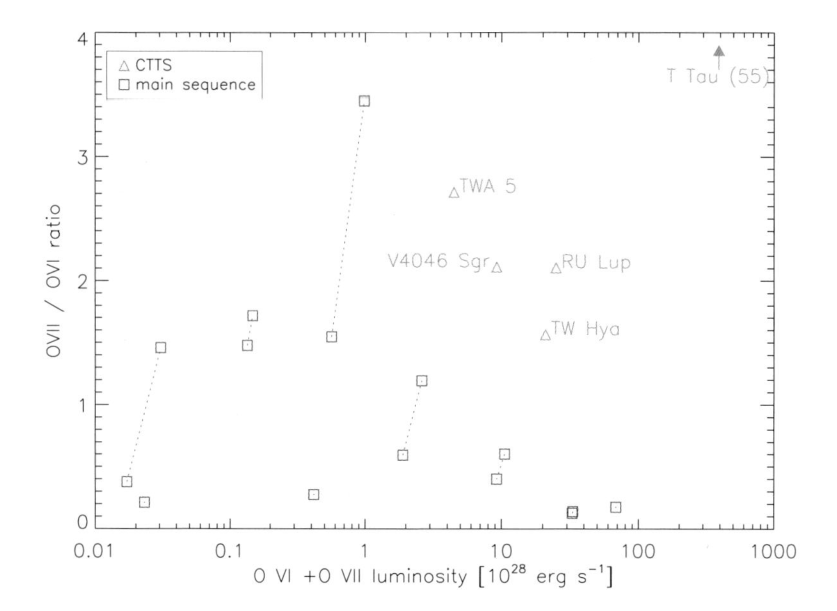


**Figure 3:** Flux ratio of O VII He-like triplet to the O VI doublet at 1032 and 1038 Å, plotted over the sum of those emission lines. For comparison purposes MS stars are shown. If two X-ray observations are available in Ness et al. (2004), the figure shows two ratios calculated from those fluxes to give an estimate of variability. Color in electronic version only.

and radius $R_*$ this is:

$$v_0 = \sqrt{\frac{2GM_*}{R_*}} = 600\sqrt{\frac{M_*}{M_\odot}}\sqrt{\frac{R_\odot}{R_*}}\frac{\text{km}}{\text{s}} \,. \tag{1}$$

Typically, CTTS have masses comparable to the Sun and radii between $R_* = 1.5\ R_\odot$ and $R_* = 4\ R_\odot$ (Muzerolle et al. 2003), because they have not yet finished their contraction. This gives infall velocities in the range 300-500 km s$^{-1}$. Not all kinetic energy is converted into thermal energy, because the total momentum needs to be conserved. In strong shocks the post-shock velocity $v_1$ is given by

$$v_1 = \frac{1}{4}v_0 \,,$$

where $v_0$ is the pre-shock velocity. Particle number conservation then demands the following relation for the pre-shock particle number density $n_0$ and the post-shock number density $n_1$:

$$n_1 = 4n_0 \,.$$

The pre-shock pressure is completely dominated by the ram pressure of the infalling material, which drops dramatically across the shock front. In a quasi-equilibrium state, this is compensated by thermal pressure on the post-shock side and this leads

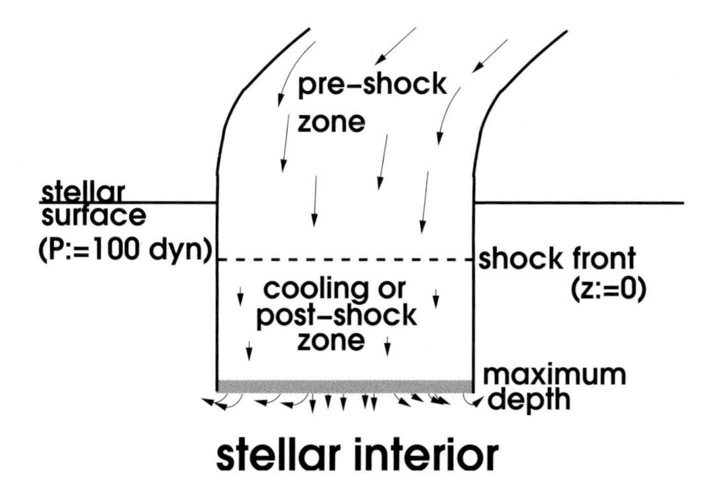


**Figure 4:** Sketch of the accretion shock geometry. This sketch is not to scale. Matter falls in from the accretion streams and close to the stellar surface an accretion shock develops. In the post-shock zone the plasma cools via radiation. The stellar surface is arbitrarily represented at pressure $p = 100$ dyn.

to an expression for the post-shock temperature $T_1$:

$$T_1 = \frac{3}{16}\frac{\mu m_{\rm H}}{k} v_0^2,$$

where $\mu$ is the particle mass number, averaged over ions and electrons, and $m_{\rm H}$ is the mass of a hydrogen atom. $k$ is the Boltzmann constant. For infall velocities close to the free-fall velocity this equation predicts extra emission at 1-2 MK. Strictly speaking, mostly the ions are heated in the strong shock, and is takes several mean-free path lengths to transfer heat to the electron gas. However, non-equilibrium ionization and temperature have a negligible effect on the final spectrum (Günther et al. 2007). After plasma passes through the accretion shock, it cools down via radiation. If the magnetic field of the accretion funnels is strong enough, all motion perpendicular to the field lines is suppressed and the problem can be simulated in 1D (sketch in figure 4). Calvet & Gullbring (1998) presented a grid of such models. They find that the radiation from the accretion shock heats the photosphere below the shock to about 20 000 K. Their model fits the hydrogen continuum in the UV. Lamzin (1998) calculated the expected X-ray flux from the post-shock cooling zone. These simulations were updated by Günther et al. (2007) to resolve individual lines, especially in the He-like triplets. CTTS definitely show signs of coronal activity (section 2.1), so observed X-ray spectra have to be fitted with a combination of models which account for accretion and coronal contribution. Drake (2005) suggested that the shock might be buried in the photosphere, so that most of the X-ray radiation is absorbed by the surrounding photospheric gas. However, the fact that we observe high densities shows that at least some radiation escapes. While the photosphere would be a gray absorber, which affects all lines of the triplets simultaneously, emission from deeper layers of the post-shock cooling zone could experience line absorption. There are some hints to optical depth effects in the X-ray spectra (Argiroffi et al. 2009), but for most stars the $G$-ratios are close to the expected values. This excludes large optical depth effects, because the optical depth of the resonance line should be several orders of magnitude larger than in the forbidden or the intercombination line.

I calculated a grid of shock model with the code of Günther et al. (2007) for infall velocities between 300 km $s^{-1}$ and 1000 km $s^{-1}$ in steps of 100 km $s^{-1}$ and infall densities between $10^{10}$ $cm^{-3}$ and $10^{14}$ $cm^{-3}$ in five logarithmic steps. The emission for all ions of C, N, O, Ne, Mg, Si, S and Fe is calculated separately from the continuum emission and the remaining elements, so that these abundances can be fit individually. However, large changes in the abundance would significantly alter the cooling function and thus the thermal structure of the shock. Simulations with different abundances found in the literature for CTTS were performed, and in this range no significant change was observed. The model grid is provided as a table model. This table model is available at http://www.hs.uni-hamburg.de/DE/Ins/Per/Guenther/shock_model/.

This basic shock interpretation has been extended by Brickhouse et al. (2010), who found a lower density and less absorption for O VII than for Ne IX in TW Hya, although the O VII triplet is formed at lower temperatures and therefore deeper in the post-shock cooling zone. They conclude, that most of the O VII emission does not originate in the accretion shock itself, but in plasma that has been heated by the shocked material. This situation can no longer be described in 1D. Simulations of the accretion shock region in more dimensions have been performed by Orlando et al. (2010). They show the flow to be well constrained for strong magnetic fields, but weaker fields cannot hold the plasma and mass flows sideways, thus changing the temperature and density profiles in the shock.

There is no a-priory reason to expect the shocks to be stable over time. Koldoba et al. (2008) and Sacco et al. (2008) both looked at this issue and predicted sub-second oscillations of the shock front, albeit at very different time scales, mostly due to the vastly different densities they assume in the shock. Observationally, this has not been found, neither in X-rays (Drake et al. 2009) nor in the optical response (Günther et al. 2010). This does not rule out fast oscillations of the shock front, it just requires the accretion along different field lines to oscillate independently with slightly different time constants.

### 2.3.2 Observations and fitting

Table 2 gives a list of X-ray observations of CTTS, where previous authors have found significant deviations from the low-density limit in the He-like triplets.

For observations split over several orbits, the day of the first exposure and the summed exposure time is given in the table. The *XMM-Newton* data was retrieved from the archive and processed with the standard *XMM-Newton* Science Analysis System (SAS) software, version 10.0 (Gabriel et al. 2004) with all standard selection criteria to filter out background contamination. *Chandra* data was retrieved from the archive and processed with CIAO 4.3 (Fruscione et al. 2006) to obtain CCD spectra; the grating spectra were taken directly from TGCat [2]. Positive and negative first-

[2] http://tgcat.mit.edu

**Table 2:** Observations

| star | satellite | exp. time [ks] | obs date | ObsID |
|---|---|---|---|---|
| TW Hya | XMM-Newton | 30 | 2001-07-09 | 0112880201 |
| TW Hya | Chandra | 490 | 2007-02-15 | 7435, 7436, 7437, 7438 |
| BP Tau | XMM-Newton | 130 | 2004-08-15 | 0200370101 |
| MP Mus | XMM-Newton | 110 | 2006-08-19 | 0406030101 |
| V4046 Sgr | Chandra | 150 | 2006-08-06 | 5423, 6265 |

order spectra were merged, then spectra from different orbits were combined, but HEG and MEG are kept separate. All CCD spectra were binned to 15 counts per bin, but grating data only to 5 counts per bin, because the density information in the triplets, which often contain only very few counts, would be destroyed by a coarser binning.

### 2.3.3 A sample of CTTS

For small count numbers a $\chi^2$ statistic is no longer applicable, instead the fit was done using the C-statistic. Still, the CCD spectra with their higher count rates tend to dominate the statistic. For *XMM-Newton* data only the MOS1 was fitted and for the *Chandra* spectra only the ACIS zeroth order of one of the available exposures. In this way, the cool shock component, which is mostly constrained by the information from the gratings is effectively fitted, otherwise the optimization would prefer small improvements in the coronal components at the prize of systematic deviations in the low energy region because the hot components cause higher CCD count rates. The C-statistic takes the proper Poisson uncertainties in bins with small count number into account, but it does not provide a goodness-of-fit.

X-ray spectra can be used to fit relative abundances of metals, but it is very difficult to obtain absolute abundances. Thus, the abundance of oxygen is fixed at solar (Grevesse & Sauval 1998) for all models in this section, i.e. all abundances are given relative to oxygen. The abundances of C, N, Ne, Mg, Si and Fe are fitted independently. The abundance of S is coupled to Fe, all other abundances are fixed at solar values. Because *Chandra*/ACIS has low effective areas at longer wavelengths, where the lines of C and N are observed, I fixed the abundance of those two elements at solar values for all *Chandra* data sets and, additionally, the absorbing column density for TW Hya in the *Chandra* data was fixed to the value found in the *XMM-Newton* observation.

Fits to the CTTS sample were obtained with XSPEC 12.6 (Arnaud 1996), fitting four model components: Three emission components (one shock model and two optically thin thermal APEC models, which represent the corona) and one cold photoabsorption component. Figure 5 shows the fitted low-resolution spectra (left panels) and also the fits to the He-like ions Ne IX (middle panels) and O VII (right panels). The results are summarized in table 3.The panels show the contribution of shock and corona independently. The corona is responsible for the hot emission, which cannot

**Table 3:** Fit results (errors are statistical only and give 90% confidence intervals)

| | TW Hya *XMM-Newton* | TW Hya *Chandra* | BP Tau *XMM-Newton* | MP Mus *XMM-Newton* | V4046 Sgr *Chandra* |
|---|---|---|---|---|---|
| | absorbing column density | | | | |
| $N_H$ [$10^{20}$ cm$^{-2}$] | $5.2^{+0.6}_{-0.4}$ | =5.2 | $14 \pm 4$ | $2 \pm 1$ | $6 \pm 3$ |
| | abundances relative to O | | | | |
| C | $1.5 \pm 0.2$ | =1 | $2.2^{+2.4}_{-1.7}$ | $1.4 \pm 0.4$ | =1 |
| N | $3.0 \pm 0.3$ | =1 | $2.2 \pm 1.2$ | $1.6 \pm 0.4$ | =1 |
| Ne | $7.6^{+3.2}_{-2.6}$ | $3.2 \pm 0.2$ | $1.7 \pm 0.5$ | $1.8 \pm 0.2$ | $3.5 \pm 0.4$ |
| Mg | $1.0 \pm 0.3$ | $0.36 \pm 0.05$ | $0.3 \pm 0.3$ | $0.8 \pm 0.2$ | $0.5 \pm 0.1$ |
| Si | $0.8 \pm 0.3$ | $0.5 \pm 0.1$ | $0.2 \pm 0.2$ | $0.6 \pm 0.2$ | $0.6 \pm 0.1$ |
| Fe | $0.9 \pm 0.1$ | $0.42 \pm 0.03$ | $0.2 \pm 0.1$ | $0.5 \pm 0.1$ | $0.46 \pm 0.06$ |
| | coronal components | | | | |
| $kT_1$ [keV] | $0.69^{+0.04}_{-0.02}$ | $0.36 \pm 0.01$ | $0.7 \pm 0.1$ | $0.63 \pm 0.03$ | $0.73 \pm 0.04$ |
| $VEM_1$ [$10^{51}$cm$^{-3}$] | $7.2 \pm 1.2$ | $27 \pm 2$ | $43 \pm 2$ | $83 \pm 15$ | $28 \pm 5$ |
| $kT_2$ [keV] | $1.7^{+0.2}_{0.1}$ | $2.0 \pm 0.05$ | $2.2 \pm 0.2$ | $2.2 \pm 0.1$ | $2.2 \pm 0.2$ |
| $VEM_2$ [$10^{51}$cm$^{-3}$] | $19 \pm 1.5$ | $37 \pm 1$ | $100 \pm 10$ | $111 \pm 5$ | $38 \pm 3$ |
| | shock properties | | | | |
| $v_0$ [km s$^{-1}$] | $500 \pm 5$ | $504 \pm 3$ | $440^{+70}_{-25}$ | $510 \pm 10$ | $500^{+10}_{-50}$ |
| $\log(n_0)$ [cm$^{-3}$] | $11.9^a$ | $13.0^a$ | $12.9^a$ | $11.1 \pm 0.2$ | $11.2^a$ |
| $A$ [cm$^2$] | $6.5 \times 10^{19}$ | $4 \times 10^{18}$ | $4 \times 10^{20}$ | $4 \times 10^{20}$ | $3 \times 10^{20}$ |
| $\dot{M}$ [$M_\odot$ yr$^{-1}$] | $7 \times 10^{-11}$ | $6 \times 10^{-11}$ | $4 \times 10^{-9}$ | $6 \times 10^{-11}$ | $6 \times 10^{-11}$ |
| | X-ray mass accretion rates from literature | | | | |
| $\dot{M}$ [$M_\odot$ yr$^{-1}$] | $2 \times 10^{-11}$ (1) | | $9 \times 10^{-10}$ (2) | $2 \times 10^{-11}$ (3) | $3 \times 10^{-11}$ (4) |
| $\dot{M}$ [$M_\odot$ yr$^{-1}$] | $2 \times 10^{-10}$ (5) | | | $7.7 \times 10^{-11}$ (6) | |

(a) The model grid uses density steps of 1 in log-space. Here, the interpolation error between two models is much larger than the statistical uncertainty. (1) Stelzer & Schmitt (2004); (2) Schmitt et al. (2005); (3) Argiroffi et al. (2007); (4) Günther & Schmitt (2007); (5) Günther et al. (2007); (6) Sacco et al. (2008)

be explained by the accretion shock, because such high temperature would require infall velocities above the free-fall velocity. The He-like triplets are predominantly formed at lower temperatures in the accretion shock. This can be immediately seen from the strong $i$ line. The corona is in the low density limit and thus its $f/i$ ratio is high in contrast to the observations. Therefore, the $i$ line has to be formed in the shock, which automatically requires the shock to contribute most of the emission in the $r$ and $f$ lines as well.

Table 3 only shows the statistical uncertainties for the fit, not any systematic model uncertainties. The plasma temperatures e.g. just represent the average plasma properties. The small statistical uncertainties on the temperature do not mean that the plasma temperature distribution is bimodal with very narrow peaks. Fits which prescribe e.g. polynomial emission measure distributions are also possible.

The mass accretion rate $\dot{M}$ is related to the accretion spot size $A$, the infall velocity $v_0$ and the density $n_0$ as:

$$\dot{M} = \mu m_H n_0 v_0 A \,. \tag{2}$$

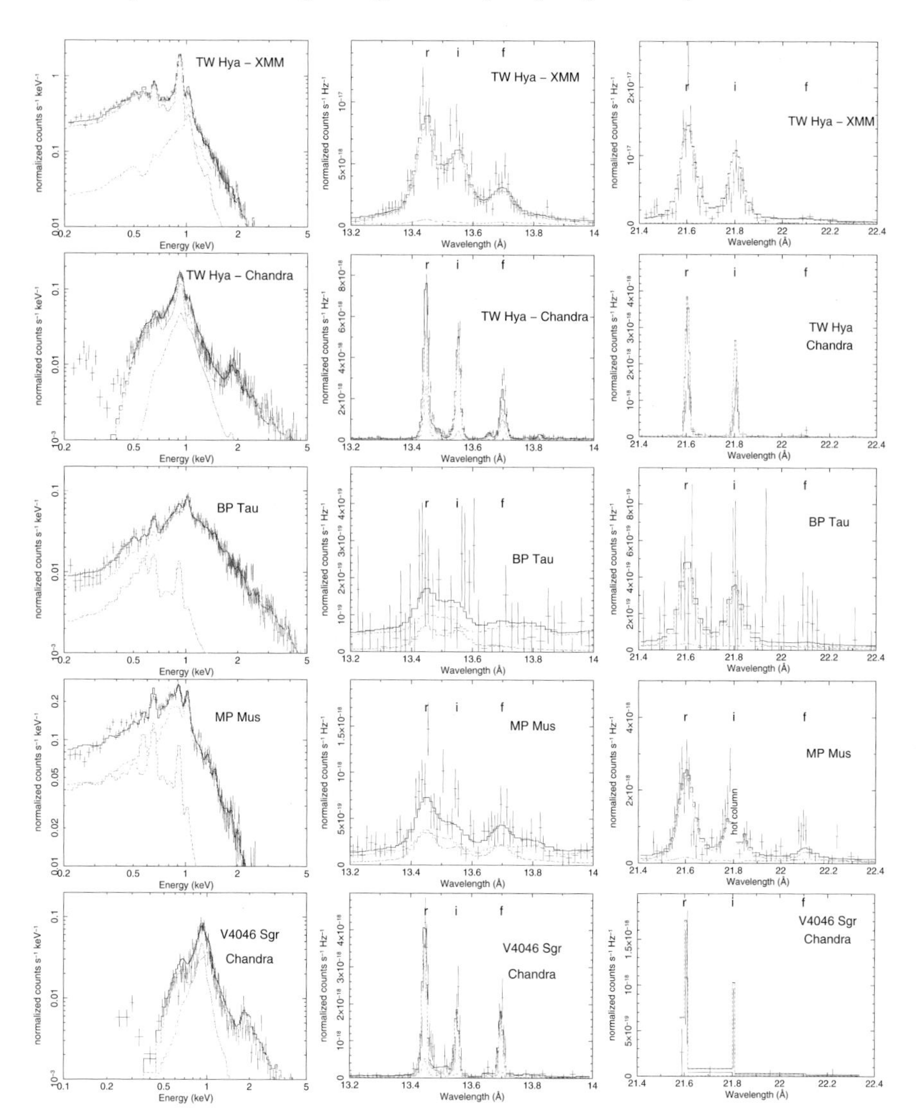


**Figure 5:** Spectra of CTTS with fitted model overplotted. The panels show spectra in CCD resolution (left) and grating data for the Ne IX (middle) and O VII (right) He-like triplets. Best-fit models are overplotted, the contribution of corona (red/gray dashed) and shock (red/gray) is shown individually. In the He-like triplets most emission comes from the shock. The wavelength of the $r$, $i$ and $f$ line is labeled. The second and fifth row show *Chandra* data, which has narrower lines in the grating spectra, but little effective area at lower wavelength in the CCD spectra.

For most stars the signal in the He-like ions is relatively weak and the statistical error on the density is large. The error in $n_0$ and $A$ is then correlated such that a small accretion spot with a high density or a large accretion spot with a low density are both possible, if they have the same mass accretion rate and thus the same total luminosity. So, while $n_0$ and $A$ are uncertain, $\dot{M}$ is still a well-determined quantity. Fitted mass accretion rates are given in table 3.

The sample was selected to contain only stars with high densities. The prime example is TW Hya, where Kastner et al. (2002) observed $10^{13}$ cm$^{-3}$, Stelzer & Schmitt (2004) the same, Raassen (2009) $10^{12}$ cm$^{-3}$ and Brickhouse et al. (2010) between $6 \times 10^{11}$ cm$^{-3}$ and $3 \times 10^{12}$ cm$^{-3}$ for O VII and Ne IX respectively. In BP Tau Schmitt et al. (2005) find $3 \times 10^{11}$ cm$^{-3}$, Günther et al. (2006) give $3 \times 10^{11} - 10^{12}$ cm$^{-3}$ for V4046 Sgr and Argiroffi et al. (2007) found $5 \times 10^{11}$ cm$^{-3}$ in MP Mus. This is largely consistent with the post-shock values expected from the pre-shock densities given in table 3.

If, however, the cool emission is formed only partly in a high-density accretion shock and partly in a low-density corona, the $f/i$ ratio in the He-like triplet can indicate medium densities. In this case, the $\dot{M}$ is overestimated. One further problem of the model chosen in this paragraph shows up in TW Hya. Brickhouse et al. (2010) found that the accretion shock is less absorbed than the corona in TW Hya and the absorbing column density differs even between Ne IX and O VII. Also, Ne IX indicates a larger density than O VII, although is should be formed higher in the accretion shock. All this is not reproduced by the simple model used here and figure 5 shows that Ne IX is fitted very well for TW Hya, while O VII is overpredicted.

Unfortunately, TW Hya is the only CTTS with such a high signal-to-noise in the spectrum that these differences to the model can be seen. Similar shortcomings as in TW Hya might exist for the other CTTS but would be hidden in the noise. The best I can do to compare the mass accretion rates for several CTTS consistently is to fit all stars in the sample with a consistent model.

The table also shows, that the Ne abundances are enhanced in all CTTS, while the abundances of Fe, Si and Mg are reduced. Stelzer & Schmitt (2004) put forward the idea that Fe, Si and Mg condense on grains and settle, but the noble gases cannot and are accreted onto the star. However, Ne also has a higher first ionization potential and in all active stars the coronal abundance of those elements is enhanced (IFIP effect) (see the review by Güdel & Nazé 2009). Even for TW Hya the signal is not strong enough to determine the Ne abundance in the shock and the corona independently, so any one or both of the above scenarios might be important.

In most cases, the corona is described by one temperature component with $kT \approx 0.7$ keV and another with $kT \approx 2$ keV; the volume emission measures $VEM$ of the corona vary over the sample. That is not surprising as several of the lightcurves contain strong flares which cause higher temperatures and higher emission measures for the duration of the flare.

The fitted values for the infall velocity in all stars are compatible with estimates for the free-fall velocity. The normalization of the shock models is proportional to the total mass flux, with is around $6 \times 10^{-11} M_\odot$ yr$^{-1}$ for TW Hya, MP Mus and V4046 Sgr. TW Hya and MP Mus are both relatively old CTTS, so a smaller mass accretion rate here is expected. The mass accretion rate of BP Tau is two orders of magnitude stronger.

### 2.3.4 Comparison of mass accretion rates

The shock models fitted in the previous section agree with other estimates in the literature, that are also based on X-ray spectra, within a factor of 2–4 (table 3). Some of those estimates are far simpler (Argiroffi et al. 2007; Schmitt et al. 2005; Stelzer & Schmitt 2004). They just use the fitted volume emission measure and density with a single value for the cooling function and some rough estimate for the depth of the post-shock cooling zone. Still, the results are comparable. In the case of TW Hya the mass accretion rates of both observations agree. Again, this shows that the fitting of the mass accretion rates is relatively robust, but the density should be determined from line fitting and not with a global fit on binned spectra.

However, all X-ray determined mass accretion rates are one or two orders of magnitude smaller than measurements obtained with the UV flux or the optical veiling (Curran et al. 2010). It is unknown what causes this effect. One possibility are inhomogeneous accretion spots. Either only a small part of the accretion stream impacts at free-fall velocity and the remaining mass accretes at a lower velocities or parts of the accretion spots are hidden by complete absorption. Although some hydrodynamical simulations of accretion streams show inhomogeneous impact velocities (Long et al. 2007; Romanova et al. 2004), it remains unclear which physical mechanism reduces the speed and where the corresponding gravitational energy is lost. Also, Argiroffi et al. (2009) find no resonant scattering in the O VII lines in TW Hya, thus is seems unlikely that absorption can explain the small mass accretion rate found.

## 2.4 Winds

Many, if not all, CTTS have outflows (a review is given by Bally et al. 2007), but the physical driving mechanism is unknown. Theoretical models propose a variety of stellar winds (Kwan & Tademaru 1988; Matt & Pudritz 2005), X-winds (Shu et al. 1994) and disc winds (Anderson et al. 2005; Blandford & Payne 1982). Winds, which could be of stellar origin or come from the disk, are observed over a wide wavelength range from radio to the UV (e.g. Alencar & Basri 2000; Beristain et al. 2001; Dupree et al. 2005a; Edwards et al. 2006; Lamzin et al. 2004). Some CTTS also have highly collimated jets, which are discussed in more detail in section 2.5. The dynamics of the gas around CTTS are measured by UV spectroscopy with *HST/GHRS*, *HST/STIS* and *FUSE* (Ardila et al. 2002a,b; Herczeg et al. 2002, 2006, 2005). The best-studied CTTS is TW Hya, where Dupree et al. (2005a) fit the asymmetric line profile of the O VI 1032 Å line with a Gaussian with the centroid matching the stellar rest-frame. They explain the missing flux on the blue side of the line by a spherically-symmetric and smoothly-accelerated hot wind. However, Johns-Krull & Herczeg (2007) argue that this model is incompatible with *HST/STIS* observations, especially for the C IV 1550 Å doublet. The existence of a hot wind in TW Hya, therefore, remains an open issue. Günther & Schmitt (2008) extracted line profiles for all CTTS in the *FUSE* archive and find line centroids to be shifted between -170 km $s^{-1}$ and +100 km $s^{-1}$. The blue-shifted emission is likely caused by shocks in outflows from the CTTS.

Figure 6 compares the gas column density as measured by X-rays with the optical reddening, which is mainly caused by dust (see Günther & Schmitt 2008, and

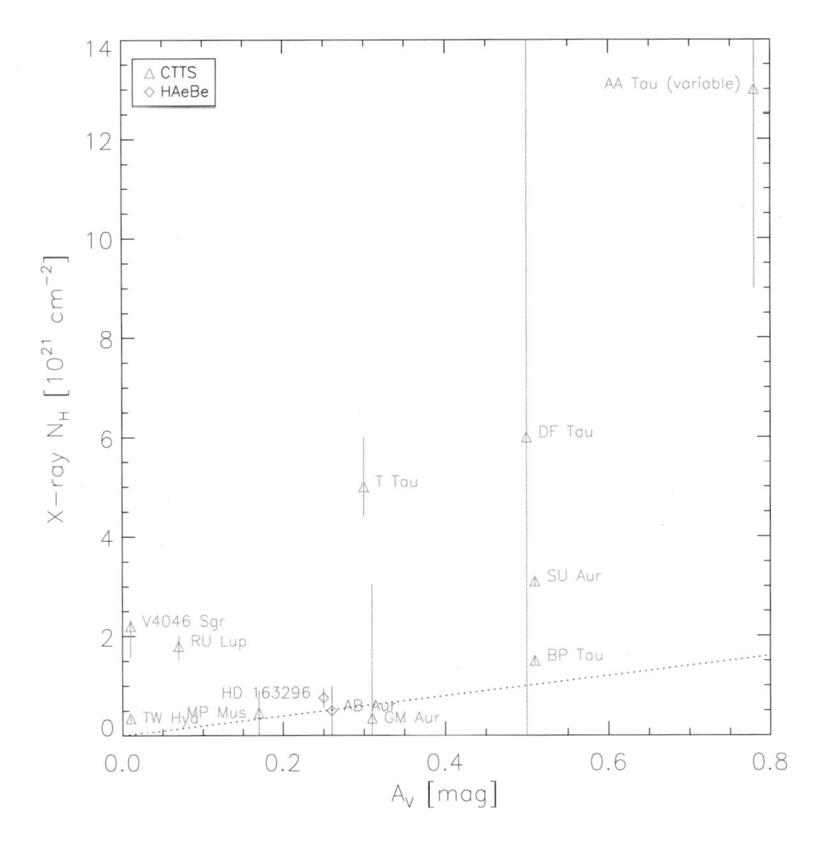


**Figure 6:** Optical reddening and X-ray absorbing column density (with 90% confidence error bars). The dotted line shows the interstellar gas-to-dust ratio. DM Tau and GM Aur have *ROSAT*/PSPC spectra only, so the error bars are much larger.

references therein for data sources). The optical reddening is notoriously difficult to measure in accreting systems, but at least in some cases, notably V4046 Sgr, RU Lup and T Tau, the gas absorption is much higher than expected from an interstellar gas-to-dust ratio (Vuong et al. 2003) and the optical reddening. These are also the stars with blue-shifted UV emission. It is possible that the same outflows, which provide the UV emission lines also act as dust-depleted absorbers for the stellar light.

## 2.5 Jets

In addition to wide-angle winds CTTS can also drive highly collimated jets (Coffey et al. 2004; Güdel et al. 2005; Rodriguez 1995), but DG Tau is the only CTTS where X-ray emission has been found from the jets. Other X-ray jets like HH 2 (Pravdo et al. 2001) or HH 154 (Bally et al. 2003; Favata et al. 2006, 2002) have younger and more embedded driving sources. Usually, the inner components of these jets are faster and denser (figure 7). Bacciotti et al. (2000) used the *HST* to resolve the jet of DG Tau in

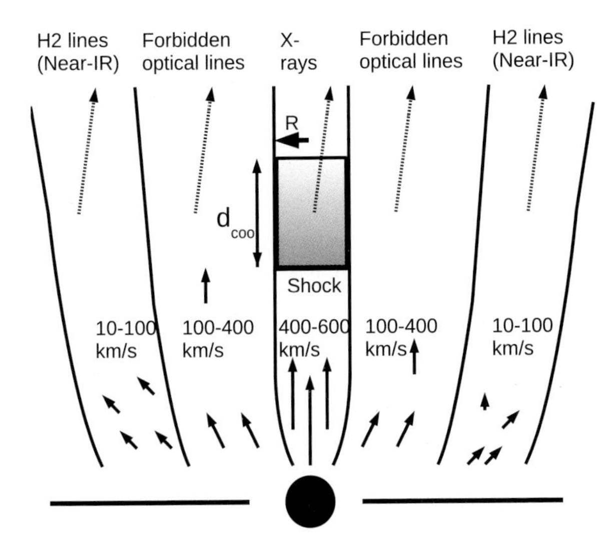


**Figure 7:** Sketch of a cut through the outflow from a CTTS (not to scale). The inner components of the outflow are more collimated and faster and thus heat to higher temperatures when shocked.

several long-slit observations. The emission of gas faster than 200 km s$^{-1}$ is mostly confined to the inner slit, corresponding to a radius of 15 AU. The ratios of [O I], [N II] and [S II] give a lower limit on the gas density of $10^4$ cm$^{-3}$. This component seems to be surrounded by slower moving gas with lower densities of the order $10^3$ – $10^4$ cm$^{-3}$. On larger scales an even cooler outflow, with an opening angle of 90° is seen in molecular hydrogen (Takami et al. 2004). Lavalley et al. (1997) estimate a mass loss rate of $6.5 \cdot 10^{-6}$ M$_\odot$ yr$^{-1}$ assuming shock heating in the gas. Hartigan et al. (1995) obtain $3 \cdot 10^{-7}$ M$_\odot$ yr$^{-1}$. The kinematics of the gas can be calculated from line shifts and proper motion, accounting for the inclination of the jet to the plane of the sky. Optical and IR lines are blue-shifted up to deprojected velocities of 600 km s$^{-1}$ (Bacciotti et al. 2000; Lavalley-Fouquet et al. 2000; Pyo et al. 2003) and the proper motion of the knots in the jets is 0.″28 yr$^{-1}$, which translates into a deprojected velocity of 300 km s$^{-1}$ (Pyo et al. 2003).

In this context, the X-ray emission from the jet can be understood as shock-heated plasma from the densest and innermost component of the outflow (Günther et al. 2009). Only $10^{-3}$ of the total mass outflow is required to shock at the velocities observed in the fastest jet component to explain the observed X-ray spectra and emission measures. Given the density from the forbidden emission lines and the temperature for the X-ray spectrum the cooling length $d_{\rm cool}$ can be calculated with a model very similar to the accretion shock model. The total intensity determines the volume emission measure of the plasma and with the density and $d_{\rm cool}$ this yields the area of the shock. For DG Tau the estimated dimensions are only a few AU, if all X-ray emission is produced in a single shock (figure 7). Because of the small dimensions, a shock in the innermost component does not necessarily disturb the flow in the outer layers, which are resolved with the *HST*. So far, this model is compatible with all available observations.

# 3 Herbig Ae/Be stars

Herbig Ae/Be stars (HAeBes) are in a similar evolutionary state as CTTS, but they have spectral type A or B. Due to their higher mass their evolution progresses faster and typically they are younger than CTTS. Like CTTS HAeBes are surrounded by a disk and they actively accrete matter, but they are not expected to have an outer convective envelope, thus they should not generate magnetic fields and coronal activity. It is unclear if they can support magnetically funneled accretion. Zinnecker & Preibisch (1994) discovered X-ray emission from many HAeBes. More recent studies with the current generation of X-ray telescopes often identify the X-rays with a co-eval companion, i.e. a CTTS, but some HAeBes still seem to generate intrinsic X-ray emission (Hamaguchi et al. 2005; Skinner et al. 2004; Stelzer et al. 2006, 2009). There are two cases of HAeBes without any evidence of binarity and an existing X-ray grating spectrum: AB Aur (Telleschi et al. 2007c) and HD 163296 (Günther & Schmitt 2009; Swartz et al. 2005). Figure 8 shows the O VII triplets for those two HAeBes. The signal for AB Aur is weak, because it falls on the edge of the detector. Both stars seem to have strong $f$ and weak $i$ lines, for HD 163296 $f/i > 2.6$ on the 90% confidence level (Günther & Schmitt 2009). This means that the emission originates in a region of low density and, given the strong UV field on the surface of A-type stars, this region is at an radius $R > 1.7\ R_*$, i.e. at least $0.7\ R_*$ above the surface. Otherwise, the UV photons would shift emission from the $f$ to the $i$ line. Given the low absorbing column density it is unlikely that the hot plasma resides in a region closer to the surface, that is somehow shielded from the stellar radiation field. HD 163296 is known to drive a powerful jet, just as the CTTS DG Tau, so a similar scenario, where the X-ray emission is caused by a shock in the jet, is possible. In fact, one knot in the jet of HD 163296 might itself be an X-ray source (Swartz et al. 2005). In addition to this soft component, HAeBes also show surprisingly hot emission with fitted temperatures around 30 MK. This cannot be caused by accretion and is a clear signature of magnetic activity. HAeBes are not expected to drive solar-like dynamos, but they might operate turbulent dynamos in the atmosphere or retain the primordial magnetic field of the interstellar cloud. If the field lines are frozen-in the field strength increases by many orders of magnitude during the collapse to the proto-star. Giardino et al. (2004) observed a large flare in V892 Tau, a system of a HAeBe with a companion. The system is not resolved but they could tentatively identify the flare with the HAeBe star. This fits the picture of a magnetically heated corona.

The absence of high-density emission and the minimum distance between O VII emission and the stellar surface shows that accretion does not contribute to the X-ray emission from HAeBes. Likely, the accretion process differs from CTTS, because the magnetic field is not strong enough to disrupt the disk at the co-rotation radius

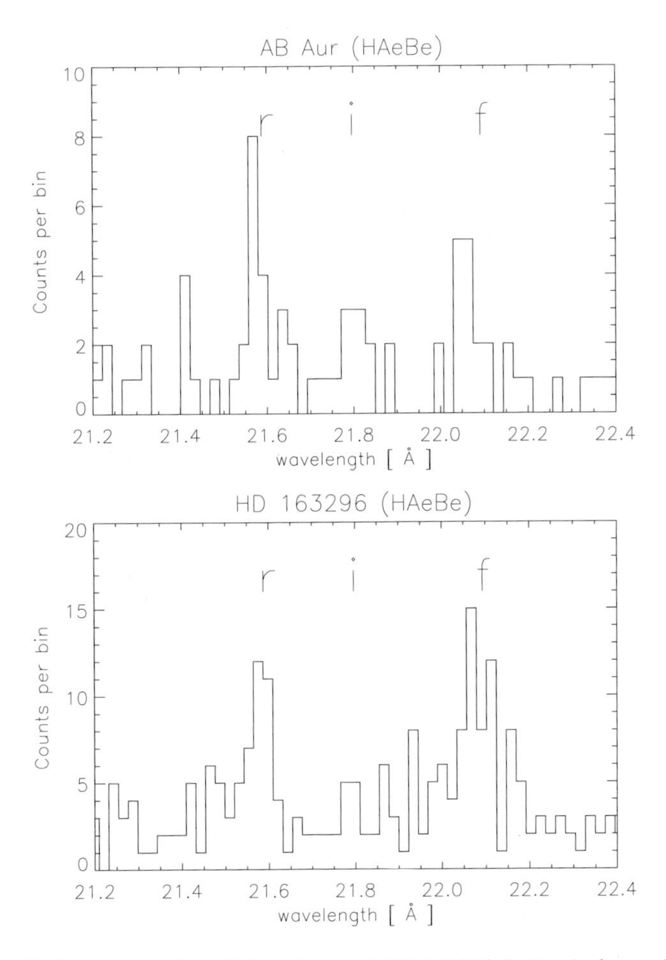


**Figure 8:** O VII triplet for AB Aur (top) and HD 163296 (bottom) observed with *XMM-Newton*/RGS. $r$, $i$ and $f$ line are labeled.

and to support magnetically funneled accretion. The mass may reach the star in the equatorial plane and form some kind of a boundary layer.

# 4 Summary

This article summarizes the results of high-resolution X-ray spectroscopy of young stars in the mass range of CTTS and HAeBes.

For two *Chandra* observations I could deblend the Mg XI triplet from the Ne Lyman series. It turns out that the Ne blend is weak, despite the enhanced Ne abundance.

At least three different processes contribute to the X-ray emission from CTTS, their importance varies between individual objects. First, there is a solar-like corona, second, the post-shock zone of the accretion shock cools radiatively in X-rays and other wavelengths and, third, internal shocks in jets can heat matter to X-ray emitting temperatures. This is studied for the individual sources with the best spectra, but there is no reason why these results should not apply to CTTS as a class. The combined models explain the observed line ratios in the He-like triplets of O VII and Ne IX very well.

The mass accretion rate required to power the X-ray emission is typically lower than values estimated from other wavelengths. This might be due to inhomogeneous spots, partial absorption or accretion streams, which impact at velocities significantly below the free-fall speed.

The mass flux in the X-ray component of the DG Tau jets is also $10^{-3}$ of the total mass loss rate. This is fully compatible with optical observations, that show only the innermost jet component to be sufficiently fast (400 to 500 km $s^{-1}$) to heat the gas in shocks to X-ray emitting temperatures. For an electron density $> 10^5$ $cm^{-3}$ all dimensions of the shock cooling zone are only a few AU, so even in optical observations this cannot be resolved.

The circumstellar environment of some CTTS differs markedly from the interstellar gas-to-dust ratio. This can be explained, if the outflows of those CTTS are dust-depleted.

Last, the X-ray emission from HAeBes is likely intrinsic and not due to an unresolved companion. It can thus be established that HAeBes have a hot emission component similar to a solar-like corona. The line ratio in the O VII He-like triplet rules out an accretion shock origin for the soft emission. Likely, a large fraction of the soft component is produced in the jet similar to DG Tau.

**Acknowledgements**

I thank the Astronomische Gesellschaft for awarding me the newly established Promotionspreis in 2010. The work, which is summarized in this article, would not have been possible without the help and advice of my colleagues at the Hamburger Sternwarte and especially my thesis adviser Prof. J. H. M. M. Schmitt. Jan-Uwe Ness helped a lot i nthe CORA fits to the Mg triplet. H. M. G. acknowledges financial support from Chandra grant GO6-7017X and from the Faculty of the European Space Astronomy Centre. This research made use of the Chandra Transmission Grating Catalog and archive (http://tgcat.mit.edu).

# References

Alcala, J. M., Krautter, J., Schmitt, J. H. M. M., et al. 1995, A&AS, 114, 109

Alcala, J. M., Terranegra, L., Wichmann, R., et al. 1996, A&AS, 119, 7

Alencar, S. H. P. & Basri, G. 2000, AJ , 119, 1881

Anderson, J. M., Li, Z.-Y., Krasnopolsky, R., & Blandford, R. D. 2005, ApJ , 630, 945

Ardila, D. R., Basri, G., Walter, F. M., Valenti, J. A., & Johns-Krull, C. M. 2002a, ApJ , 566, 1100

Ardila, D. R., Basri, G., Walter, F. M., Valenti, J. A., & Johns-Krull, C. M. 2002b, ApJ , 567, 1013

Argiroffi, C., Maggio, A., & Peres, G. 2007, A&A, 465, L5

Argiroffi, C., Maggio, A., Peres, G., et al. 2009, A&A, 507, 939

Argiroffi, C., Maggio, A., Peres, G., Stelzer, B., & Neuhäuser, R. 2005, A&A, 439, 1149

Arnaud, K. A. 1996, in ASP Conf. Ser. 101: Astronomical Data Analysis Software and Systems V, 17–+

Bacciotti, F., Mundt, R., Ray, T. P., et al. 2000, ApJ , 537, L49

Bally, J., Feigelson, E., & Reipurth, B. 2003, ApJ , 584, 843

Bally, J., Reipurth, B., & Davis, C. J. 2007, in Protostars and Planets V, ed. B. Reipurth, D. Jewitt, & K. Keil, 215–230

Balucinska-Church, M. & McCammon, D. 1992, ApJ , 400, 699

Beristain, G., Edwards, S., & Kwan, J. 2001, ApJ , 551, 1037

Bertout, C., Basri, G., & Bouvier, J. 1988, ApJ , 330, 350

Blandford, R. D. & Payne, D. G. 1982, MNRAS , 199, 883

Brickhouse, N. S., Cranmer, S. R., Dupree, A. K., Luna, G. J. M., & Wolk, S. 2010, ApJ , 710, 1835

Calvet, N. & Gullbring, E. 1998, ApJ , 509, 802

Cardelli, J. A., Clayton, G. C., & Mathis, J. S. 1989, ApJ , 345, 245

Coffey, D., Bacciotti, F., Woitas, J., Ray, T. P., & Eislöffel, J. 2004, ApJ , 604, 758

Curran, R. L., Argiroffi, C., Sacco, G. G., et al. 2010, ArXiv e-prints

Dere, K. P., Landi, E., Mason, H. E., Fossi, B. C. M., & Young, P. R. 1998, in ASP Conf. Ser. 143: The Scientific Impact of the Goddard High Resolution Spectrograph, 390–+

Donati, J.-F., Jardine, M. M., Gregory, S. G., et al. 2007, MNRAS , 380, 1297

Donati, J.-F., Jardine, M. M., Gregory, S. G., et al. 2008, MNRAS , 386, 1234

Drake, J. J. 2005, in 13th Cambidge Workshop on Cool Stars, Stellar Systems and the Sun, 519–523

Drake, J. J., Ratzlaff, P. W., Laming, J. M., & Raymond, J. 2009, ApJ , 703, 1224

Dupree, A. K., Brickhouse, N. S., Smith, G. H., & Strader, J. 2005a, ApJ , 625, L131

Dupree, A. K., Lobel, A., Young, P. R., et al. 2005b, ApJ , 622, 629

Edwards, S., Fischer, W., Hillenbrand, L., & Kwan, J. 2006, ApJ , 646, 319

Eisner, J. A., Chiang, E. I., & Hillenbrand, L. A. 2006, ApJ , 637, L133

Fang, M., van Boekel, R., Wang, W., et al. 2009, A&A, 504, 461

Favata, F., Bonito, R., Micela, G., et al. 2006, A&A, 450, L17

Favata, F., Fridlund, C. V. M., Micela, G., Sciortino, S., & Kaas, A. A. 2002, A&A, 386, 204

Feigelson, E. D. & Decampli, W. M. 1981, ApJ , 243, L89

Feigelson, E. D. & Montmerle, T. 1999, ARA&A , 37, 363

Fruscione, A., McDowell, J. C., Allen, G. E., et al. 2006, in Society of Photo-Optical Instrumentation Engineers (SPIE) Conference Series, Vol. 6270, Society of Photo-Optical Instrumentation Engineers (SPIE) Conference Series

Gabriel, A. H. & Jordan, C. 1969, MNRAS , 145, 241

Gabriel, C., Denby, M., Fyfe, D. J., et al. 2004, in Astronomical Society of the Pacific Conference Series, Vol. 314, Astronomical Data Analysis Software and Systems (ADASS) XIII, ed. F. Ochsenbein, M. G. Allen, & D. Egret, 759–+

Getman, K. V., Feigelson, E. D., Broos, P. S., Micela, G., & Garmire, G. P. 2008a, ApJ , 688, 418

Getman, K. V., Feigelson, E. D., Micela, G., et al. 2008b, ApJ , 688, 437

Getman, K. V., Flaccomio, E., Broos, P. S., et al. 2005, ApJS , 160, 319

Giardino, G., Favata, F., Micela, G., & Reale, F. 2004, A&A, 413, 669

Grevesse, N. & Sauval, A. J. 1998, Space Science Reviews, 85, 161

Güdel, M. & Nazé, Y. 2009, A&A Rev., 17, 309

Güdel, M., Skinner, S. L., Briggs, K. R., et al. 2005, ApJ , 626, L53

Güdel, M., Skinner, S. L., Mel'Nikov, S. Y., et al. 2007, A&A, 468, 529

Güdel, M. & Telleschi, A. 2007, A&A, 474, L25

Gullbring, E., Calvet, N., Muzerolle, J., & Hartmann, L. 2000, ApJ , 544, 927

Günther, H. M., Lewandowska, N., Hundertmark, M. P. G., et al. 2010, A&A, 518, A54+

Günther, H. M., Liefke, C., Schmitt, J. H. M. M., Robrade, J., & Ness, J.-U. 2006, A&A, 459, L29

Günther, H. M., Matt, S. P., & Li, Z.-Y. 2009, A&A, 493, 579

Günther, H. M. & Schmitt, J. H. M. M. 2007, Memorie della Societa Astronomica Italiana, 78, 359

Günther, H. M. & Schmitt, J. H. M. M. 2008, A&A, 481, 735

Günther, H. M. & Schmitt, J. H. M. M. 2009, A&A, 494, 1041

Günther, H. M., Schmitt, J. H. M. M., Robrade, J., & Liefke, C. 2007, A&A, 466, 1111

Hamaguchi, K., Yamauchi, S., & Koyama, K. 2005, ApJ , 618, 360

Hartigan, P., Edwards, S., & Ghandour, L. 1995, ApJ , 452, 736

Herczeg, G. J., Linsky, J. L., Valenti, J. A., Johns-Krull, C. M., & Wood, B. E. 2002, ApJ , 572, 310

Herczeg, G. J., Linsky, J. L., Walter, F. M., Gahm, G. F., & Johns-Krull, C. M. 2006, ApJS , 165, 256

Herczeg, G. J., Walter, F. M., Linsky, J. L., et al. 2005, AJ , 129, 2777

Huenemoerder, D. P., Kastner, J. H., Testa, P., Schulz, N. S., & Weintraub, D. A. 2007, ApJ , 671, 592

Johns-Krull, C. M. & Herczeg, G. J. 2007, ApJ , 655, 345

Kastner, J. H., Huenemoerder, D. P., Schulz, N. S., et al. 2004, ApJ , 605, L49

Kastner, J. H., Huenemoerder, D. P., Schulz, N. S., Canizares, C. R., & Weintraub, D. A. 2002, ApJ , 567, 434

Koenigl, A. 1991, ApJ , 370, L39

Koldoba, A. V., Ustyugova, G. V., Romanova, M. M., & Lovelace, R. V. E. 2008, MNRAS , 388, 357

Kwan, J. & Tademaru, E. 1988, ApJ , 332, L41

Lamzin, S. A. 1998, Astronomy Reports, 42, 322

Lamzin, S. A., Kravtsova, A. S., Romanova, M. M., & Batalha, C. 2004, Astronomy Letters, 30, 413

Landi, E., Del Zanna, G., Young, P. R., et al. 2006, ApJS , 162, 261

Lavalley, C., Cabrit, S., Dougados, C., Ferruit, P., & Bacon, R. 1997, A&A, 327, 671

Lavalley-Fouquet, C., Cabrit, S., & Dougados, C. 2000, A&A, 356, L41

Long, M., Romanova, M. M., & Lovelace, R. V. E. 2007, MNRAS , 374, 436

Matt, S. & Pudritz, R. E. 2005, ApJ , 632, L135

Muzerolle, J., Calvet, N., & Hartmann, L. 1998a, ApJ , 492, 743

Muzerolle, J., Calvet, N., Hartmann, L., & D'Alessio, P. 2003, ApJ , 597, L149

Muzerolle, J., Hartmann, L., & Calvet, N. 1998b, AJ , 116, 2965

Ness, J.-U., Güdel, M., Schmitt, J. H. M. M., Audard, M., & Telleschi, A. 2004, A&A, 427, 667

Ness, J.-U. & Wichmann, R. 2002, Astronomische Nachrichten, 323, 129

Neuhäuser, R., Sterzik, M. F., Schmitt, J. H. M. M., Wichmann, R., & Krautter, J. 1995, A&A, 297, 391

Orlando, S., Sacco, G. G., Argiroffi, C., et al. 2010, A&A, 510, A71+

Padgett, D. L., Cieza, L., Stapelfeldt, K. R., et al. 2006, ApJ , 645, 1283

Porquet, D., Mewe, R., Dubau, J., Raassen, A. J. J., & Kaastra, J. S. 2001, A&A, 376, 1113

Pravdo, S. H., Feigelson, E. D., Garmire, G., et al. 2001, Nature, 413, 708

Preibisch, T., Kim, Y.-C., Favata, F., et al. 2005, ApJS , 160, 401

Pyo, T.-S., Kobayashi, N., Hayashi, M., et al. 2003, ApJ , 590, 340

Raassen, A. J. J. 2009, A&A, 505, 755

Redfield, S., Linsky, J. L., Ake, T. B., et al. 2002, ApJ , 581, 626

Robrade, J. & Schmitt, J. H. M. M. 2007, A&A, 473, 229

Rodriguez, L. F. 1995, in Revista Mexicana de Astronomia y Astrofisica Conference Series, ed. S. Lizano & J. M. Torrelles, 1–+

Romanova, M. M., Ustyugova, G. V., Koldoba, A. V., & Lovelace, R. V. E. 2004, ApJ , 610, 920

Sacco, G. G., Argiroffi, C., Orlando, S., et al. 2008, A&A, 491, L17

Schmitt, J. H. M. M., Robrade, J., Ness, J.-U., Favata, F., & Stelzer, B. 2005, A&A, 432, L35

Shu, F., Najita, J., Ostriker, E., et al. 1994, ApJ , 429, 781

Skinner, S. L., Güdel, M., Audard, M., & Smith, K. 2004, ApJ , 614, 221

Stelzer, B., Flaccomio, E., Briggs, K., et al. 2007, A&A, 468, 463

Stelzer, B., Micela, G., Hamaguchi, K., & Schmitt, J. H. M. M. 2006, A&A, 457, 223

Stelzer, B. & Neuhäuser, R. 2001, A&A, 377, 538

Stelzer, B., Robrade, J., Schmitt, J. H. M. M., & Bouvier, J. 2009, A&A, 493, 1109

Stelzer, B. & Schmitt, J. H. M. M. 2004, A&A, 418, 687

Strassmeier, K. G., Rice, J. B., Ritter, A., et al. 2005, A&A, 440, 1105

Swartz, D. A., Drake, J. J., Elsner, R. F., et al. 2005, ApJ , 628, 811

Takami, M., Chrysostomou, A., Ray, T. P., et al. 2004, A&A, 416, 213

Telleschi, A., Güdel, M., Briggs, K. R., Audard, M., & Palla, F. 2007a, A&A, 468, 425

Telleschi, A., Güdel, M., Briggs, K. R., Audard, M., & Scelsi, L. 2007b, A&A, 468, 443

Telleschi, A., Güdel, M., Briggs, K. R., et al. 2007c, A&A, 468, 541

Testa, P., Drake, J. J., & Peres, G. 2004, ApJ , 617, 508

Uchida, Y. & Shibata, K. 1984, PASJ, 36, 105

Unruh, Y. C., Collier Cameron, A., & Guenther, E. 1998, MNRAS , 295, 781

Vuong, M. H., Montmerle, T., Grosso, N., et al. 2003, A&A, 408, 581

Zinnecker, H. & Preibisch, T. 1994, A&A, 292, 152

# The physics and astrophysics of supernova explosions

Wolfgang Hillebrandt

Max-Planck-Institut für Astrophysik
Karl-Schwarzschild-Str. 1
D-85748 Garching, Germany

wfh@mpa-garching.mpg.de

**Abstract**

*Because of their complexity models of supernova explosions have always been a challenge. In the case of core collapse models, which utilize gravitational binding for the explosion, the problem is not so much an energy but rather a momentum problem. There is plenty of energy available, in principle, and only a small fraction of it has to be converted into outward momentum of the stellar envelope, but this is a non-trivial problem. Thermonuclear explosion models, on the other hand, suffer from the fact that once nuclear burning is ignited the progenitor star, presumably a white dwarf, tends to expand and cool. It has been shown that an explosion results only if the burning front propagates with a velocity much larger than the laminar speed of nuclear flame in degenerate matter. Since for most supernova models hydrodynamic instabilities play a key role numerical experiments to prove or disprove certain ideas have to be 2- or 3-dimensional and, in fact, they have become feasible in recent years. Here we discuss the present status of 2- and 3-D numerical simulations of supernova explosions. In the case of explosions triggered by gravitational energy release, the crucial problem is the transport of neutrinos in convectively unstable matter, and for explosions powered by nuclear energy release, the problem of turbulent combustion has to be addressed as well as the diversity of potential progenitor channels. Robust predictions of the various models are confronted with observations.*

## 1 Introduction – Some observational facts

In general, supernovae are classified according to their maximum-light spectra. Those showing Balmer lines of H are called Type II's, and all the others are of Type I. Those of Type I which show at maximum light a strong Si II absorption feature near 6100 Å are named Type Ia, and the others are Ib's or Ic's, depending on whether or no they have also He I features in their spectra (Wheeler & Harkness 1990). At later times, several months after the explosion, when the supernova ejecta become optically thin, Type II spectra are dominated by emission lines of H, O, and Ca, whereas Type Ia's have no O, but Fe and Co. Type Ib,c's, on the other hand side, show emission lines of O and Ca, just like the Type II's (Filippenko 1989).

*Reviews in Modern Astronomy 23: Zoomimg in: The Cosmos at High Resolution.* First Edition.
Edited by Regina von Berlepsch.

In contrast to SNe II (and, to a certain extend, SN Ib,c), which have rather different light curves in general, SNe Ia are quite similar, making them good candidates for standard candles to measure cosmic distances. Moreover, since they are the brightest among all supernovae, they can be observed even at high redshifts and, in fact, Type Ia supernovae out to a redshift of about 2 have been observed in the past few years (Riess et al. 2004). Although their absolute peak luminosity may vary by more than a magnitude, an observed correlation between the luminosity and light-curve shape (the brighter ones have broader lightcurves, Phillips (1993)) allows one to correct for the differences. So recently observations of Type Ia's have become a powerful tool in attempts to determine cosmological parameters (Schmidt et al. 1998; Riess et al. 1998; Perlmutter et al. 1999; Garnavich et al. 1998; Tonry et al. 2003; Astier et al. 2006; Kowalski et al. 2008).

Observational information on the physics of supernovae is in general less solid although due to a number of very active transient and supernova searches, followed by detailed photometric and spectroscopic follow-ups, the situation has improved considerably (e.g., Foley et al. (2009) for ESSENCE, Nordin et al. (2011) and references therein for SDSS-II, Ganeshalingam (2010) for the Lick Observatory Supernova Search). However, with the exception of a few cases, including SN 1987A, we do not have direct information on the properties of the progenitors, the energetics, nor the masses of the ejecta and of the (compact) remnants, if there are any (see, e.g., Smartt (2009) for a discussion of Type II-P SNe progenitors). Indirect information on those properties can be obtained from lightcurves and spectra leading, however, to an "inverse" problem as far as the theoretical interpretation is concerned (Nadyozhin 2003; Pastorello et al. 2006; Sahu et al. 2006; Tsvetkov et al. 2006; Utrobin 2007; Mazzali et al. 2007; Tominaga et al. 2008; Foley et al. 2009; Taubenberger et al. 2009; Scalzo et al. 2010). However, these attempts did not lead to fully conclusive results yet.

## 2 Physical classification

The physical classification of supernovae is usually done according to the suspected progenitor stars and the explosion mechanism, and it dates back to a classic paper of Fred Hoyle and Willy Fowler in 1960 (Hoyle & Fowler 1960). Based on very few observational facts available at that time they postulated that Type II supernovae are the consequence of an implosion of non-degenerate stars, whereas Type I's are the result of the ignition of nuclear fuel in degenerate stars, and today this is still believed to be true in general, if "Type I's" are substituted by "Type Ia's." It is now generally believed that Type II and Ib,c supernovae stem from collapsing massive stars with or without hydrogen envelopes, respectively, and that SN Ia originate from thermonuclear explosions of white dwarfs.

The classification with respect to the energetics is more complex, and a variety of possible models is available but no unique and well accepted answer. For example, in the case of thermonuclear explosions, it is not clear what the mode of propagation of the burning front is, and observations supply poor constraints only. A (fast) deflagration wave in a Chandrasekhar-mass white dwarf can explain the spectra and light curves of Type Ia's, but so can deflagrations changing into detonations at low densities with or without pulsations (so-called delayed detonations), or even pure detonations in stars with low enough densities are not ruled out. As far as the physics of core collapse supernovae is concerned, the situation is not significantly better. Whether or not rotation is important, whether or not magnetic fields play any role, whether neutrinos are diffusively or convectively transported from the cooling protoneutron star to the stellar mantle, and many other questions, are largely unanswered.

# 3 Numerical simulations

Of course, a simple way to try to improve the rather unsatisfactory situation outlined in the previous section is to call for more observations. Since computing realistic theoretical lightcurves and synthetic spectra for a given supernova model has become feasible recently (although computing reliable spectra for Type Ia's still requires novel techniques) high quality data might help to rule out certain possibilities. But as long as most theoretical models contain a large number of more or less free parameters one may doubt the success of this approach.

A few examples may serve as illustrations. As long as for a thermonuclear explosion the propagation velocity of the burning front and the degree of mixing can still be treated as free functions, it is easy to fit a given light curve and spectrum of a Type Ia supernova (see, e.g., Nomoto et al. 1984; Branch et al. 1995), but one does not prove that the model is correct. Similarly, in the core collapse scenario, neutrinos can transfer momentum to the mantle of a star and cause a supernova explosion, provided their luminosity is sufficiently high. Whether this luminosity results from low opacity or from convective motions does not matter. Only thousands of neutrino events from a single (galactic) supernova seen in neutrino detectors could possibly tell the difference. Finally, similar momentum transfer can, in principle, be mediated by appropriate combinations of rotation and magnetic fields. Again, a lucky incident of a galactic supernova may supply enough data, including detections of gravitational waves, to eliminate several models, but only for one special event and not for the entire class.

Therefore, it appears to be more promising to search for ways to reduce the freedom one still has in building theoretical models. An obvious possibility is to replace parameter studies by first principle calculations, whenever this is possible. The numerical methods that are widely used in this context are finite volume schemes based on Godunov-type algorithms or, more recently, also smoothed particle hydrodynamic codes, coupled to either a neutrino transport scheme (in the case of core collapse supernovae) or to the reaction kinetics (for thermonuclear ones).

## 3.1 Core collapse models

Again, a few examples are given to demonstrate that there has been significant progress but that there is plenty of room for improvements still, both in the physics and the numerics of core collapse supernova models. A first and obvious example is again neutrino transport. One will note that still rather primitive (and likely incorrect) approximations for neutrino interactions with dense nuclear matter are commonly used in all hydrodynamic simulations, although some attempts have been made to compute those cross sections on the basis of microscopic theories (Raffelt et al. 1995; Raffelt & Strobel 1997; Reddy et al. 1998; Chandra et al. 2002; Reddy et al. 2003; Lykasov et al. 2008; Bacca et al. 2009). Similarly, well-developed methods to calculate the properties of dense nuclear matter are available, but the equations of state used in numerical studies are still computed ignoring nucleon-nucleon correlations, despite the fact that the stiffness of the equation of state is very important ingredient. In the following subsections we will discuss a few of these points in a bit more detail (see also Janka et al. (2007b) for a recent review).

### 3.1.1 Basic input physics

In constructing a (core-collapse) supernova model one has to solve the hydrodynamic equations, some field equations describing gravity, rate equations for composition changes, and transport equations for particle numbers and energy for a given set of initial conditions (densities, entropies, velocities, composition variables, etc.) and material functions (equation of state, reaction rates, interaction cross-sections, etc.). It is obvious that this set of equations cannot be solved in full generality and that many approximations are necessary in order to make the problem tractable.

Concerning numerical methods for solving the hydro-equations great progress has been made during the last couple of years, provided Newtonian mechanics and gravity is a valid prescription (e.g., Kifonidis et al. (2003)). However, in stellar collapse peak velocities approach several tenths of the velocity of light and general relativistic effects are not negligibly small, in particular at core bounce and during the early cooling and deleptonization phase of the newly born neutron star. The use of the Newtonian approximation, therefore, is questionable. For general relativistic hydrodynamics, on the other hand, the numerical techniques are much less advanced, in particular for realistic microphysics and neutrino transport and multidimensional simulations which seem to be necessary (Shibata & Sekiguchi 2004, 2005; Dimmelmeier 2005; Cerdà-Duràn et al. 2005; Ott et al. 2007a, 2007b; Dimmelmeier et al. 2007; Müller et al. 2010). One has to keep this situation in mind when discussing uncertainties in and implications from microphysics input data.

### 3.1.2 The nuclear equation of state

One of the most important ingredients is the equation of state (EOS) and, consequently, also a major fraction of the uncertainties results from our incomplete knowledge of it. Also, there is little hope that we can learn much about the EOS from laboratory experiments, such as heavy ion collisions, because they test the nuclear EOS under rather different conditions. Therefore we have to rely on theoretical models or, possibly, on interpretations of astrophysical observations.

At rather low densities ($\varrho \lesssim 10^{-2}\varrho_0$; $\varrho_0 \sim 3 \times 10^{14}$ g/cm$^3$) the EOS can in principle be calculated from a Boltzmann-gas approximation for nucleons and nuclei, provided nuclear binding energies and partition functions are known. However, most nuclei present in the interior of a collapsing stellar core or in the outer layers of a neutron star would be very short-lived under laboratory conditions, and most of them have not even been synthesized yet by experiments. Therefore one has to rely on extrapolations from the properties of stable and "mildly" unstable nuclei.

At higher densities the Boltzmann-gas approach to the nuclear part of the EOS is no longer valid. This happens because the nuclear radius becomes comparable to the Coulomb interaction radius. Therefore, at those densities (above about $10^{12}$g/cm$^3$) self-consistent models have to be used, and at even higher densities, $\varrho \gtrsim 0.1\varrho_0$, similar arguments show that also nucleon-nucleon interactions have to be included in such a self-consistent model. The most advanced method which has been applied to the supernova problem so far is the temperature dependent Hartree-Fock method but very little progress has been made in the past years.

Even more problematic is the EOS beyond nuclear saturation density. In the deep interior of a newly born neutron star "nuclei" dissolve into a homogeneous fluid of free neutrons and protons once the density exceeds $\varrho_0$. Consequently nucleon-nucleon interactions in a dense Fermi fluid dominate the EOS. Phenomenologically determined nucleon-nucleon forces gradually lose reliability with increasing density and it is therefore not surprising that up to now the EOS at densities above, say, twice nuclear saturation density is still subject to considerable dispute, but it might be crucial for the neutrino luminosity during the first seconds past core-bounce, thought to trigger the ejection of the stellar envelope. Moreover, there may be exotic states of nuclear matter such as pion and kaon condensates realized, hyperons may be important, or even a phase transition to a quark-gluon plasma may happen (Fischer et al. 2010), which would affect the equation of state considerably.

However, most EOS in use in core-collapse simulations are simple, and include only part of the important physics. They are either phenomenologically motivated (e.g., Lattimer & Swesty (1991))), or are based on mean field approaches (e.g., Shen et al. 1998), or on effective potentials (Hillebrandt et al. 1984, see also Hillebrandt (1994)).

### 3.1.3 Weak interaction rates

Next we want to mention briefly some of the uncertainties entering through our incomplete knowledge of weak interaction rates. It is well known that during most of core collapse and during the early cooling phase of newly born neutron stars typical weak interaction timescales are of the same order as the dynamical or evolutionary time scale of the stellar core or star, respectively. So in contrast to strong and also electromagnetic interactions weak rates have to be known explicitly. Moreover, because in some cases neutrino energy distributions are not in equilibrium, it is not even sufficient to calculate energy-averaged rates. Fortunately, in this field considerable progress has been made recently. The best rates available to date are based on shell model wave functions or use the quasi-particle random phase approximation, but from the sensitivity of those results to details of the nuclear model one may still conclude that the calculated rates are uncertain to within a factor of two on average, and possibly by an order of magnitude in some particular cases (Langanke & Martinez-Pinedo 2000, 2003; Langanke et al. 2003).

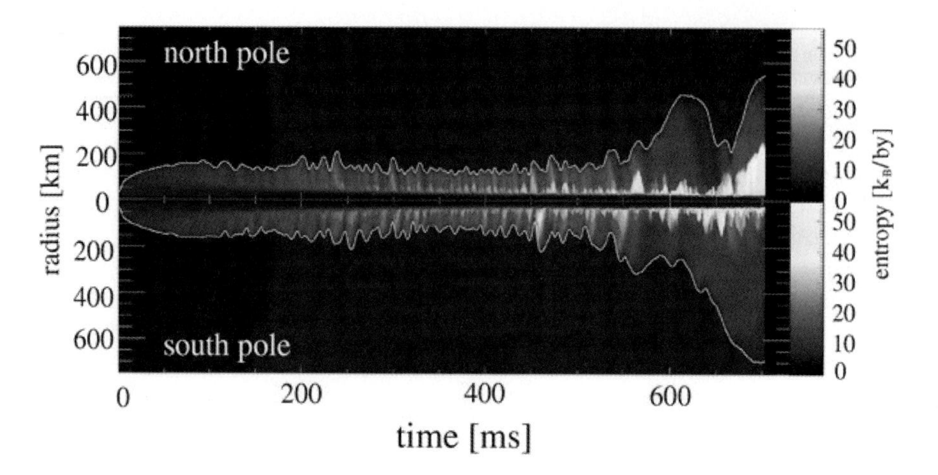


**Figure 1:** Radial positions of the shock near the north and south poles of of a rotating 15 $M_\odot$ model as functions of post-bounce time (white lines). The color coding represents the entropy per nucleon of the stellar gas. The quasi-periodic, bi-polar shock expansion and contraction due to the standing accretion shock instability can clearly be seen (Marek & Janka 2009).

Other important weak interaction processes include neutrino-absorption by free neutrons, neutrino-electron scattering, neutrino-nucleus coherent scattering and neutrino-neutron scattering. These cross sections and reaction rates can, in principle, be computed numerically exactly but in most numerical studies only approximate values are used.

### 3.1.4 Neutrino transport

During collapse and after core-bounce we will always find regions in the star where neutrinos are either free streaming or diffusing outward. So, in principle, we have to solve the Boltzmann transport equation. This transport equation, however, is a set of complicated partial integro-differential equations and, therefore, has only very recently been used in computations of core-collapse supernovae, but most models still use approximations to it.

Generally speaking, neutrino transport calculations for supernovae face two major problems. Firstly, at densities below $10^{12}$g/cm$^3$ neutrinos are not in thermal equilibrium and, secondly, also the diffusion approximation to the Boltzmann equation breaks down at the neutrino-sphere, where the mean free path $\bar{\lambda}$ becomes comparable to the stellar radius. The second problem is usually circumvented by introducing a so-called flux-limiter which guarantees that for $\bar{\lambda} \gg \Delta r$ the free streaming limit is obtained. The first problem can only be solved by non-equilibrium transport models such as two-fluid models, multi-group flux-limited diffusion approximation, variable Eddington-factor methods, Monte-Carlo transport, or even direct integrations of the Boltzmann equation, the latter being feasible in 1-D and 2-D only (Rampp & Janka

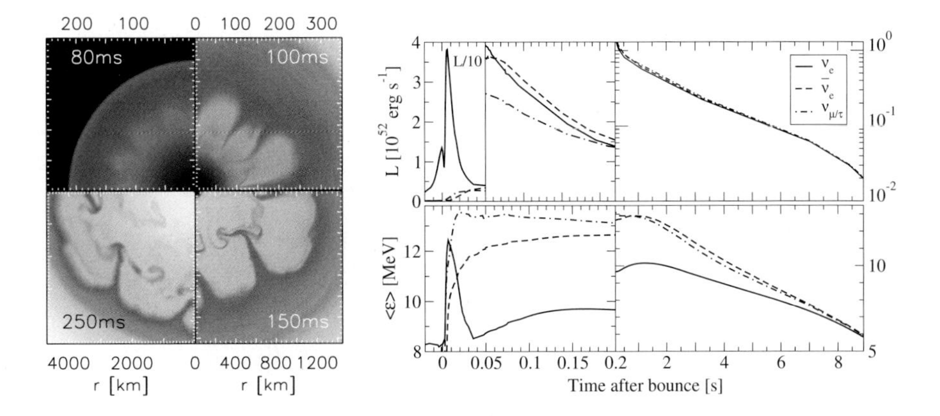


**Figure 2:** Four snapshots of the explosion of a8.8 $M_\odot$ core-collapse supernova model (upper panel; the color coding gives the entropy per nucleon with black corresponding to values of $< 7$ $k_B$, red to 10-15 $k_B$, orange to 15-20 $k_B$, and white to about 25 $k_B$; time runs clock-wise) and predicted neutrino luminosities and mean energies observed at infinity for this model (lower panel) (Janka et al. 2007a; Hüdepohl et al. 2010).

2002; Buras et al. 2003; Keil et al. 2003; Liebendörfer et al. 2003, 2004, 2005, 2009; Marek et al. 2006; Scheck et al. 2008; Müller et al. 2010; Brandt et al. 2011).

### 3.1.5 Recent Models of Core-collapse Supernovae

Since most of the recent numerical simulations are reviewed in Janka et al. (2007b) we will not go into details here but rather summarize the successes and some of the main problems.

As far as numerical simulations of stellar collapse and the subsequent supernova explosion are concerned, the most crucial phase seems to be when the newly born neutron star looses its leptons and thermal energy by neutrinos of all flavors. During this phase, both the proto-neutron star itself and the matter behind the stalled shock in the mantle are found to be unstable to entropy and/or lepton-number driven buoyancy instabilities which can increase the neutrino luminosity considerably. It is commonly believed that this increase in luminosity is required in order to revive the stalled shock by neutrino heating (Hernant et al. 1994; Burrows et al. 1995; Fryer 1999; Fryer & Warren 2002, 2004; Keil et al. 1996; Blondin et al. 2003; Dessart et al. 2006; Burras et al. 2006a,b; Scheck et al. 2008; Marek & Janka 2009; Nordhaus et al. 2010; Brandt et al. 2011; Bruenn et al. 2010).

Although none of the recent simulations give quite the desired results, namely an envelope ejection with a typical energy of about $10^{51}$erg, leaving behind a neutron of about 1.4 $M_\odot$, "hope is left in Pandora's box". In fact, the simulations give weak explosions or are close to explosions, and minor changes in the input physics and/or the numerical treatment might change the outcome. An obvious example is the neutrino luminosity of the newly-born neutron star which seems to be a bit too low for success in some cases, but can be changed during the contraction and cooling phase by different EOS's or neutrino-transport properties of hot and dense nuclear matter. A second example are hydrodynamic instabilities of the neutrino-heated convectively unstable matter behind the stalled shock which, again, can change the energy (and momentum) transport. Finally, general relativistic effects have not yet been studied in great detail in models based on realistic micro-physics input.

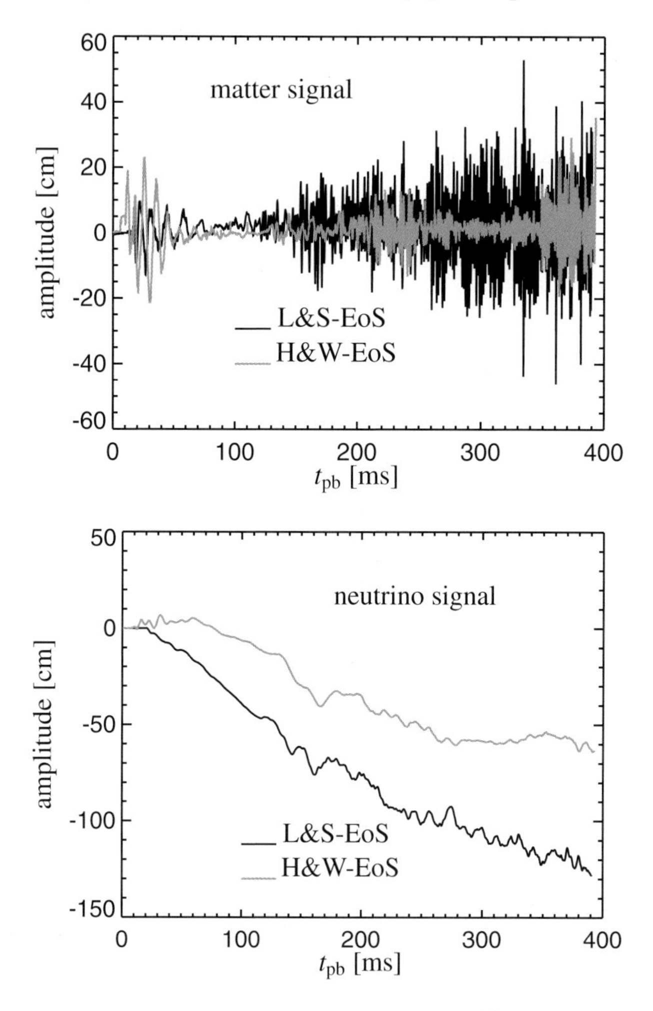


**Figure 3:** Gravitational-wave quadrupole amplitudes $A_{20}^{E2}$ as functions of post-bounce time (associated with mass motions (*top*) and anisotropic neutrino ($\nu_e + \bar{\nu}_e$ plus all heavy lepton neutrinos and antineutrinos) emission (*bottom*) for a distant observer in the equatorial plane of the axisymmetric source for a $15\,M_\odot$ stellar model and the EOS of Lattimer & Swesty (1991)(L&S) and Hillebrandt & Wolff (1984)(H&W)(from Marek et al. (2009).

### 3.1.6 A few observable predictions

Since models of core-collapse supernovae are not yet predicting the observed explosion energy correctly it would not make much sense to compute lightcurves and spectra, and to compare them with data, although models based on rather low-mass progenitors ($M \simeq 8 - 10\,M_\odot$) seem to be able to explain the sub-luminous Type II-P supernovae and the Crab supernova SN 1054 (Kitaura et al. 2006; Marek & Janka 2009). However, observable predictions related to the properties of the newly born neutron star are more robust.

An obvious example are neutrino luminosity and spectra. Here detailed simulations performed in the past few years with state-of-the-art stellar models and realistic input physics and neutrino transport make clear predictions (Burrows & Lattimer 1986; Keil & Janka 1995; Totani et al. 1998; Pons et al. 1999; Keil et al. 2003; see also Hüdepohl et al. (2010) for a recent study of lower-mass core-collapse supernovae and further references, Fig. 2). The standard prediction are monotonically decreasing neutrino energies after no more than a short ($\simeq$ 100 ms) period of increase and spectral hardening over 2–5 s before turning over to cooling. Characteristic phases of neutrino emission of successfully exploding models are: (i) A rising luminosity during collapse. (ii) A shock breakout burst. (iii) An accretion phase, ending a fraction of a second past bounce when neutrino heating reverses the infall. (iv) Kelvin-Helmholtz cooling of the hot proto-neutron star with a duration of typically 10 s or more (e.g., Hüdepohl et al. (2010)). Future detectors and/or a local (galactic) core-collapse supernova will allow to confirm (disprove) these predictions.

A second example are gravitational waves emitted from the newly born neutron star after bounce. Several recent paper deal with this question (Müller et al. 2004; Cardall 2005; Kotake et al. 2006, 2007, 2009; Ott et al. 2006; Murphy et al. 2009; Marek et al. 2009). For instance, Marek et al. (2009), by running 2-dimensional axisymmetric neutrino-hydrodynamic simulations of the long-time accretion phase of a $15\,M_\odot$ progenitor star, find that the predicted gravitational-wave (as well as the neutrino) emission depends on the compactness of the newly born neutron star and, thus, on the nuclear equation of state. In general terms, they predict that the non-radial mass motions in the supernova core impose a time variability on the neutrino and gravitational-wave signals with larger amplitudes, as well as higher frequencies in the case of a more compact nascent neutron star resulting from a softer equation of state. Right after bounce such non-radial mass motions occur due to prompt post-shock convection and contribute mostly to the early wave production around 100 Hz. Later, the gravitational-wave power peaks at about 300 to 800 Hz, connected to changes in the mass quadrupole moment on a timescale of milliseconds, caused by the so-called standing accretion shock instability and various convective activities which occur above and below the neutrino sphere. It is interesting to note that, independent of the equation of state, at late times, i.e. a few tenths of a second after core-bounce, the gravitational-wave amplitude is dominated by the anisotropic neutrino emission rather than the matter signal and produces low-frequency emission (below 100–200 Hz). Future gravitational wave detectors should take this effect into account (see also Fig. 3).

## 3.2 Thermonuclear explosions

Next, we will review thermonuclear explosions. The most popular progenitor model for the average type Ia supernova is a massive white dwarf, consisting of carbon and oxygen, which approaches the Chandrasekhar mass ($M_{Chan} \simeq 1.4\,M_{\odot}$) by a yet unknown mechanism, presumably accretion from a companion star, and is disrupted by a thermonuclear explosion (see, e.g., Hillebrandt & Niemeyer (2000) for a review). At high densities explosive carbon burning mostly produces radioactive $^{56}$Ni. At lower densities intermediate-mass nuclei, like $^{28}$Si, are produced. These elements give rise to the typical observed spectra of SNe Ia, which are dominated by lines of Fe, Si and S.

The general picture of such an explosion is that first carbon burns rather quietly in the core of the contracting white dwarf. Because this core is convectively unstable temperature fluctuations will be present, and they may locally reach run-away values. After ignition, the flame is thought to propagate through the star as a sub-sonic turbulent deflagration wave which may or may not change into a detonation at low densities (around $10^7$ g/cm$^3$), disrupting the star in the end in both cases.

Early attempts to model SNe Ia were based on one-dimensional numerical simulations. Such models gave valuable insight into the basic mechanism of the explosions. However, their predictive power was limited due to the fact that underlying physical processes enter the models in a parametrized way. In particular, the velocity of the thermonuclear flame front entered as a free parameter. However, recent models are in 3D and try to avoid such parameters.

### 3.2.1 The physics of turbulent thermonuclear combustion in white dwarfs

Due to the strong temperature dependence of the carbon-fusion reaction rates nuclear burning during the explosion of a white dwarf is confined to microscopically thin layers that propagate either conductively as subsonic deflagrations ("flames") or by shock compression as supersonic detonations. Both modes are hydrodynamically unstable to spatial perturbations. The best studied and probably most important hydrodynamical effect for modeling SN Ia explosions is the Rayleigh-Taylor (RT) instability resulting from the buoyancy of hot, burned fluid with respect to the dense, unburned material (Niemeyer & Hillebrandt 1995). Subject to the RT instability, small surface perturbations grow until they form bubbles (or "mushrooms") that begin to float upward while spikes of dense fluid fall down. In the nonlinear regime, bubbles of various sizes interact and create a foamy RT mixing layer whose vertical extent grows with time. Secondary instabilities related to the velocity shear along the bubble surfaces (Niemeyer & Hillebrandt 1995) quickly lead to the production of turbulent velocity fluctuations that cascade from the size of the largest bubbles ($\approx 10^7$ cm) down to the microscopic Kolmogorov scale, $l_{\rm k} \approx 10^{-4}$ cm, where they are dissipated. Since no computer is capable of resolving this range of scales, one has to resort to statistical or scaling approximations of those length scales that are not properly resolved. The most prominent scaling relation in turbulence research is Kolmogorov's law for the cascade of velocity fluctuations, stating that in the case of isotropy and statistical stationarity, the mean velocity $v$ of turbulent eddies with size $l$ scales as $v \sim l^{1/3}$.

Given the velocity of large eddies, e.g. from computer simulations, one can use this relation to extrapolate the eddy velocity distribution down to smaller scales under the assumption of isotropic, fully developed turbulence. Turbulence wrinkles and deforms the flame. These wrinkles increase the flame surface area and therefore the total energy generation rate of the turbulent front. In other words, the turbulent flame speed, defined as the mean overall propagation velocity of the turbulent flame front, becomes larger than the laminar speed. If the turbulence is sufficiently strong the turbulent flame speed becomes independent of the laminar speed, and therefore of the microphysics of burning and diffusion, and scales only with the velocity of the largest turbulent eddy (Clavin 1994).

As the density of the white dwarf material declines and the laminar flamelets become slower and thicker, it is plausible that at some point turbulence significantly alters the thermal flame structure (Niemeyer & Woosley 1997). So far, modeling this so-called distributed burning regime in exploding white dwarfs has not been attempted explicitly since neither nuclear burning and diffusion nor turbulent mixing can be properly described by simplified prescriptions. However, it is this regime where the transition from deflagration to detonation is assumed to happen in certain phenomenological models.

### 3.2.2 Numerical models for turbulent combustion

Numerical methods to handle turbulent combustion are presently developed for combustion in engines and for reactor safety (see, e.g., Smilianovski et al. (1997), and references therein), and there is no principle problem in applying them also to supernovae. Moreover, simple microscopic models for turbulence can be calibrated to laboratory combustion experiments, which again will help to reduce the number of parameters (see Kerstein (1991, 1999) for a recent approach). The aim is always to model numerically unresolved scales and their impact on the resolved large scales.

Fortunately, this problem is not as complicated as it may seem to be. At high densities, near the center of the white dwarf, the thermal width of the thermonuclear combustion front is small and the turbulence intensity is moderate. The flame is said to be in the "flamelet"-regime and can be modelled by means of front capturing/tracking schemes (Smiljanovski et al., 1997), coupled to a turbulence model (Niemeyer & Hillebrandt, 1995; Schmidt et al. 2006). In this approach the front is tracked by a level set, and the thermodynamic properties of the matter before and after the front are reconstructed from the jump conditions for mass, energy, and momentum. This is possible if the normal velocity of the burning front is known which, in turn, can be obtained from the turbulence model, because in the limit of high propagation speed, the front propagates with the turbulent velocity and is independent of the laminar burning velocity.

### 3.2.3 Chandrasekhar-mass explosion models

We start by describing Chandrasekhar-mass explosion models (e.g., Röpke et al. (2007)). Alternative scenarios and models will be discussed in the next subsection. A typical evolution of such a SN Ia explosion modeled as described above is shown in Fig. 4. Starting from an ignition in multiple sparks the flame propagates outward. The mushroom-shaped features due to the buoyancy instability are clearly visible. Subsequently, the flame becomes increasingly corrugated and is accelerated by its interaction with turbulence. Therefore, it burns through a large fraction of the white dwarf material. The snapshot at $t = 0.6\,\mathrm{s}$ shows the flame evolution around the peak of energy production due to nuclear burning. Up to this point, the burning terminated in nuclear statistical equilibrium (NSE) and the carbon/oxygen material was primarily converted to iron group elements.

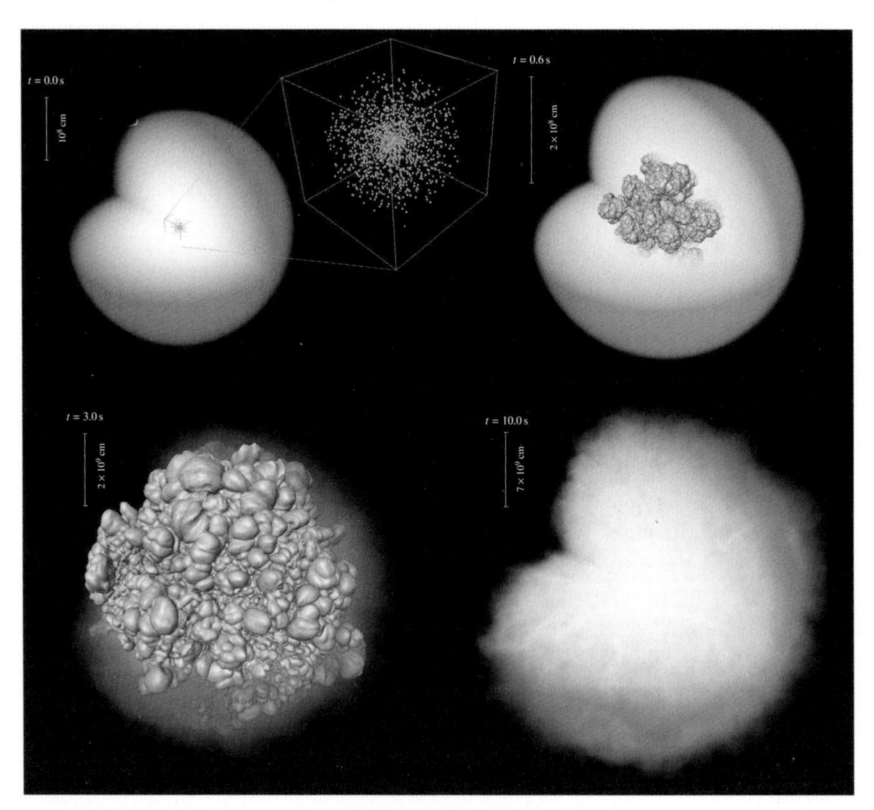


**Figure 4:** Snapshots from a full-star SN Ia pure-deflagation simulation starting from a multi-spot ignition scenario. The logarithm of the density is volume rendered indicating the extend of the WD star and the (blue) isosurface corresponds to the thermonuclear flame. The upper left panel shows the initial set up and the close-up illustrates the chosen flame ignition configuration. The last snapshot marks the end of the simulation and is not on scale with the earlier snapshots (Röpke et al. 2007).

The expansion of the WD decreases the fuel density steadily and once it falls below $5 \times 10^7$ g cm$^{-3}$ nuclear burning becomes incomplete and produces mainly intermediate mass elements. About 2 s after ignition, expansion quenches the burning and the following evolution is characterized by the relaxation to homologous expansion of the ejecta, which is reached to a reasonable accuracy about 10s after ignition (Röpke 2005).

As we shall discuss below, pure-deflagration Chandrasekhar-mass explosion models do not fit the lightcurves and spectra of "normal" SNe Ia well. A modification that brings the models closer to the observations is a transition of the mode of thermonuclear burning from a subsonic deflagration to a supersonic detonation in the course of the explosion, called the "delayed-detonation" scenario (Blinnikov & Khokhlov 1986; Khokhlov 1991). Since here the detonation propagates through a pre-expanded star that is brought out of hydrostatic equilibrium by the preceding deflagration phase, the detonation produces a layer of intermediate-mass elements that encompasses almost all of the outer ejecta, in agreement with observations (Röpke & Niemeyer 2007; Mazzali et al. 2007).

An open question, however, is whether or not such deflagration-to-detonation transitions (DDTs) occur in SNe Ia. Recent studies (Röpke 2007; Woosley 2007; Woosley et al. 2009) indicate that they may be possible. Numerical simulations have been carried out in which the DDTs are parametrized. In 1-D models (Hoeflich & Khokhlov 1996), a pre-defined transition density is assumed. Multi-D simulations, however, allow for a physically better motivated choice of parameters taking into account the strength of turbulent velocity fluctuations. Although DDT models are able to reproduce most of the properties of normal SNe Ia well, a potential problem appears to be that rate of formation of near-Chandrasekhar-mass white dwarfs is too low to explain the SN Ia rate (e.g., Ruiter et al. (2009)). A second problem are "subluminous" and "super-luminous" SNe Ia which have been discovered recently. The mass of radioactive Ni that is required in order to explain their luminosity is clearly in conflict with the predictions of all Chandrasekharmass models (see next subsection).

### 3.2.4 Non-Chandrasekhar mass explosion models

Non-Chandrasekhar mass explosion models come in two flavors. In sub-Chandrasekhar mass models, a white dwarf accretes helium from a binary companion (which may be a non-gegenerate He star or a He white dwarf). Above the C+O core, a He layer builds up and when it becomes sufficiently massive, a detonation triggers by compression. This detonation burns the He layer and drives a shock wave into the core. If strong enough, the shock wave may initiate a secondary detonation in the C+O core (either at the interface between the core and the He shell or at the center of the core). Sub-Chandrasekhar mass models were first studied by Woosley & Weaver (1994) and Livne & Arnett (1995) but were then kind of "forgotten" because the SNe Ia class appeared to be too homogeneous for them and, even more important, lightcurves and spectra predicted by those models seemed to contradict observations. From a population synthesis point of view, however, sub-Chandrasekhar mass scenarios could perhaps even account for the bulk of normal SNe Ia (Ruiter et al. 2009). In fact, recent multi-dimensional simulations show that once the He shell detonates,

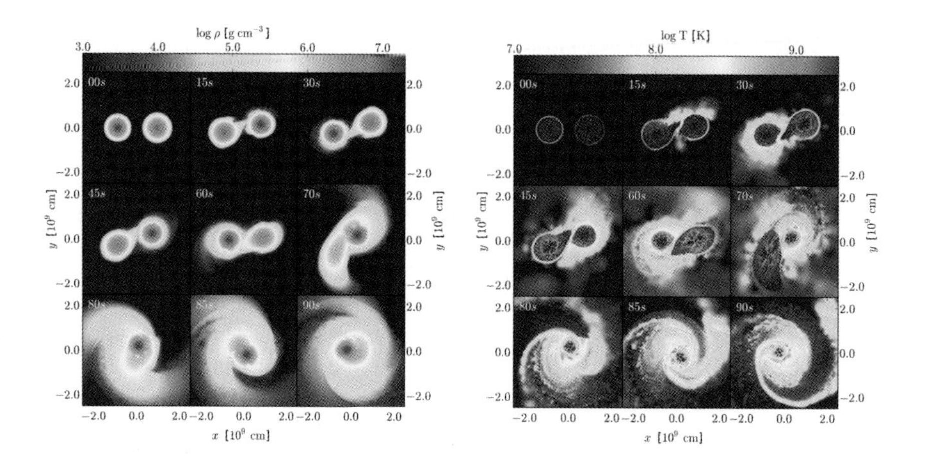


**Figure 5:** Snapshots of the evolution of a white-dwarf binary system during the inspiral. The system consists of two WDs of 0.90 and 0.81 $M_\odot$, respectively. Color-coded is the logarithm of the density (left panel) and of the temperature (right panel) (Pakmor et al. 2011).

the initiation of a core detonation is a very robust phenomenon, relatively independent of the He shell ignition geometry (Fink et al. 2007).

In contrast, super-Chandrasekhar mass explosions can result if two C+O white dwarfs merge. If the masses of the white dwarfs are unequal, the merger proceeds in a very asymmetric way. The lighter star is disrupted by tidal forces and may form an accretion disk around the heavier companion. Such WD-WD mergers are not expected to lead to thermonuclear explosions but rather to the formation of a neutron star by burning from an initial C+O to an O+Ne white dwarf and subsequent gravitational collapse (Saio et al.1985). The parameter space for WD-WD mergers, however, is rich and has not been fully explored yet.

In any case, explosions are more likely in the special case of the merger of two nearly equal mass white dwarfs. Due to the symmetry in the initial setup, a break-up of the lighter star is avoided. Instead, the stars merge violently within a few orbits. In the violent merger scenario (Pakmor et al. 2010, 2011), this leads to a thermonuclear explosion and disruption of both white dwarfs (see Fig. 5).

### 3.2.5 A few observable predictions

Again, we start with a discussion of Chandrasekhar-mass models. Apart from the initial conditions simulations as described above contain no free parameters. Therefore we may ask whether such models are capable of reproducing observations without any fitting. The explosion energies achievable in the outlined scenarios reach up to $\sim 8 \times 10^{50}$ erg for pure-deflagrations and can be up to 50% higher for delayed-detonation models, and the models produce up to $0.4\,M_\odot$ of $^{56}$Ni in case of pure deflagrations and up to about $1\,M_\odot$ for DDTs. This falls into the range of obser-

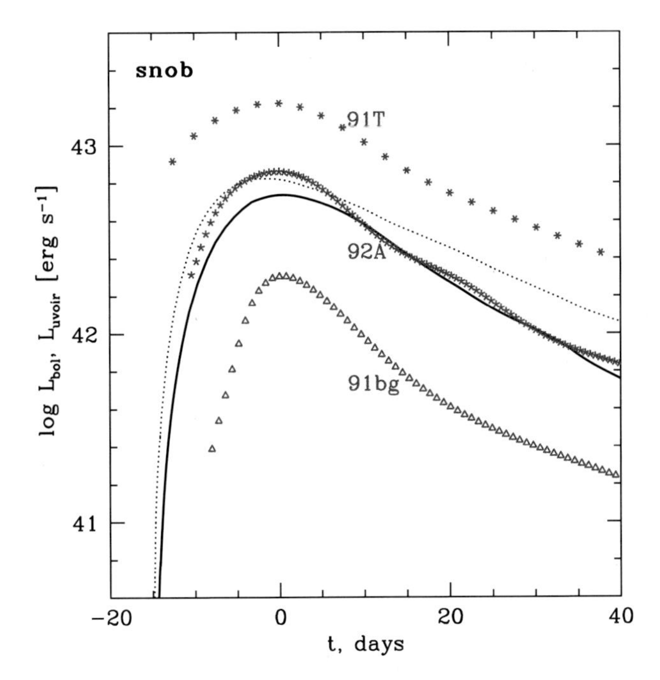


**Figure 6:** Bolometric light curve derived for the deflagration model of Fig. 4 (black curves; solid is the "UVOIR-bolometric" light curve and the complete bolometric light curve is dotted.) The blue dotted curves correspond to observed bolometric light curves from Stritzinger et al. (2006). SN 1991bg was the first example of a sub-luminous SN Ia, and SN 1991T is an example of a bright, but not "super-luminous" SN Ia. SN 1992A was a rather "normal" explosion (Röpke et al. 2007).

vational expectations (Contardo et al. 2000; Stritzinger et al. 2006). However, both classes of models cannot explain the sub-luminous nor the super-luminous SNe Ia. Sub-luminous explosions typically have Ni-masses around 0.1 $M_\odot$, about a factor of two below the minimum value obtained from Chandrasekhar-mass models. In contrast, in super-luminous explosions the Ni-mass may even exceed $M_{Chan}$ (e.g., SN 2009dc (Taubenberger et al. 2011; Silverman et al. 2011)).

Nonetheless, synthetic lightcurves derived from models of these class fit the observations in the $B$ and $V$ bands around maximum luminosity rather well (Sorokina & Blinnikov 2003; Blinnikov et al. 2006; Röpke et al. 2007; Woosley et al. 2007; Sim et al. 2007; Kasen, et al. 2007, 2008, 2009)(see Fig. 6). A much harder constraint on the explosion model is posed by spectral observations, since spectra are particularly sensitive to the chemical composition of the ejecta. Kozma et al. (2005) pointed out a potential problem of deflagration SN Ia models. In late time "nebular" spectra, unburned material (transported towards the center in down drafts due to the large-scale buoyancy-unstable flame pattern) gives rise to a strong oxygen line of low-velocity material which has never been observed. Clearly, delayed-detonation models do not have this problem. In more general terms, DDT models are able to fit both lightcurves and spectra of "normal" SNe Ia quite well, and even their observed width-luminosity relation (Kasen et al. 2009), whereas pure-deflagration models can only fit the low-luminosity end of the "normal" SN Ia well (Sim et al. 2011).

Detailed spectral observations also allow to determine the chemical composition of the ejecta in velocity space (Stehle et al. 2005). The mixed composition of the ejecta observed points to a deflagration phase being a significant contribution to most SN Ia explosions. The central parts are found to be clearly dominated by iron group elements, which are mixed out to velocities of about 12 000 km/s. Intermediate mass elements are distributed over a wide range in radii, and no unburned material is found at velocities below 5000 km/s. Again, high-resolution full-star $M_{Chan}$ SN Ia simulations can get close to these observational constraints as can do DDT models (Röpke et al. 2007; Tanaka et al. 2011)).

Recently, alternative progenitor channels have received renewed attention, mainly because the observed SNe Ia rate seems to be in conflict with population-synthesis models predicting a $M_{Chan}$ C+O white dwarf formation rate about a factor of 10 too low (Ruiter et al. 2009). In contrast, the rate of double-degenerate mergers as well as of He accreting sub- $M_{Chan}$ C+O white dwarfs and merging pairs of C+O and He white dwarfs can account for the rate more easily.

As far as sub-$M_{Chan}$ explosions are concerned, recent multi-dimensional simulations show that once the He shell detonates, the initiation of a core detonation is a very robust phenomenon for C+O core masses above, say, $0.8\,M_{\odot}$. It is relatively independent of the He shell ignition geometry (Fink et al. 2007, 2010). As in previous 1-D models, however, the nucleosynthesis expected from the detonation of a massive He shell above a C+O core is inconsistent with the observations (Hoeflich et al. 1996; Nugent et al. 1997: Kromer et al. 2010). This is mainly due to the fact that previous models predicted a substantial $^{56}$Ni production in this layer. Recently, models with larger core masses (around $1\,M_{\odot}$) have been studied. They have the advantage that a thin He shell may trigger a detonation (Bildsten et al 2007). Although the He shell detonation then does not produce significant amounts of $^{56}$Ni, other iron-group isotopes are still synthesized (Fink et al. (2010); see, however, Woosley & Kasen (2010)) in multi-dimensional simulations. In particular, absorption by Cr and Ti leads to strong flux redistribution towards the red parts of the spectrum making the models inconsistent with observations. This may cause a problem but it may be avoided if the He shell is polluted by carbon (Kromer et al. 2010), for instance by instabilities in the accretion or by He burning prior to the onset of the detonation. A carbon enrichment reduces the mass number of alpha elements that can be synthesized and with a $\sim$33% admixture, the spectrum agrees very well with the observations (see Fig. 7).

In contrast to sub-$M_{Chan}$ models which, in principle, can explain the full range of SN Ia (leaving aside the super-luminous ones for which no convincing model exists) violent mergers of two C+O white dwarfs of roughly equal mass can explain the class of sub-luminous events only, as was shown recently by Pakmor et al. (2010, 2011), at least for masses around $1\,M_{\odot}$. More massive white dwarf mergers may give rise to more luminous explosions, but they are so rare that they would not account for the bulk of the SNe Ia.

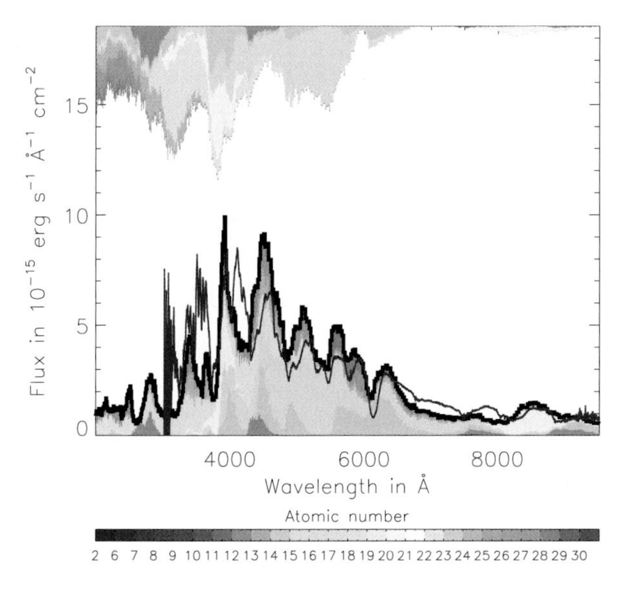


**Figure 7:** Angle-averaged (thick black line) spectra at three days before $B$-band maximum for a sub-$M_{Chan}$ model with $M_{core} = 1.025\,M_\odot$ and $M_{shell} = 0.055\,M_\odot$ and carbon enriched to 33%. For comparison the blue line shows the de-redshifted and de-reddened spectrum of SN 2004eo (Pastorello et al. 2006) at the corresponding epoch. The color coding indicates the element(s) responsible for both bound-bound emission and absorption of quanta in the Monte Carlo simulation. The region below the synthetic spectrum is color coded to indicate the fraction of escaping quanta in each wavelength bin which last interacted with a particular element (the associated atomic numbers are illustrated in the color bar). Similarly, the colored regions above the spectra indicate which elements were last responsible for removing quanta from the wavelength bin (either by absorption or scattering/fluorescence)(Kromer et al. 2010).

The reason for this finding is simple. Even under the optimistic assumption that the merged object is completely disrupted by a prompt detonation (which, however, seems to be a reasonable one (Pakmor et al. 2010, 2011)) nuclear burning happens at rather low density such that only a small fraction of the matter is incinerated to Fe-group elements, but most of it is burned to intermediate-mass elements or is left unburned. Consequently, the mass of radioactive Ni is low, of the order of $0.1\,M_\odot$, in agreement with the sub-luminous class. Moreover, because the total mass is rather large, about $2\,M_\odot$, despite of the high kinetic energy the ejecta velocity stays fairly low, again in agreement with observations. Finally, also the abundance tomography, carried out for the faint SN Ia 2005bl, strongly supports this conclusion (Hachinger et al. 2009).

# 4 Summary and conclusions

The physics of supernovae as well as the present status of numerical simulations of the explosion process have been reviewed. By means of a few examples it was demonstrated that multi-dimensional models are needed and that, to a certain extend, they have been carried out in 2-D with rather good physics input for core-collapse supernovae and in 3-D for thermonuclear explosions with most of the relevant physics included. 3-D core-collapse models with full neutrino transport are still out of reach for present supercomputers, but may become feasible in the near future. The bad news is that these very "expensive" computations have not solved the supernova problem yet, neither for explosions driven by stellar collapse nor for thermonuclear ones. In both cases the limiting factor does not seem to be the numerics only, but also still existing uncertainties in the relevant physics.

Since in case of core collapse supernovae it is likely that with the next generation of supercomputers one will be able to carry out fully resolved simulations in 3-D, most of the work has to go into improving the micro-physics and the neutrino transport. In contrast, for thermonuclear explosions one will never be able to resolve all relevant scales numerically and, therefore, developing new and clever tools to overcome this difficulty is a must. Here, in particular, the physics of the (desired) transition from a deflagration to a detonation in Chandrasekhar-mass C+O white dwarfs is crucial, but also the question which of the potential progenitor channels contributes how much to the observed population of SNe Ia is important, in particular for their use as cosmic distance indicators.

As far as observational constraints for the models are concerned, the theory of thermonuclear supernovae is certainly more advanced than that of explosions triggered by a collapsing stellar core. In fact, some SNe Ia models are able to reproduce lightcurves and spectra amazingly well, given the fact, that these models have almost no tunable (non-physical) parameters. The problem is, however, that rather different models do this equally well. Core-collapse models, in contrast, have to wait for future neutrino and gravitational wave detectors, or a galactic supernova, to test their predictions, because their lightcurves and spectra reflect more the physics in the outer parts of the star and not so much what is causing the explosion.

# References

Astier P., et al.: 2006, A&A 447, 31

Bacca S., Hally K., Pethick C. J., Schwenk A.: 2009, Phys. Rev. C 80, 2802

Bildsten L., Shen K. J., Weinberg N. N., Nelemans, G.: 2007, ApJ 662, L95

Blinnikov S. I., Khokhlov A. M.: 1986, Sov. Astron. Lett. 12, 131

Blinnikov S. I., et al.: 2006, A&A 453, 229

Blondin J. M., Mezzacappa A., DeMarino C.: 2003, ApJ 584, 971

Branch D. et al.: 1995, PASP 107, 1019

Brandt T. D., Burrows A., Ott, C. D., Livne, E.: 2011, ApJ 728, 8

Bruenn, S. W., et al.: 2010, e-print 2010arXiv1002.4914B

Buras R., Rampp M., Janka H.-T., Kifonidis K.: 2003, Phys. Rev. Lett. 90, id. 241101

Buras R., Rampp M., Janka H.-Th., Kifonidis K.: 2006a, A&A 447, 1049

Buras R., Janka H.-Th., Rampp M., Kifonidis K.: 2006b, A&A 457, 281

Burrows A., Lattimer, J. M.: 1986, ApJ 307, 178

Burrows, A., Hayes J., Fryxell B. A.: 1995, ApJ 450, 830

Cardall C. Y.: 2005, Nuc. Phys. B 138, 436

Chandra D., Goyal A., Goswami K.: 2002, Phys. Rev. D 65, 3003

Cerdà-Duràn P., et al.: 2005, A&A 439, 1033

Clavin P.: 1994, Annual Rev. Fluid Mech. 26, 321

Contardo G., Leibundgut B., Vacca W. D.: 2000, A&A 359, 876

Dessart L., et al.: 2006, ApJ 644, 1063

Dimmelmeier H., Novak J., Font J. A., Ibanez J. M., Müller, E.: 2005, Phys. Rev. D 71, 4023

Dimmelmeier H., Ott C. D., Janka H.-T., Marek A., Müller E.: 2007, Phys. Rev. Lett. 98, id. 251101

Filippenko A.V.: 1988, ApJ 96, 1941

Fink M., Hillebrandt W., Röpke F. K.: 2007, A&A 476, 1133

Fink M., et al.: 2010, A&A 514, 53

Fryer, C. L.: 1999, ApJ 522, 413

Fryer, C. L., Warren M. S.: 2002, ApJ 574, 65

Fryer, C. L., Warren M. S.: 2004, ApJ 601, 391

Garnavich P.M. et al.: 1998, ApJ 509, 74

Filippenko, A.V.: 1989, Proc. Conf. Particle Astrophysics: Forefront Experimental Issues, Berkeley; Eric B. Norman (ed.), World Scientific Publ., p.177

Fischer T., et al.: 2010, Classical and Quantum Gravity 27, 114102

Foley R. J., et al.: 2009, AJ 137, 3731

Ganeshalingam M., et al.: 2010, ApJS 190, 418

Hachinger S., et al.: 2009, MNRAS 399, 1238

Herant M., Benz W., Hix W. R., Fryer C. L., Colgate S. A.: 1994, ApJ 435, 339

Hillebrandt W., Nomoto K., Wolff R. G.: 1984, A&A 133, 175

Hillebrandt W.: 1994, in Supernovae; NATO Advanced Science Institutes (ASI) Series, eds. S.Bludman, R. Mochkovitch & J. Zinn-Justin; Elsevier Science, North-Holland, p.251

Hillebrandt W., Niemeyer J. C.: 2000, ARAA 38,191

Hoeflich P., Khokhlov A.: 1996, ApJ 457, 500

Hoeflich P., et al.: 1996, ApJ 472, 81

Hoyle F., Fowler W.A.: 1960, ApJ 132, 565

Hüdepohl L., et al.: 2010, Phys. Rev. Lett. 104, id. 251101

Janka H.-Th., Kitaura, F., Marek, A.: 2007a: in Supernova 1987A: 20 Years After, AIP Conference Proceedings, 937, p. 144

Janka H.-Th., Langanke K., Marek A., Martínez-Pinedo G., Müller B.: 2007b, Phys. Rep. 442, 38

Kasen D., Woosley S. E., Nugent P., Röpke F. K.: 2007, Journal of Physics 78, 12037

Kasen D., Thomas, R. C., Röpke F. K., Woosley S. E.: 2008, Journal of Physics 125, 12007

Kasen D., Röpke F. K., Woosley S. E.: 2009, Nature 462, 624

Keil W., Janka H.-T.: 1995, A&A 296, 145

Keil W., Janka H.-Th., Müller E.: 1996, ApJ 473, L111

Keil W., Raffelt G., Janka H.-Th.: 2003, ApJ 590, 971

Kerstein A.R.: 1991, J. Fluid Mech. 231, 361

Kerstein A.R.: 1999, J. Fluid Mech. 392, 277

Khokhlov A. M.: 1991, A&A 245, 114

Kifonidis K., Plewa T., Janka H.-Th., Müller E.: 2003, A&A 408, 621

Kitaura F. S., Janka H.-Th., Hillebrandt W.: 2006, A&A 450, 345

Kotake K., Sato K., Takahashi K.: 2006, Rep. Progr. Phys. 69, 971

Kotake K. Ohnishi N., Yamada S.: 2007, ApJ 655, 406

Kotake K., Iwakami W., Ohnishi N., Yamada S.: 2009, ApJ 697, 133

Kowalski, M., et al.: 2008, ApJ 686, 749

Kozma C., et al.: A&A 437, 983

Kromer, M., et al.: 2010, ApJ 719, 1067

Langanke K., Martinez-Pinedo G.: 2000, Nuc. Phys. A 673, 481

Langanke K., Martinez-Pinedo G.: 2003, Rev. Mod. Phys. 75, 819

Langanke K., et al.: 2003, Pys. Rev. Lett. 90, id. 241102

Lattimer J. M., Swesty D. F.: 1991, Nuc. Phys. A 535, 331

Liebendörfer, M., et al.: 2003, Nuc. Phys. A 719, 144

Liebendörfer, M., et al.: 2004, ApJs 150, 263

Liebendörfer, M., Rampp M., Janka H.-Th., Mezzacappa A.: 2005, ApJ 620, 840

Liebendörfer, M., et al.: 2009, Nuc. Phys. A 827, 573

Livne E., Arnett D.: 1995, ApJ 452, 62

Lykasov G. I., Pethick C. J., Schwenk, A.: 2008, Phys.Rev. C 78, 5803

Marek A., et al.: 2006, A&A 445, 273

Marek A., Janka H.-Th.: 2009, ApJ, 694, 664

Marek A., Janka H.-Th., Müller E.: 2009, A&A 496, 475

Mazzali P. A., Röpke F. K., Benetti S., Hillebrandt W.: 2007, Science 315, 825

Mazzali P. A., et al.: 2007, ApJ 670, 592

Messer B., Mezzacappa, A., Bruenn, S., Guidry, M.: 1998, e-print, astro-ph/9805276

Müller, B. Janka, H.-T., Dimmelmeier, H.: 2010, ApJS 189. 104

Müller E., et al.: 2004, ApJ 603, 221

Murphy J. W., Ott C. D., Burrows A.: 2009, ApJ 707, 1173

Nadyozhin D. K.: 2003, MNRAS 346, 97

Niemeyer J.C., Hillebrandt W.: 1995, ApJ 452, 769

Niemeyer. J. C., Woosley, S. E.: 1997, ApJ 475, 740

Nomoto K., Thielemann F.-K., Yokoi, K.: 1984, ApJ 286, 644

Nordhaus J., Burrows A., Almgren A., Bell J.: 2010, ApJ 720, 694

Nordin, J. et al.: 2011, A&A 526, 119

Nugent, P., et al.: 1997, ApJ 485, 812

Ott C. D., Burrows A., Dessart L., Livne E.: 2006, Phys. Rev. Lett 96, id. 201102

Ott C. D., et al.: 2007a, Classical and Quantum Gravity 24, 139

Ott C. D., et al.: 2007b, Phys. Rev. Lett. 98, id. 261101

Pakmor R., et al.: 2010, Nature 463, 61

Pakmor R., Hachinger S., Röpke F. K., Hillebrandt W.: 2011, A&A, in print

Pastorello A., et al.: 2006, MNRAS 370, 1752

Perlmutter, S., et al.: 1999, ApJ 517, 565

Phillips, M. M.: 1993, ApJ 413, 105

Pons J. A., et al.: 1999, ApJ 513, 780

Raffelt G., Seckel D., Sigl G.: 1995, Phys. Rev. D 54, 2784

Raffelt G., Strobel T.: 1997, Phys. Rev. D 55, 523

Rampp M., Janka H.-T.: 2002, A&A 396, 361

Reddy S., Prakash M., Lattimer J.M.: 1998, Phys. Rev. D 58, 3009

Reddy S., Sadzikowski M., Tachibana M.: 2003, Nuc. Phys. A 714, 337

Reinecke M., Hillebrandt W., Niemeyer J.C.: 1998, Proc. 16th Int. Conf on Numerical Methods in Fluid Dynamics; Arcachon, France, Springer, Lecture Notes in Physics 515, 542

Riess A. G., et al.: 1998, AJ 116, 1009

Riess A. G., et al.: 2004, ApJ 607, 665

Röpke, F. K.: 2005, A&A 432, 969

Röpke, F. K.: 2007, ApJ 668, 1103

Röpke, F. K., Niemeyer, J. C.: 2007, A&A 464, 683

Röpke, F. K., et al.: 2007, ApJ 668, 1132

Ruiter, A. J., Belczynski, K., Fryer, C.: 2009, ApJ 699, 2026

Sahu D. K., Anupama G. C., Srividya S., Muneer, S.: 2006, MNRAS 372, 1315

Saio H., Nomoto K.: 1985, A&A 150, L21

Scalzo R. A., et al.: 2010, ApJ 713, 1073

Scheck L., Janka H.-Th., Foglizzo T., Kifonidis K.: 2008, A&A 477, 931

Schmidt B. P., et al.: 1998, ApJ 507, 46
Shibata M., Sekiguchi Y.: 2004, Phys. Rev. D 69, 4024
Shibata M., Sekiguchi Y.: 2005, Phys. Rev. D 71, 4014
Shen H., Toki H., Oyamatsu K., Sumiyoshi K.: 1998, Nuc. Phys. A 637, 435
Silverman J. M., et al.: 2011, MNRAS 410, 585
Sim S. A., Sauer D. N., Röpke F. K., Hillebrandt W.: 2007, MNRAS 378, 2
Sim S. A., et al.: 2011, to be submitted
Smartt S. J.: 2009, ARA&A 47, 63
Smiljanovski V., Moser V., Klein R.: 1997, Combust. Theory Modelling 1, 183
Sorokina E., Blinnikov S.: 2003, e-print arXiv:astro-ph/0309146
Stritzinger M., Leibundgut B., Walch S., Contardo G.: 2006, A&A 450, 241
Tanaka M., et al.: 2011, MNRAS 410, 1725
Taubenberger S., et al.: 2009, MNRAS 397, 677
Taubenberger S., et al.: 2011, e-print arXiv:1011.5665, MNRAS, in print
Tominaga N., et al.: 2008, ApJ 687, 1208
Tonry J. L., et al.: 2003, ApJ 594, 1
Totani T., Sato K., Dalhed H. E., Wilson J. R.: 1998, ApJ 496, 216
Tsvetkov D. Yu., et al.: 2006, A&A 460, 769
Utrobin V. P.: 2007, A&A 461, 233
Wheeler J.C., Harknesss R.P.: 1990, Rep. Prog. Phys. 55, 1467
Woosley S. E.: 2007, ApJ 668, 1109
Woosley S. E., Weaver, T. A.: 1994, ApJ 423, 371
Woosley S. E., Kasen D.: 2010, e-print arXiv:1010.5292
Woosley S. E., Kasen D., Blinnikov S., Sorokina E.: 2007, ApJ 662, 487
Woosley S. E., et. al: 2009, ApJ 704, 255

# The Facility for Antiproton and Ion Research: A new era for supernova dynamics and nucleosynthesis

Karlheinz Langanke

GSI Helmholtzzentrum für Schwerionenforschung
Darmstadt, Germany

k.langanke@gsi.de

Technische Universität Darmstadt, Germany

and

Frankfurt Institute for Advanced Studies
Frankfurt, Germany

**Abstract**

*In the next years the Facility for Antiproton and Ion Research FAIR will be constructed at the GSI Helmholtzzentrum für Schwerionenforschung in Darmstadt, Germany. This new accelerator complex will allow for unprecedented and path-breaking research in hadronic, nuclear, and atomic physics as well as in applied sciences. This manuscript will discuss some of these research opportunities, with a focus on supernova dynamics and nucleosynthesis.*

## 1 The Facility for Antiproton and Ion Research FAIR

FAIR is the next generation facility for fundamental and applied research with antiproton and ion beams (Henning *et al.* 2001). It will provide worldwide unique accelerator and experimental facilities, allowing for a large variety of unprecedented fore-front research in physics and applied sciences. FAIR is an international project with 16 partner countries and more than 2500 scientists and engineers involved in the planning and construction of the accelerators and associated experiments. FAIR will be realized in a stepwise approach. The Modularized Start Version, comprising the Heavy-Ion Synchroton SIS100, the antiproton facility, the Superconducting Fragment Separator and experimental areas and novel detectors for atomic, hadron, heavy-ion, nuclear, and plasma physics and applications in material sciences and biophysics, is expected to be operational in 2016. Completion of FAIR, including the synchroton SIS300, to follow soon after.

*Reviews in Modern Astronomy 23: Zoomimg in: The Cosmos at High Resolution.* First Edition.
Edited by Regina von Berlepsch.

The main thrust of FAIR research focuses on the structure and evolution of matter on both a microscopic and on a cosmic scale – bringing our Universe into one laboratory. In particular, FAIR will expand our knowledge in various scientific fields beyond current frontiers addressing:

- the investigation of the properties and the role of the strong (nuclear) force in shaping basic building blocks of the visible world around us and its role in the evolution of the universe;
- tests of symmetries and predictions of the standard model and search for physics beyond in the electro-weak sector and in the domain of the strong interaction;
- the properties of matter under extreme conditions, both at the subatomic as well as the macroscopic scale of matter;
- applications of high-intensity, high-quality ion and antiproton beams in research areas that provide the basis for, or directly addresses, issues of applied sciences and technology.

The basis for the realization of these wide-spread and pathbreaking research activities are worldwide unique accelerator and experimental facilities. FAIR will allow, in comparison to the existing GSI facilities, to increase the intensities of secondary beams by factors 100 to 10000 and the beam energies by factors 15-20. Moreover, the use of beam cooling techniques enables the production of antiproton and ion beams of highest quality, i.e. with very precise energy and extremely fine profile. Upon completion, the FAIR accelerator complex can support up to 5 experimental programs simultaneously with beams of different ion species in genuine parallel operation. This unique feature is made possible by an optimal balance of the usage of the various accumulator, collector and experimental storage rings.

In this manuscript we will discuss some of the outstanding research opportunities at FAIR, concentrating on forefront questions relevant to supernova dynamics and nucleosynthesis, as they are part of the program of the NuSTAR collaboration. This gives only a restricted glimpse on the research program at FAIR, omitting important aspects. For example,

- the CBM/HADES experiment will explore the phase diagram of Quantum-chromodynamics (QCD) in the region of very high baryon densities and moderate temperatures, supplementing related programs at the Relativistic Hadron Collider and the Large Hadron Collider, which probe the phase diagram at high temperatures and almost zero net baryon densities. The aim of CBM/HADES is to study the nuclear matter equation of state, to search for the predicted first order phase transition between hadronic and partonic matter, and to search for the QCD critical point and the chiral phase transition;
- the highest-quality antiproton beams and the extreme energy resolution of the PANDA detector allow to produce and unravel the structure of exotic hadronic states made of quarks and gluons testing the predictions of QCD and bringing

forward our understanding of the generation of mass in the world of the strong interaction. The PANDA experiment will also extend the nuclear chart into the third dimension of strangeness by producing many new hypernuclei. This will also include $\Lambda\Lambda$ hypernuclei opening here the possibility to experimentally study the interaction of two $\Lambda$ particles and the $N\Lambda\Lambda$ interaction which both are important requirements for the study of neutron star structure;

- the experimental program of the FLAIR collaboration with low-energy antiprotons focuses on the production of novel configurations of antimatter like anti-hydrogen atoms and molecules and anti-helium ultimately using these physical systems for tests of matter-antimatter symmetry;
- for plasma physics the availability of high-energy, high-intensity ion-beams enables the investigation of High-Energy Dense Matter in regimes of temperature, density and pressure, as they exist insides of stars and large planets, but which have not been experimentally accessible so far;
- experiments by the APPA collaboration with stored relativistic heavy ions, like uranium, which are stripped of most of their electrons and have a helium- or lithium-like configuration will study electron correlations in extreme electromagnetic fields like those expected on the surface of neutron stars and perform test of the fundamental symmetries of QED;
- the study of the biological effectiveness of high-energy and high-intensity beams is a prerequisite to estimate the expected radiation damage induced by cosmic rays on Moon and Mars missions and to solve related protection issues.

# 2 Introductory remarks to nuclear astrophysics

Nuclear astrophysics has developed in the last twenty years into one of the most important subfields of 'applied' nuclear physics. It is a truly interdisciplinary field, concentrating on primordial and stellar nucleosynthesis, stellar evolution, and the interpretation of cataclysmic stellar events like novae and supernovae.

The field has been tremendously stimulated by recent developments in laboratory and observational techniques. In the laboratory the development of radioactive ion beam facilities as well as low-energy underground facilities have allowed to remove some of the most crucial ambiguities in nuclear astrophysics arising from nuclear physics input parameters. This work has been accompanied by significant progress in nuclear theory which makes it now possible to derive some of the input at stellar conditions based on microscopic models. Nevertheless, much of the required nuclear input is still insufficiently known. Here, decisive progress is expected once radioactive ion beam facilities of the next generation, like FAIR, are operational. The nuclear progress goes hand-in-hand with tremendous advances in observational data arising from satellite observations of intense galactic gamma-sources, from observation and analysis of isotopic and elemental abundances in deep convective Red Giant and Asymptotic Giant Branch stars, and abundance and dynamical studies of nova ejecta and supernova remnants. Recent breakthroughs have also been obtained for measuring the solar neutrino flux, giving clear evidence for neutrino oscillations and confirming the solar models. Also, the latest developments in modeling stars, novae, x-ray bursts, type I supernovae, and the identification of the neutrino wind driven shock in type II supernovae as a possible site for the r-process allow now much better predictions from nucleosynthesis calculations to be compared with the observational data.

It is impossible to present all these exciting developments in this manuscript. We will rather focus on the nuclear processes important for the dynamics of core-collapse supernovae and their associated explosive nucleosynthesis.

# 3 Electron capture in core-collapse supernovae

Massive stars end their lifes as core-collapse supernovae, triggered by a collapse of their central iron core with a mass of more than 1 $M_{\odot}$ (Bethe 1990). Despite of improved description of the microphysics entering the simulations and sophisticated neutrino transport treatment spherically symmetric simulations of core-collapse supernovae fail to explode (Buras *et al.* 2003, Janka and Rampp 2000, Mezzacappa *et al.* 2001) (however, see (Sagert *et al.* 2009) who consider a quark-hadron phase transition at relatively low densities); i.e. the energy transported to the stalled shock, which triggers the explosion, by absorption of neutrinos on free neutrons and protons is not sufficient to revive the shock wave which has run out of energy by dissociating the matter it traverses into free nucleons. However, the revival is successful in two-dimensional simulations as convection supported by hydrodynamical instabilities increases the efficiency by which neutrinos carry energy to the shock region. Successful two-dimensional simulations are reported in (Burrows *et al.* 2006, Marek and Janka 2009, Hix *et al.* 2009).

# 4 Electron capture in core-collapse supernovae

During most of the collapse the equation of state is given by that of a degenerate relativistic electron gas (Bethe 1990); i.e. the pressure against the gravitational contraction arises from the degeneracy of the electrons in the stellar core. However, the electron chemical potential at core densities in excess of $10^8$ g/cm$^3$ is of order MeV and higher, thus making electron captures on nuclei energetically favorable. As these electron captures occur at rather small momentum transfer, the process is dominated by Gamow-Teller transitions; i.e. by the $\mathrm{GT}_+$ transitions, in which a proton is changed into a neutron.

When the electron chemical potential $\mu_e$ (which grows with density like $\rho^{1/3}$) is of the same order as the nuclear $Q$-value, the electron capture rates are very sensitive to phase space and require a description of the detailed $\mathrm{GT}_+$ distribution of the nuclei involved which is as accurate as possible. Furthermore, the finite temperature in the star requires the implicit consideration of capture on excited nuclear states, for which the $\mathrm{GT}_+$ distribution can be different than for the ground state. It has been demonstrated (Caurier *et al.* 1999, Langanke *et al.* 2000) that modern shell model calculations are capable to describe $GT_+$ distributions rather well (Frekers 2005) (an example is shown in Figure 1) and are therefore the appropriate tool to calculate the weak-interaction rates for those nuclei ($A \sim 50 - 65$) which are relevant at such densities (Langanke and Martinez-Pinedo 2003). At higher densities, when $\mu_e$ is sufficiently larger than the respective nuclear $Q$ values, the capture rate becomes less sensitive to the detailed $GT_+$ distribution and depends practically only on the total GT strength. Thus, less sophisticated nuclear models might be sufficient. However, one is facing a nuclear structure problem which has been overcome only very recently. Once the matter has become sufficiently neutronrich, nuclei with proton numbers $Z < 40$ and neutron numbers $N > 40$ will be quite abundant in the core. For such nuclei, Gamow-Teller transitions would be Pauli forbidden (Fuller 1982) ($GT_+$ transitions change a proton into a neutron in the same harmonic oscillator shell) were it not for nuclear correlation and finite temperature effects which move nucleons from the $pf$ shell into the $gds$ shell. To describe such effects in an appropriately large model space (e.g. the complete $fpgds$ shell) is currently only possible by means of the Shell Model Monte Carlo approach (SMMC) (Langanke *et al.* 1995, Koonin, Dean and Langanke 1997). In (Langanke *et al.* 2003) SMMC-based electron capture rates have been calculated for more than 100 nuclei.

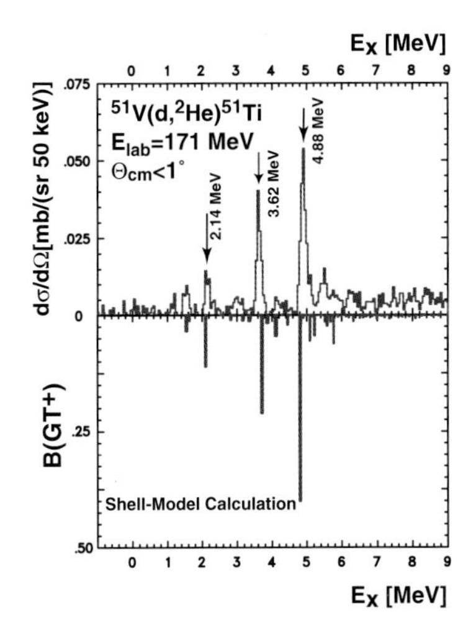


**Figure 1:** Comparison of the measured $^{51}V(d,^2He)^{51}Ti$ cross section at forward angles (which is proportional to the $GT_+$ strength) with the shell model GT distribution in $^{51}V$ (from (Bäumer *et al.* 2003)).

Recently the $GT_+$ strength for $^{76}Se$ (with $Z = 34$ and $N = 42$) has been measured (Grewe *et al.* 2008). Indeed it is experimentally observed that cross-shell correlations induce a non-vanishing $GT_+$ strength which is well reproduced by shell model calculations (Sieja and Nowacki 2010). These studies consider the $(pf)$ orbitals for protons and the $(p, f_{5/2}, g_{9/2})$ orbitals for neutrons. The calculations also reproduce the experimentally determined orbital occupation numbers (Kay *et al.*

2008), predicting about 3.5 neutron holes in the $(pf)$ shell which make $GT_+$ transitions possible. Figure 2 compares the electron capture rates determined from the shell model and experimental GT ground state distributions. These rates do not include contributions from thermally excited nuclear states and hence correspond to stellar capture rates at finite temperatures only if one assumes that the $GT_+$ distributions are the same for all parent states (often called Brink's hypothesis). For comparison also the SMMC/RPA capture rates are shown. These rates are smaller than the others at smaller temperatures indicating that the model is not completely capable to resolve the strength distribution at low excitation energies. The agreement gets significantly improved at the larger temperatures (and densities) at which the SMMC/RPA model is used to predict stellar capture rates.

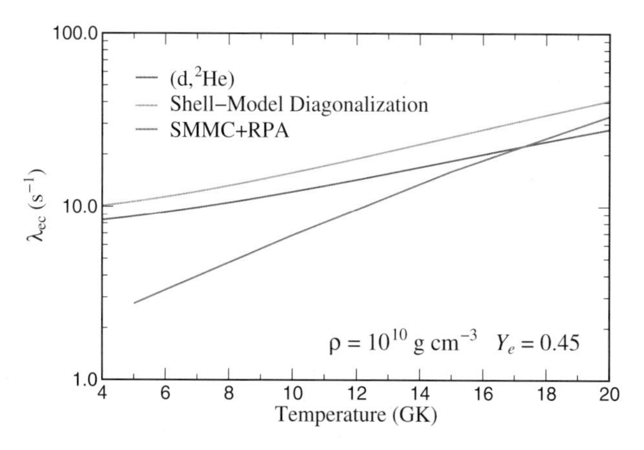


**Figure 2:** Electron capture rates on $^{76}$Se as function of temperature (in units of $10^9$ K) derived from the experimental Grewe, E.W. et al. (2008) and shell model (Sieja, K., Nowacki, F.) $GT_+$ strength distributions. A density of $10^{10}$ g/cm$^3$ and an electron-to-nucleon ratio $Y_e = 0.45$ are assumed. For comparison also the SMMC/RPA rates are shown.

The shell model results for $fd$ shell nuclei (Langanke *et al.* 2000) and the SMMC/RPA results for heavier nuclei (Langanke *et al.* 2003) have been combined for a compilation of electron rates which covers the composition during the early collapse phase well (until densities to a few $10^{10}$ g/cm$^3$). However, at even higher densities the continuous electron capture drives the matter composition to more neutron-rich and heavier nuclei than considered in (Langanke *et al.* 2003). The neglect of these nuclei could lead to a systematic overestimate of the capture rates as the neglected nuclei have larger $Q$ values and enhanced Pauli blocking due to increased neutron excess than the nuclei considered. Unfortunately an SMMC evaluation of the several thousands of nuclei present in the matter composition during the late phase of collapse before neutrino trapping is numerically not feasible. Hence a simpler approach has to be adopted. This is based on the observation that the single particle occupation numbers obtained in the SMMC calculations can be well approximated by a parametrized Fermi-Dirac distribution. By adjusting the parameters of this distribution to the SMMC results for about 250 nuclei, occupation numbers were derived for nearly 3000 nuclei and the respective individual rates have been calculated within an RPA calculation based on these partial occupations. It is found that this simplified approach reproduces the SMMC results quite well at electron chemical potential $\mu_e > 15$ MeV (Juodagalvis *et al.* 2010). This corresponds to the density regime where the previously neglected nuclei become abundant in the matter composition. At smaller electron chemical potentials (i.e. at lower densities) details of the GT distribution are important which are not recovered by the simple parametrized approach. However, at these conditions electron capture is dominated by nuclei for which individual shell model rates exist.

Figure 3 compares the capture rate derived for the pool of more than 3000 nuclei (i.e. combining the rates from shell model diagonalization, SMMC, and from the parametrized approach) with those obtained purely on the basis of the shell model results. While the agreement is excellent at small electron chemical potentials (here the shell model rates dominate), the rates for the large pool are slightly smaller at higher $\mu_e$ values due the presence of neutron-rich heavy nuclei with smaller individual rates. Furthermore the new rates also include plasma screening effects which lead to an increase of the effective $Q$ values and a reduction of the electron chemical potential, which both reduce the electron capture rates. The effect is rather mild and does not alter the conclusion that electron capture on nuclei dominates over capture on protons during the collapse.

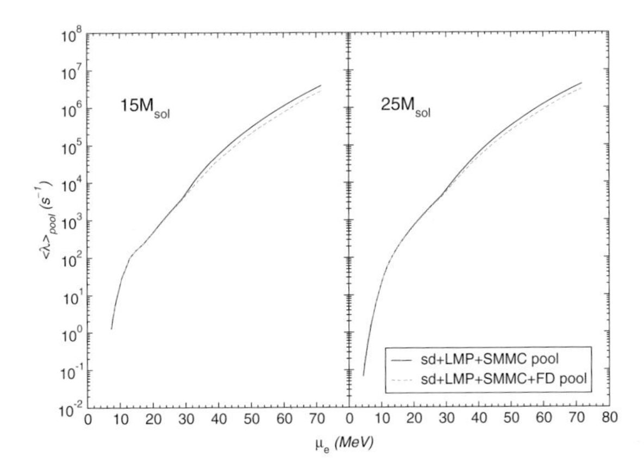


**Figure 3:** Comparison of NSE-averaged electron capture rates calculated for about 3000 individual nuclei (solid, see text) with those obtained for the restricted set of nuclei (dashed) considered in (Langanke *et al.* 2003) (from (Juodagalvis *et al.* 2010)).

Shell model capture rates have significant impact on collapse simulations (see Figs. 4,5). In the presupernova phase ($\rho < 10^{10}$ g/cm$^3$) the captures proceed slower than assumed before and for a short period during silicon burning $\beta$-decays can compete (Heger *et al.* 2001a, Heger *et al.* 2001b). As a consequence, the core is cooler, more massive and less neutronrich before the final collapse. However, until recently

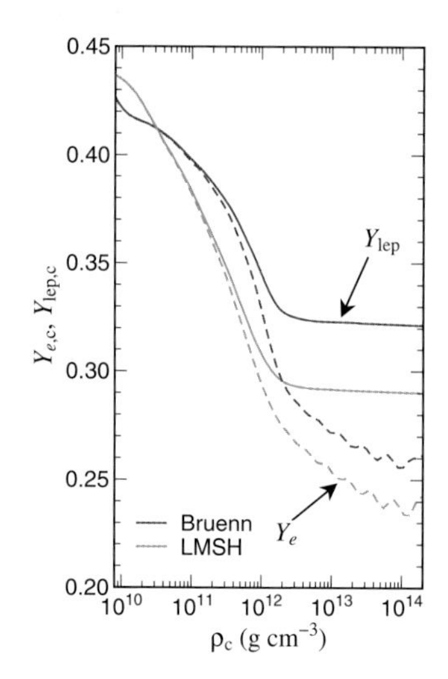


**Figure 4:** The electron and lepton fraction in the center during the collapse phase. The thin lines are a simulation using the Bruenn parameterization (Bruenn 1985), while the thick lines are for a simulation using the shell model rates. (courtesy of Hans-Thomas Janka)

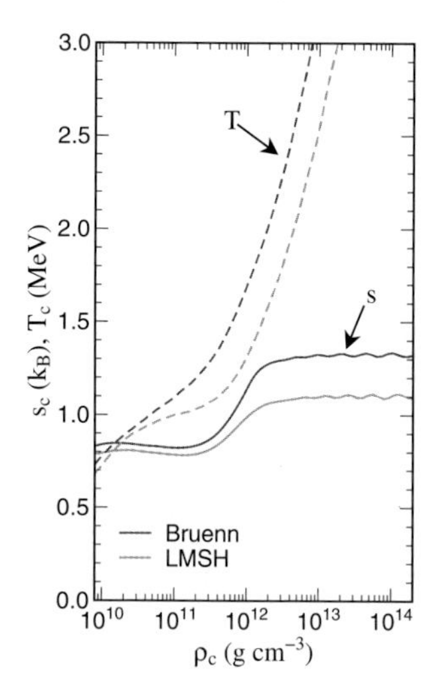


**Figure 5:** The temperature and entropy in the center during the collapse phase. The thin lines are a simulation using the Bruenn parameterization (Bruenn 1985), while the thick lines are for a simulation using the shell model rates. (courtesy of Hans-Thomas Janka)

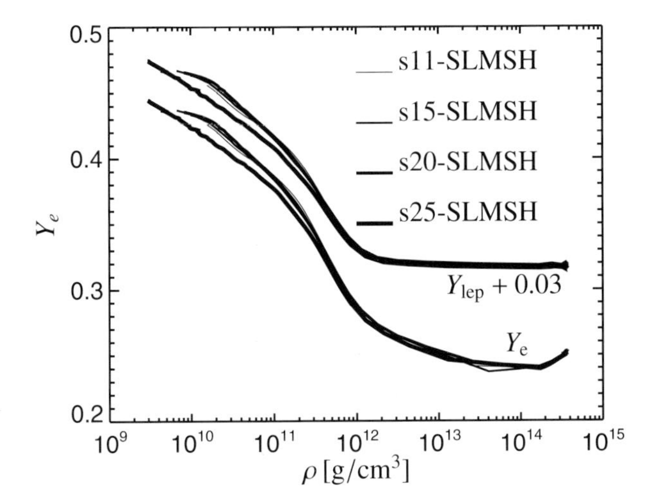


**Figure 6:** The evolution of the electron-to-baryon ratio $Y_e$ and the lepton-to-baryon ratio $Y_{lep}$ as function of central density for stars with masses between 11 and 25 $M_\odot$. (from (Janka *et al.* 2007)).

simulations of this final collapse assumed that electron captures on nuclei are prohibited by the Pauli blocking mechanism, mentioned above. However, based on the SMMC calculations it has been shown in (Langanke *et al.* 2003) that capture on nuclei dominates over capture on free protons. The changes compared to the previous simulations are significant (Langanke *et al.* 2003, Hix *et al.* 2003, Janka *et al.* 2007). Importantly the shock is now created at a smaller radius with more infalling material to traverse, but also the density, temperature and entropy profiles are strongly modified (Hix *et al.* 2003).

The possibility to capture electrons on nuclei and free protons leads to a strong self-regulation during the collapse and it is observed that the collapse approaches the same core trajectory for stars between 11 and 25 solar masses before neutrino trapping and thermalization is reached at densities around $10^{12}$ g/cm$^3$ (Fig. 6, (Janka *et al.* 2007)). Thus it is expected that supernovae, independent of the stellar mass, have a quite similar neutrino burst spectrum. This spectrum, however, has also significantly changed after inelastic neutrino-nucleus reactions have been included in the simulations (Fig. 7, Langanke *et al.* 2008).

## 5 Supernova nucleosynthesis

In a successful explosion the shock heats the matter it traverses, inducing an explosive nuclear burning on short time-scales. This explosive nucleosynthesis can alter the elemental abundance distributions in the inner (silicon, oxygen) shells. Recently explosive nucleosynthesis has been investigated consistently within supernova simulations. These studies found that in the early phase after the bounce the

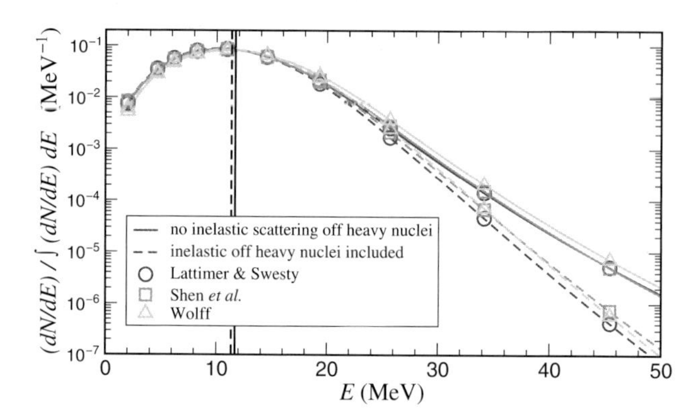


**Figure 7:** Comparison of the normalized neutrino spectra, arising from the $\nu_e$ burst shortly after bounce, without (solid) and with (dashed) consideration of inelastic neutrino-nucleus scattering in the supernova simulations. (from (Langanke *et al.* 2008)).

ejected matter is protonrich (Thielemann *et al.* 2003, Pruet *et al.* 2005, Fröhlich *et al.* 2006a), giving rise to a novel nucleosynthesis process (the $\nu$p-process (Fröhlich *et al.* 2006b)). In later stages, the matter becomes neutronrich (Wanajo 2006) allowing for the r-process to occur.

## 5.1 The $\nu$p process

The studies presented in (Pruet *et al.* 2006, Fröhlich *et al.* 2006b, Wanajo 2006) show that matter with $Y_e$ larger than 0.5 will always be present in core-collapse supernovae explosions with ejected matter irradiated by a strong neutrino flux, independently of the details of the explosion. As this proton-rich matter expands and cools, nuclei can form resulting in a composition dominated by $N = Z$ nuclei, mainly $^{56}$Ni and $^{4}$He, and protons. Without the further inclusion of neutrino and antineutrino reactions the composition of this matter will finally consist of protons, alpha-particles, and heavy (Fe-group) nuclei (in nucleosynthesis terms a proton- and alpha-rich freeze-out), with enhanced abundances of $^{45}$Sc, $^{49}$Ti, and $^{64}$Zn (Pruet *et al.* 2005, Fröhlich *et al.* 2006a). In these calculations the matter flow stops at $^{64}$Ge with a small proton capture probability and a beta-decay half-life (64 s) that is much longer than the expansion time scale ($\sim$ 10 s) (Pruet *et al.* 2005, Pruet *et al.* 2006).

As noted by Martinez-Pinedo and explored in (Fröhlich *et al.* 2006b) and in (Pruet *et al.* 2006) the synthesis of nuclei with $A > 64$ can be obtained, if one explores the previously neglected effect of neutrino interactions on the nucleosynthesis of heavy nuclei. $N \sim Z$ nuclei are practically inert to neutrino capture (converting a neutron into a proton), because such reactions are endoergic for neutron-deficient nuclei located away from the valley of stability. The situation is different for antineutrinos that are captured in a typical time of a few seconds, both on protons and nuclei, at the distances at which nuclei form ($\sim$ 1000 km). This time scale is much shorter than the beta-decay half-life of the most abundant heavy nuclei reached without neu-

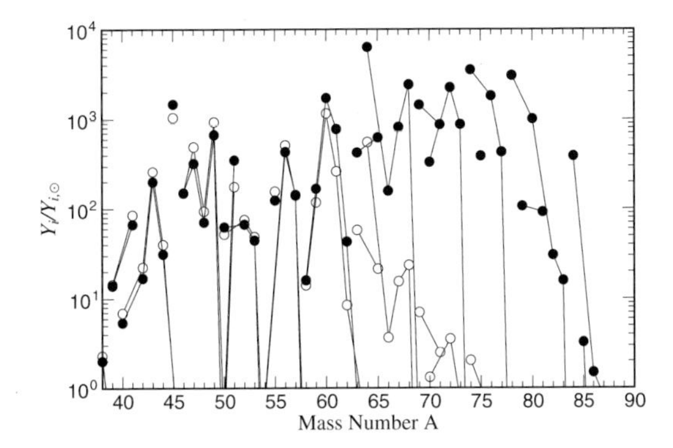


**Figure 8:** Elemental abundance yields (normalized to solar) for elements produced in the protonrich environment shortly after the supernova shock formation. The matter flow stops at nuclei like $^{56}$Ni and $^{64}$Ge (open circles), but can proceed to heavier elements if neutrino reactions are included during the network (full circles) (from (Fröhlich *et al.* 2006b).

trino interactions (e.g. $^{56}$Ni, $^{64}$Ge). As protons are more abundant than heavy nuclei, antineutrino capture occurs predominantly on protons, causing a residual density of free neutrons of $10^{14}$–$10^{15}$ cm$^{-3}$ for several seconds, when the temperatures are in the range 1–3 GK. The neutrons produced via antineutrino absorption on protons can easily be captured by neutron-deficient $N \sim Z$ nuclei (for example $^{64}$Ge), which have large neutron capture cross sections. The amount of nuclei with $A > 64$ produced is then directly proportional to the number of antineutrinos captured. While proton capture, $(p, \gamma)$, on $^{64}$Ge takes too long, the $(n, p)$ reaction dominates (with a lifetime of 0.25 s at a temperature of 2 GK), permitting the matter flow to continue to nuclei heavier than $^{64}$Ge via subsequent proton captures.

Fröhlich *et al.* (2006b) argue that all core-collapse supernovae will eject hot, explosively processed matter subject to neutrino irradiation and that this novel nucleosynthesis process (called $\nu$p process) will operate in the innermost ejected layers producing neutron-deficient nuclei above $A > 64$. However, how far the mass flow within the $\nu$p process can proceed, strongly depends on the environment conditions, most noteably on the $Y_e$ value of the matter. Obviously the larger $Y_e$, the larger the abundance of free protons which can be transformed into neutrons by antineutrino absorption. Figure 9 shows the dependence of the $\nu$p-process abundances as a function of the $Y_e$ value of the ejected matter. Nuclei heavier than $A = 64$ are only produced for $Y_e > 0.5$, showing a very strong dependence on $Y_e$ in the range 0.5–0.6. A clear increase in the production of the light $p$-nuclei, $^{92,94}$Mo and $^{96,98}$Ru, is observed as $Y_e$ gets larger. Thus the $\nu p$ process offers the explanation for the production of these light $p$-nuclei, which was yet unknown.

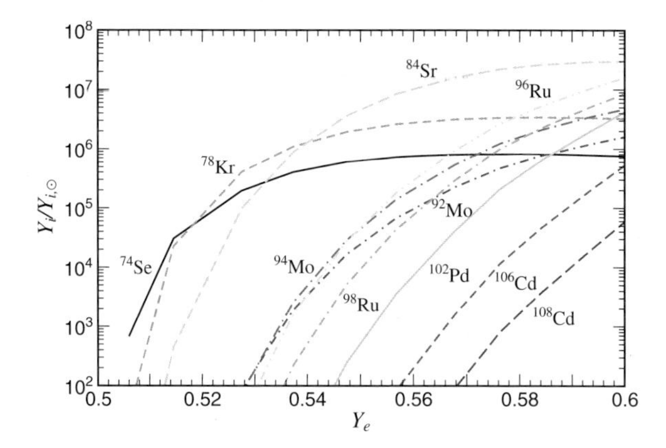


**Figure 9:** Light $p$-nuclei abundances in comparison to solar abundances as a function of $Y_e$. The $Y_e$-values given are the ones obtained at a temperature of 3 GK that corresponds to the moment when nuclei are just formed and the $\nu p$-process starts to act (from (Fröhlich *et al.* 2006b).

## 5.2 The r-process

About half of the elements heavier than mass number $A \sim 60$ are made within the r-process, a sequence of rapid neutron captures and $\beta$ decays (Burbidge *et al.* 1957, Cameron 1957). The process occurs in environments with extremely high neutron densities (Cowan, Thielemann and Truran 1991). Then neutron captures are much faster than the competing decays and the r-process path runs through very neutron-rich, unstable nuclei. Once the neutron source ceases, the process stops and the produced nuclides decay towards stability producing the neutronrich heavier elements.

Despite many promising attempts the actual site of the r-process has not been identified yet. However, parameter studies have given clear evidence that the observed r-process abundances cannot be reproduced at one site with constant temperature and neutron density (Kratz *et al.* 1993). Thus the abundances require a superposition of several (at least three) r-process components. This likely implies a dynamical r-process in an environment in which the conditions change during the duration of the process. The currently favored r-process sites (core-collapse supernovae (Woosley *et al.* 1994) and neutron-star mergers (Freiburghaus, Rosswog and Thielemann 1999)) offer such dynamical scenarios. However, recent meteoritic clues might even point to more than one distinct site for our solar r-process abundance (Wasserburg *et al.* 2006). The same conclusion can be derived from the observation of r-process abundances in low-metallicity stars (Sneden *et al.* 2003), a milestone of r-process research.

The r-process path runs through such extremely neutron-rich nuclei that most of their properties (i.e mass, lifetime and neutron capture cross sections) are experimentally unknown and have to be modelled, based on experimental guidance. The most important nuclear ingredient in r-process simulations are the nuclear masses as they determine the flow-path. They are traditionally modelled by empirical mass formulae parametrized to the known masses (Möller, Nix and Kratz 1997, Möller *et al.* 1995, Aboussir *et al.* 1995, Borsov, Goriely and Pearson 1997). A new era has been opened very recently, as for the first time, nuclear mass tables have been derived on the basis of nuclear many-body theory (Hartree-Fock-Bogoliobov model) (Goriely *et al.* 2005, Goriely *et al.* 2009, Stone 2005) rather than by parameter fit to data.

The nuclear halflives strongly influence the relative r-process abundances. In a simple $\beta$-flow equilibrium picture the elemental abundance is proportional to the halflive, with some corrections for $\beta$-delayed neutron emission (Kratz *et al.* 1988, Martinez-Pinedo and Langanke 1999). As r-process halflives are longest for the magic nuclei, these waiting point nuclei determine the minimal r-process duration time; i.e. the time needed to build up the r-process peak around $A \sim 200$. We note, however, that this time depends also crucially on the r-process path and can be as short as a few 100 milliseconds if the r-process path runs close to the neutron dripline.

There are a few milestone halflife measurements including the N=50 waiting point nuclei $^{80}$Zn, $^{79}$Cu, $^{78}$Ni and the N=82 waiting point nuclei $^{130}$Cd and $^{129}$Ag (Kratz *et al.* 1986, Hosmer *et al.* 2005, Lettry *et al.* 1998). Although no halflives for $N = 126$ waiting points have yet been determined, there has been decisive progress towards this goal recently (Kurtukian-Nieto *et al.* 2009). These data play crucial roles in constraining and testing nuclear models which are still necessary to predict the bulk of halflives required in r-process simulations. It is generally assumed that the halflives are dominated by allowed Gamow-Teller (GT) transitions, with forbidden transitions contributing noticeably for the heavier r-process nuclei (Borzov 2003, Kurtukian-Nieto *et al.* 2009).

If the r-process occurs in strong neutrino fluences, different neutrino-induced charged-current (e.g. $(\nu_e, e^-)$) and neutral-current (e.g. $(\nu, \nu')$) reactions, which are often accompanied by the emission of one or several neutrons (Fuller and Meyer 1995, Qian *et al.* 1997, Haxton *et al.* 1997), have to be modelled and included as well. Recently neutrino-induced fission has been suggested to explain the robust r-process pattern observed in old, metalpoor stars (Qian 2002). Respective fission rates and yield distributions have been calculated on the the basis of the RPA model (Kolbe, Langanke and Fuller 2004, Kelic *et al.* 2006).

Recently Martinez-Pinedo and Petermann have developed an r-process code which consistently couples an extended nuclear network with trajectories describing the dynamics of the astrophysical environment. The network explicitly accounts for all relevant reactions including the various fission processes (spontaneous, neutron-, beta- and neutrino-induced fission) and their yield distributions (Benlliure *et al.* 1998). First applications studied the r-process in a supernova environment. In this neutrino-driven wind scenario (Woosley *et al.* 1994, Qian and Woosley, 1996, Hoffman, Woosley and Qian 1997) matter is ejected as free protons and neutrons. Interactions with the extreme neutrino fluence produced by neutron star cooling makes the matter ejected a few seconds after bounce presumably neutron-rich. Upon reaching cooler regions with increasing distance from the neutron star surface, nucleons can be combined into nuclei. This nucleosynthesis includes reactions involving protons and $\alpha$-particles, but at a certain temperature charged-particle reaction rates are too slow compared to the matter expansion rate. The abundance distribution of heavier nuclei at this freeze-out becomes the seed for continuous captures of the remaining free neutrons which, together with eventual $\beta$-decays, constitutes r-process nucleosynthesis. Modern supernova simulations show that the matter is ejected with rather high velocities, until it runs into the slower moving matter of the outer shells, where it is decelerated. As temperature is a crucial parameter for r-process nucleosynthesis it is important whether matter reaches the deceleration phase fast and is kept hence relatively long at high temperatures. We will in the following assume such a 'hot environment'; i.e. we have chosen a time evolution of temperature where, after a fast exponential decrease, matter is kept at $T \approx 10^9$ K for a few seconds. Furthermore we have adopted the matter parameters such to allow for sufficiently large neutron-to-seed ratios so that the r-process reaches the range of fissioning nuclei (Petermann 2010). The studies focussed on three interesting topics.

### 5.2.1 Robustness of r-process abundances

Figure 10 shows the elemental abundances obtained in r-process simulations with various numbers for the neutron-to-seed (n/s) ratio and using the ETFSI nuclear input. In all cases the n/s value is large enough that nuclei in the actinide mass range and beyond are reached during the process. We observe the striking result that the elemental abundances do basically not change if the n/s value is varied. This is true for the total and relative abundances of the r-process elements (for $Z > 40$). Only small variations are observed for elements at and just before the lead peak. Indeed the n/s values adopted in these studies are large enough that fission and fission cycling play a role in determining the final abundances. Further studies are planned to investigate

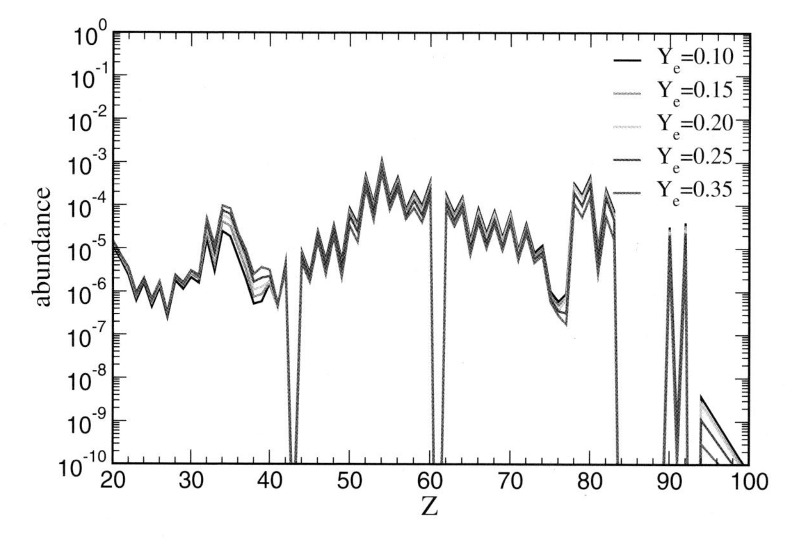


**Figure 10:** Elemental abundances for r-process simulations in the 'hot environment' using different neutron-to-seed ratios. The calculations used nuclear input based on the ETFSI model.

whether and under which conditions fission might explain the robust abundance patterns between the second and third r-process peaks as observed in metal-poor stars.

We have repeated the r-process calculations, however, assuming a 'cold environment'; i.e. the matter temperature during the deceleration period was set to $10^8$ K. Under such conditions charged-particle reactions are too slow with the consequence that the continuous production of fresh seeds from lighter nuclei during this later r-process phase is suppressed, in contrast to the hot environment. A second significant difference between r-process nucleosynthesis in the hot and cold environments is that a $(\gamma, n) \leftrightarrow (n, \gamma)$ equilibrium is established during most of the nucleosynthesis in the hot scenario (as temperatures are hot enough to induce sufficiently large photodissociation rates) while this is not the case in the cold environment. Nevertheless we find quite similar elemental r-process abundances for both scenarios (see Fig. 11).

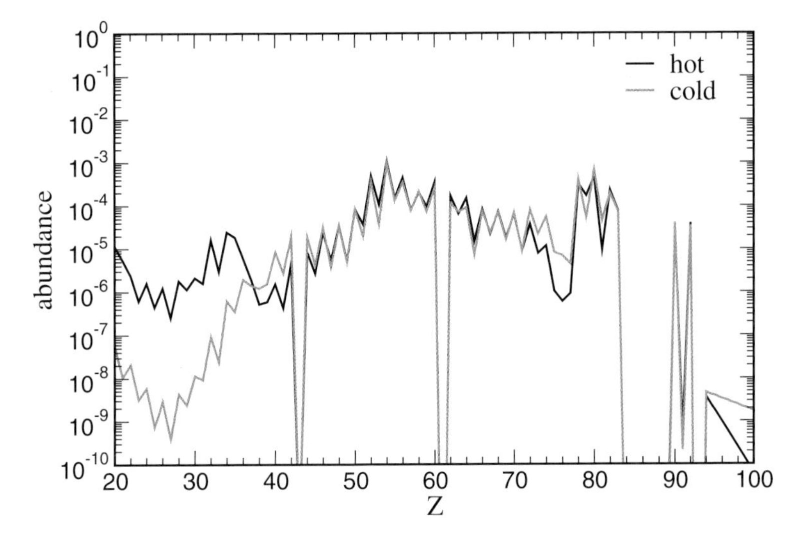


**Figure 11:** Comparison of elemental abundances for r-process simulations in hot and cold environments. The calculations used nuclear input based on the ETFSI model.

We note two interesting further observations of our studies. We also find robust elemental abundance pattern when using the other nuclear input sets and varying the environment parameters in a range that the r-process reaches the nuclear chart region of fissioning nuclei. However, the abundance patterns obtained for different nuclear input are distinct from each other indicating the sensitivity of the r-process to the nuclear ingredients. Secondly, we observe that isotopic r-process abundance distributions show a noticeably larger sensitivity to variation of the environmental parameters than do elemental abundances. Hence, observation of isotope abundances in metal-poor stars might tell us how robust the r-process really is.

### 5.2.2 Making longlived superheavies by r-process

Due to nuclear models the existence of long-lived superheavy elements with charge numbers around $Z \sim 120$ is conceivable. While the production of such elements is the goal of experimental attempts at different laboratories, Nature might have chosen the r-process to produce such nuclides (Panov, Kornnev and Thielemann 2009). Figure 12 shows that such a proposal is interesting as indeed the r-process mass-flow in our simulation reaches nuclides in excess of mass number $A = 300$. However, such nuclides are extremely neutronrich surpassing the next predicted magic neutron number $N = 184$ (which acts like those at $N = 82$ and 126 as an obstacle to the mass-flow). Once the neutron supply ceases, these short-lived neutronrich r-process progenitor nuclei will decay. The progenitor decay might produce long-lived superheavies (expected around $Z = 120, N = 184$) if the entire decay path is dominated by $\beta$-decays. If it, however, runs through regions in the nuclear chart where fission rates are faster than the competing $\beta$-decays (or $\alpha$-decays), the progenitors split into medium-mass fragments. Using the ETFSI-based nuclear input, we have indicated in Fig. 13 which of the four decay mechanisms (spontaneous and $\beta$-delayed fission, $beta$- and $\alpha$-decay) is predicted to be fastest. We find that for this model input no superheavies would be generated (even if they existed) as the decay path leads through regions of strong (spontaneous) fission dominance. We find similar results for the FRDM and HFM nuclear inputs, where again the decay path has to traverse regios in the nuclear chart where fission is predicted to dominate and hence would prohibit the production of superheavies. Nevertheless these results do not rule out production of superheavies by the r-process as the currently available nuclear input is too uncertain.

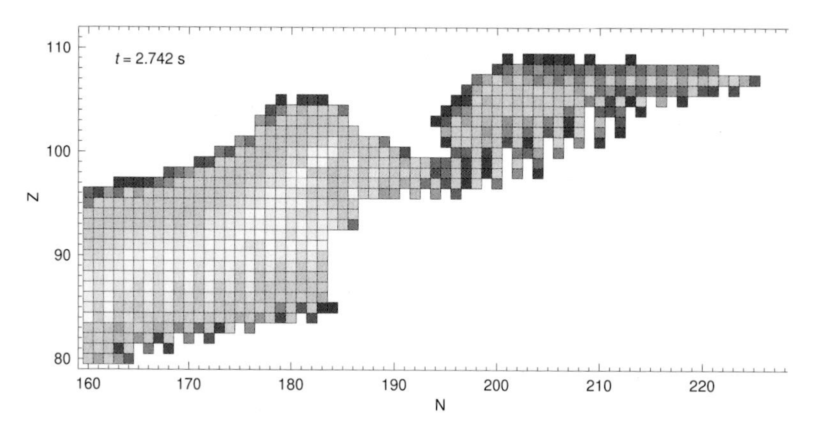


**Figure 12:** Abundance distribution of the heaviest nuclei in an r-process simulation using HFB-based nuclear input.

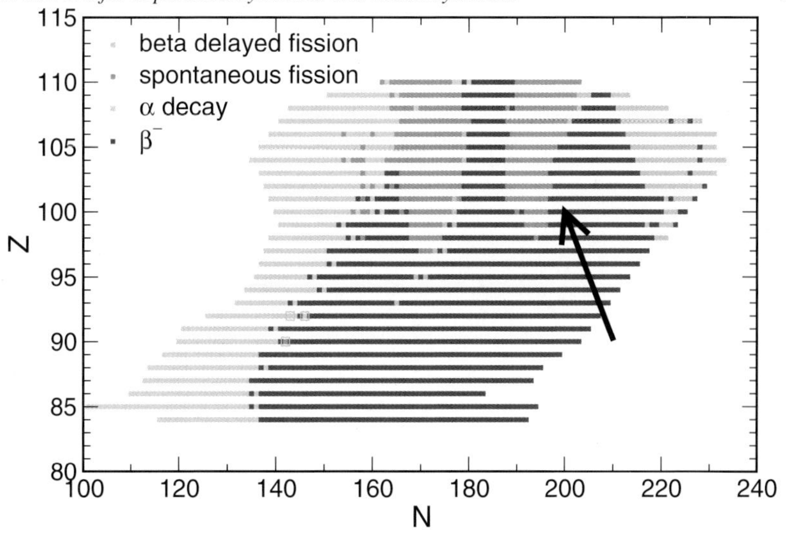


**Figure 13:** Dominant decay rates for heavy nuclei, as determined from nuclear input based on the ETFSI model. The arrow indicates the decay path of r-process progenitors towards the region of expected long-lived superheavies.

### 5.2.3 U/Th abundance ratio as cosmic chronometer

The simultaneous observation of U- and Th-lines in the atmospheres of very old metal-poor stars in our galactical halo was a milestone of recent r-process achievements. The lines presumably correspond to the long-lived isotopes $^{232}$Th and $^{238}$U with half-lives of several billion years which makes them attractive as cosmic chronometers. Such age determinations are performed on the basis of ratios of abundances assuming that many nuclear and astrophysical uncertainties cancel in the required prediction of the initial abundance ratio (Cowan *et al.* 1999, Goriely and Arnould 2001, Dauphas 2005, Schatz *et al.* 2002). This expectation is particularly true for the U/Th pair being neighbors in the nuclear chart and hence should be co-produced in the r-process.

Our calculation follows the time evolution of abundances of several thousand nuclei, including their decay to stability after the neutron supply has dried out. Figure 14 shows the abundance evolution for several long-lived isotopes relevant for the production of $^{232}$Th and $^{238}$U, performed with nuclear input derived on the basis of the FRDM. One observes that $^{232}$Th is dominantly produced by $\alpha$-decays of $^{236}$U and $^{244}$Pu with half-lives of a few $10^7$ y, while the $^{238}$U abundance is strongly boosted from the decay of $^{242}$Pu. Hence the 'initial' abundance ratio of $^{232}$Th and $^{238}$U before onset of their radioactive decay involves a larger range of the nuclear chart including also Pu and Np isotopes. This introduces a larger sensitivity to nuclear model input. Indeed calculations performed with input based on the ETFSI and HFB models predict a quite different history of the initial abundance ratio (Fig. 15).

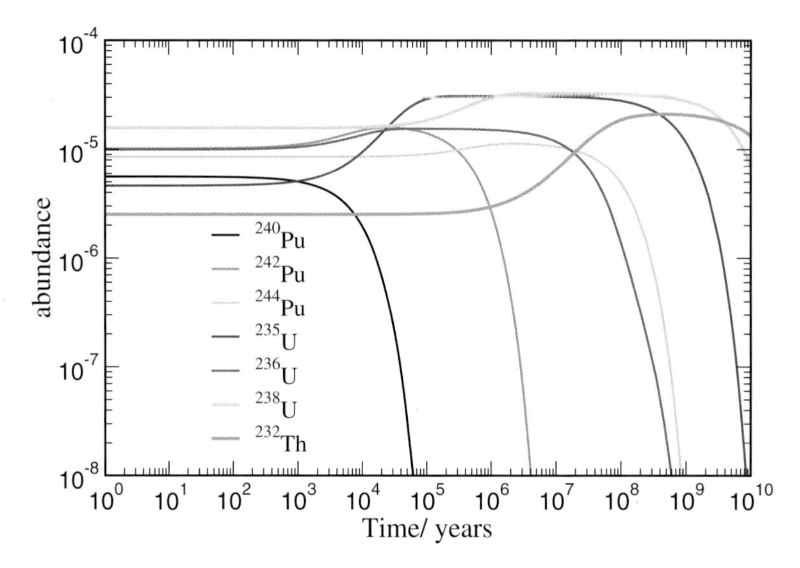


**Figure 14:** Time evolution of abundances of long-lived isotopes calculated with input based on the FRDM.

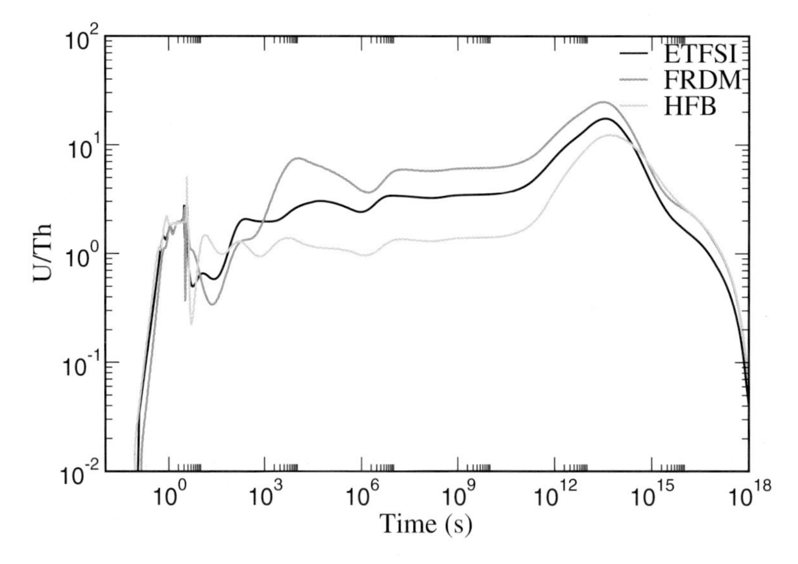


**Figure 15:** Time evolution of the U/Th abundance ratio for three different nuclear input sets.

## 5.3 Neutrino nucleosynthesis

When neutrinos, produced in the hot supernova core, pass through the outer shells of the star, they can induce nuclear reactions and in this way contribute to the elementsynthesis (the $\nu$-process, (Woosley *et al.* 1990)). For example, the nuclides $^{11}$B and $^{19}$F are produced by $(\nu, \nu'n)$ and $(\nu, \nu'p)$ reactions on the quite abundant nuclei $^{12}$C and $^{20}$Ne. These reactions are dominantly induced by $\nu_\mu$ and $\nu_\tau$ neutrinos and their antiparticles (combined called $\nu_x$ neutrinos) which have larger average energies (about 20 MeV) than $\nu_e$ and $\bar{\nu}_e$ neutrinos. As found in detailed stellar evolution studies (Heger *et al.* 2005) the rare odd-odd nuclides $^{138}$La and $^{180}$Ta are mainly made by the charged-current reaction $^{138}$Ba$(\nu_e, e^-)^{138}$La and $^{180}$Hf$(\nu_e, e^-)^{180}$Ta. Hence, the $\nu$-process is potentially sensitive to the spectra and luminosity of $\nu_e$ and $\nu_x$ neutrinos, which are the neutrino types not observed from SN1987a. Both charged-current cross sections are dominated by the low-energy tail of the GT$_-$ strength distribution which is notoriously difficult to simulate. As large-scale shell model calculations are yet not possible for $^{138}$Ba or $^{180}$Hf, the charged-current cross sections had been evaluated within the random phase approximation (Heger *et al.* 2005). As a major improvement it has been possible last year to measure the GT$_-$ strengths on $^{138}$Ba and $^{180}$Hf below the particle thresholds and to convert these data into the relevant $(\nu_e, e^-)$ cross sections (Byelikov *et al.* 2007). It is found that the new cross sections are somewhat larger than the RPA predictions. Upon using these results in stellar evolution models, slightly larger $^{138}$La and $^{180}$Hf abundances are found, which agree reasonably well with observation (Byelikov *et al.* 2007).

# 6 Summary

Advances in nuclear modelling have led to important progress in simulations of supernova dynamics and nucleosynthesis. Using two variants of shell model approaches it has been possible to derive electron capture rates for many of the nuclei present in the collapsing core of a massive star. The calculations for medium-mass nuclei in the mass range $A \sim 45 - 65$, based on the diagonalization shell model, have been validated by impressive Gamow-Teller strength distribution measurements mainly performed at the KVI Groningen. As a major result the SMMC-based studies predict that blocking of GT$_+$ transitions for nuclei with $Z < 40$ and $N > 40$ is overcome by correlations, a fact which has now been experimentally verified by measurements for $^{76}$Se. As a consequence electron captures occur predominantly on nuclei rather than free protons during the entire collapse, in contrast to previous assumptions. Significant progress has also been achieved in calculating neutrino-induced nuclear reactions. This includes inelastic neutrino scattering off nuclei, which, however, has only mild effects on the supernova dynamics.

After bounce matter is ejected as free nucleons from the surface of the hot neutron star. In the early phase, the ejected matter is proton-rich giving rize to the recently discovered $\nu$p process. At later times ejected matter might presumably be neutron-rich leading then to r-process nucleosynthesis. The $\nu$p process might be the origin for the synthesis of the light p nuclides $^{92,94}$Mo and $^{96,98}$Ru, as indicated in parameter studies. Improving the nuclear input to r-process simulations is a challenging effort. Unfortunately most of the nuclei on the r-process path have yet not been experimentally produced so that theory is needed to predict their properties. Despite noticeable progress global predictions of the masses, halflives and fission rates and yields of the very neutron-rich nuclei involved in r-process nucleosynthesis is still quite uncertain. Decisive progress is expected to come from next-generation radioactive ion-beam facilities like FAIR in Germany, FRIB in the US or FRIB in Japan. Despite these uncertainties simulations of the r-process in the dynamical environment of the supernova neutrino-driven wind scenario show interesting opportunities to explore the use of the U/Th clock as chronometer for old stars or the possibility of production of superheavy elements within the r-process.

The authors acknowledge support of the Helmholtz Alliance EMMI, the SFB 634 of the Deutsche Forschungsgemeinschaft and of the Helmholtz International Center for FAIR. The collaboration with A. Heger, W.R. Hix, Thomas Janka, Andrius Juodagalvis, Ilka Petermann, Friedel Thielemann, Stan Woosley and, in particular, Gabriel Martinez-Pinedo is appreciated.

## References

Aboussir, Y. *et al.*: 1995, At. Data Nucl. Data Tables **61** 127

Bäumer, C. *et al.*: 2003, Phys. Rev. **C68** 031303(R)

Benlliure, J. *et al.*: 1998, Nucl. Phys. **A628** 458

Bethe, H.A.: 1990, Rev. Mod. Phys. **62** 801

Borsov, I.N., Goriely, S., Pearson, J.M.: 1997, Nucl. Phys. **A621** 307c

Borzov, I.N.: 2003, Phys. Rev. **C67** 025802

Bruenn, S.W.: 1985, Astr. J. Suppl. **58** 771

Buras, R., Rampp, M., Janka, H.-Th., Kifonidis, K.: 2003, Phys. Rev. Lett. 90 241101

Burbidge, E.M., Burbidge, G.R., Fowler, W.A., Hoyle, F.: 1957, Rev. Mod. Phys. **29** 547

Burrows, A. *et al.*: 2006, Astr. j. **640** 878

Byelikov, A. *et al.*: 2007, Phys. Rev. Lett. **98** 082501

Cameron, A.G.W.: 1957, Chalk River Report CRL-41

Caurier, E., Langanke, K., Martínez-Pinedo, G., Nowacki, F.: 1999, Nucl. Phys. A **653** 439.

Cowan, J.J., Thielemann, F.-K., Truran, J.W.: 1991,

Cowan, J.J. *et al.*: 1999, Astro. J. **521** 194

Dauphas, N.: 2005, Nature **435** 03645

Freiburghaus, C., Rosswog, S., Thielemann, F.-K.: 1999, Astr. J. **525** L121

Frekers, D.: 2006, Prog. Part. Nucl. Phys. **57** 217

Fröhlich, C. *et al.*: 2006a, Astr. J. **637** 415

Fröhlich, C. *et al.*: 2006b, Phys. Rev. Lett. **96** 142502

Fuller, G.M.: 1982, Astr. J. **252** 741

Fuller, G.M., Meyer, B.S.: 1995, Astrophys.J. 453 792

Goriely, S., Arnould, M.: 2001, A&A **379** 1113

Goriely, S., Samyn, M., Pearson, J.M., Onsi, M.: 2005, Nucl. Phys. **A750** 425

Goriely, S. *et al.*: 2009, Phys. Rev. **C79** 024612

Grewe, E.W. *et al.*: 2008, Phys. Rev. **C78** 044301

Haxton, W.C., Langanke, K., Qian Y.Z., Vogel, P.: 1997, Phys. Rev. Lett. **78** 2694

Heger, A. *et al.*: 2001a, Phys. Rev. Lett. **86** 1678

Heger, A. *et al.*: 2001b, Astr. J. **560** 307

Heger, A. *et al.*: 2005, Phys. Lett. **B606** 258

Henning, W.F. *et al.*: 2001, An *International Accelerator Facility for Beams of Ions and Antiprotons*, GSI Darmstadt, http://www.gsi.de/GSI Future/cdr/

Hix, R.W. *et al.*: 2003, Phys. Rev. Lett. **91** 210102

Hix, W.R. *et al.*: 2009, Nucl. Phys. **A834** 602c

Hoffman, R.D., Woosley, S.E., Qian, Y.Z.: 1997, Astr. J. **482** 951

Hosmer, P.T. *et al.*: 2005, Phys. Rev. Lett. **94** 112501

Janka, H.-Th., Rampp, M.: 2000, Astrophys. J. **539** L33

Janka, H.-Th. *et al.*: 2007, Phys. Rep. **442** 38

Juodagalvis, A. *et al.*: 2005, Nucl. Phys. **A747** 87

Juodagalvis, A. *et al*: 2010, Nucl. Phys. **848**, 454

Kay, B.P. *et al.*: 2009, Phys. Rev. **C79** 021301

Kelic, A. *et al.*: 2006, Phys. Lett. **B616** 48

Kolbe, E., Langanke, K., Fuller, G.M.: 2004, Phys. Rev. Lett. **92** 111101

Koonin, S.E., Dean D.J., Langanke, K.: 1997, Phys. Rep. **278** 2

Kratz, K.L. *et al.*: 1986, Z. Phys. **A325** 489

Kratz, K.L. *et al.*: 1988, J. Phys. **G24** S331

Kratz, K.L. *et al.*: 1993, Astrophys.J.. **402** 216

Kurtukian-Nieto, T. *et al.*: 2009, Nucl. Phys. **A 829** 587c

Langanke, K. *et al.*: 1995, Phys. Rev. **C52** 718

Langanke, K., Martínez-Pinedo, G.: 2000, Nucl. Phys. **A673** 481

Langanke, K., Martinez-Pinedo, G.: 2003, Rev. Mod. Phys. **75** 819

Langanke, K. *et al.*: 2003, Phys. Rev. Lett. 90 241102

Langanke, K. *et al.*: 2004, Phys. Rev. Lett. **93** 202501

Langanke, K. *et al.*: 2008, Phys. Rev. Lett. **100** 011101

Lettry, J. *et al.*: 1998, Rev. Sci. Instr. **69** 761

Marek, A., Janka, H.-Th.: 2009, Astr. J. **694** 664

Martinez-Pinedo, G., Langanke, K.: 1999, Phys. Rev. Lett. **83** 4502

Mezzacappa, A. *et al.*,: 2001, Phys. Rev. Lett. **86** 1935

Möller, P., Nix, J.R., Kratz, K.L.: 1997, At. Data Nucl. Data Tables **66** 131

Möller, P. *et al.*: 1995, At. Data Nucl. Data Tables **59** 185

Panov, I.V., Korneev, I.Yu., Thielemann, F.-K.: 2009, Phys. Atom. Nucl. **72** 1026

Petermann, I.: 2010, Ph.D. thesis, Darmstadt

Pruet, J *et al.*: 2005, Astr. J. **623** 1

Pruet, J, Hoffman, R., Woosley S.E., Janka, H.-Th.: 2006, Astr. J. **644** 1028

Qian, Y.Z., Woosley, S.E.: 1996, Astr. J. **471** 331

Qian, Y.Z., Haxton, W.C., Langanke, K., Vogel, P.: 1997, Phys. Rev. **C55** 1532

Qian, Y.Z.: 2002, Astr. J. **569** L103

Sagert, I. *et al.*: 2009, Phys. Rev. Lett. **102** 081101

Sampaio, J. *et al.*: 2002, Phys. Lett. **B529** 19

Schatz, H. *et al.*: 2002, Astr. J. **579** 626

Sieja, K., Nowacki, F., to be published

Sneden, C. *et al.*: 2003, Astr. J. **591** 936

Stone, J.R.: 2005, J. Phys. **G 31** 211

Thielemann, F.-K. *et al.*: 2003, Nucl. Phys. **A718** 139c

Wanajo, S.: 2006, Astr. J. **647** 1323

Wasserburg, G.J., Busso, M., Gallino, R., Nollett, K.M.: 2006, Nucl. Phys. **A777** 5

Woosley, S.E. *et al.*: 1990, Astr. J. **356** 272

Woosley, S.E. *et al.*: 1994, Astr. J. **399** 229

*Highlight talk Astronomische Gesellschaft 2010*

# The Bar and Spiral Structure Legacy (BeSSeL) survey: Mapping the Milky Way with VLBI astrometry[1]

Andreas Brunthaler[1], Mark J. Reid[2], Karl M. Menten[1], Xing-Wu Zheng[3], Anna Bartkiewicz[4], Yoon K. Choi[1], Tom Dame[2], Kazuya Hachisuka[5], Katharina Immer[1,2], George Moellenbrock[6], Luca Moscadelli[7], Kazi L. J. Rygl[1,8], Alberto Sanna[1], Mayumi Sato[9], Yuanwei Wu[10], Ye Xu[10], and Bo Zhang[1]

[1]Max-Planck-Institut für Radioastronomie, Auf dem Hügel 69, 53121 Bonn, Germany

brunthal@mpifr-bonn.mpg.de

[2]Harvard-Smithsonian Center for Astrophysics, 60 Garden Street, Cambridge, MA 02138, USA

[3]Department of Astronomy, Nanjing University, Nanjing 210093, China

[4]Torun Centre for Astronomy, Nicolaus Copernicus University, Gagarina 11, 87-100 Torun, Poland

[5]Shanghai Astronomical Observatory, 80 Nandan Road, Shanghai, China

[6]National Radio Astronomy Observatory, Socorro, NM, USA

[7]INAF, Osservatorio Astrofisico di Arcetri, Largo E. Fermi 5, 50125 Firenze, Italy

[8]Istituto di Fisica dello Spazio Interplanetario (IFSI-INAF), Via del Fosso del Cavaliere 100, 00133 Roma, Italy

[9]Department of Astronomy, Graduate School of Science, The University of Tokyo, Tokyo 113 0033, Japan

[10]Purple Mountain Observatory, Chinese Academy of Sciences, Nanjing 210093, China

**Abstract**

*Astrometric Very Long Baseline Interferometry (VLBI) observations of maser sources in the Milky Way are used to map the spiral structure of our galaxy and*

[1]This article has already appeared in Astron. Nachr./AN 332, no. 5 (2011).

*Reviews in Modern Astronomy 23: Zoomimg in: The Cosmos at High Resolution.* First Edition.
Edited by Regina von Berlepsch.

*to determine fundamental parameters such as the rotation velocity ($\Theta_0$) and curve and the distance to the Galactic center ($R_0$). Here, we present an update on our first results, implementing a recent change in the knowledge about the Solar motion. It seems unavoidable that the IAU recommended values for $R_0$ and $\Theta_0$ need a substantial revision. In particular the combination of 8.5 kpc and 220* $\mathrm{km\,s^{-1}}$ *can be ruled out with high confidence. Combining the maser data with the distance to the Galactic center from stellar orbits and the proper motion of SgrA* gives best values of* $R_0 = 8.3 \pm 0.23$ *kpc and* $\Theta_0 = 239$ *or* $246 \pm 7\ \mathrm{km\,s^{-1}}$, *for Solar motions of* $V_\odot = 12.23$ *and* $5.25\ \mathrm{km\,s^{-1}}$, *respectively. Finally, we give an outlook to future observations in the Bar and Spiral Structure Legacy (BeSSeL) survey.*

# 1 Introduction

The Milky Way is a barred spiral galaxy, as seen from observations of CO and H I gas (e.g. Burton 1988; Dame et al. 2001), and star counts (e.g. Benjamin et al. 2005). However, our location in the galaxy makes it difficult to determine the number and positions of spiral arms, the length and position of the central bar, and the rotation curve. As a result, even the most fundamental parameters of the Milky Way, such as the distance to the Galactic center $R_0$ and the rotation speed $\Theta_0$ are still not known with high accuracy. However, these values are not only important for Galactic astronomy, but also for a wide range of different fields. This includes the interpretation of the proper motions of the Magellanic Clouds (Besla et al. 2010; Diaz & Bekki 2011; Peebles 2010; Ružička et al. 2010; Shattow & Loeb 2009) and galaxies in the Andromeda subgroup (Brunthaler et al. 2005, 2007; van der Marel & Guhathakurta 2008), the motion of the Sun relative to the cosmic microwave background (e.g. Loeb & Narayan 2008), and even on the interpretation of dark matter direct detection experiments (Foot 2010; McCabe 2010).

In recent years, many large scale surveys have covered the Galactic Plane in all wave bands from radio to gamma rays. However, almost all of these surveys are all only two dimensional, and using them to construct a three dimensional model of the Milky Way is not trivial, mainly due to large uncertainties in distance measurements (Georgelin & Georgelin 1976; Hou et al. 2009). These also affect the interpretation of these surveys since most astrophysical quantities, such as linear size, mass, and luminosities, strongly depend on the distance to the object.

# 2 Galactic distances

In Galactic astronomy, the *kinematic distance* is very commonly used. Here, a distance is deduced from a measured line-of-sight velocity, a rotation model of the Milky Way, and the assumption that the object has no peculiar motion. Apart from a distance ambiguity in the inner galaxy (near and far kinematic distance), errors in the rotation model and peculiar motions can lead to very large errors in the kinematic distance. One extreme example is the star forming region G9.62+0.20. It is located at a distance $5.2 \pm 0.6$ kpc (Sanna et al. 2009), while the near and far kinematic distances place the source at 0.5 and 16 kpc, respectively. This source shows a very large peculiar motion which may be induced by the central bar and is responsible for the large error in the kinematic distance.

Certainly, the trigonometric parallax is the most fundamental method, since it is based on pure geometry without any astrophysical assumptions except the well known orbit of the Earth about the Sun. However, the measurement of an accurate trigonometric parallax requires extremely high astrometric precision. Friedrich Wilhelm Bessel was able to measure the first stellar parallax of the star 61 Cygni (Bessel 1838a,b). A huge step forward came with the ESA's Hipparcos satellite (Perryman et al. 1997). It provided astrometric accuracies of the order of 1 milliarcsecond, which allows distance estimates in the Solar neighborhood out to 100 pc with 10 % accuracy, i.e. a very small portion of the Milky Way. Following the footsteps of Hipparcos, ESA's new Gaia mission (Perryman et al. 2001) will be launched in late 2012 to provide astrometry of up to 1 billion stars in the Milky Way with parallax accuracies up to 7 $\mu$as if it achieves specifications, a factor 100 better than Hipparcos. Although Gaia will revolutionize our view of the Milky Way, the strong extinction by gas and dust in the Galactic plane and in particular the spiral arms will prevent Gaia from making significant progress on the spiral structure of the Milky Way. Since radio waves are not affected by interstellar extinction, radio astronomy can come to the rescue and fill the gaps.

# 3 VLBI astrometry

Recently developed calibration techniques for Very Long Baseline Interferometry (VLBI) have improved the accuracy of astrometric VLBI observations significantly. When these techniques are applied to VLBI networks like the NRAO Very Long Baseline Array (VLBA) in the US, the European VLBI Network (EVN) in Europe and China, or the VLBI Exploration of Radio Astrometry (VERA) array in Japan, it is now possible to measure parallaxes of radio sources with accuracies of the order of 10 $\mu$as, comparable to the expected accuracy of Gaia.

Possible target objects for VLBI astrometry are either radio continuum sources or strong maser emission. Most radio stars are relatively weak, and the sensitivity of current instruments limits the detection to sources within a few hundred parsec (e.g. Loinard et al. 2007; Menten et al. 2007; Torres et al. 2007). However, sensitivity upgrades of existing VLBI instruments (e.g. Ulvestad et al. 2010) and eventually the Square Kilometer Array (Fomalont & Reid 2004) will extend the accessible range to even weaker and more distant sources.

Molecules with maser lines suitable for astrometry are hydroxyl (OH), methanol ($CH_3OH$), water ($H_2O$), and silicon monoxide (SiO). Since OH masers at a low frequency of 1.6 GHz are strongly affected by interstellar scattering and ionospheric effects, they are not the optimal targets for VLBI astrometry. SiO masers at 43 GHz are less affected by scattering and the ionosphere, but these observations are more vulnerable to tropospheric phase errors and require excellent weather conditions. Nevertheless, both masers have been used to measure accurate distances to evolved stars (e.g. Choi et al. 2008; van Langevelde et al. 2000).

The most suitable targets are methanol (6.7 and 12.2 GHz) and water (22 GHz) masers. They are very strong and are found in high mass star forming regions (HMSFRs) mainly in the spiral arms of the galaxy. The catalog of 6.7 GHz methanol masers by Pestalozzi et al. (2005) lists more than 500 sources, and many more were found recently with Arecibo (Pandian et al. 2007), Effelsberg (Xu et al. 2008), and in particular in the methanol multi beam survey on the Parkes telescope (Caswell et al. 2010; Green et al. 2010). Water masers can be even stronger and Valdettaro et al. (2001) list more than 1000 sources in the galaxy.

Parallaxes using the 6.7 GHz methanol transition have been reported so far only with the EVN (Rygl et al. 2010a,b). while the VLBA can currently observe only the 12.2 GHz methanol and 22 GHz water maser lines. One example is the distance to W3(OH), a high mass star forming region in the Perseus spiral arm of the Milky Way. Two independent parallax measurements with the VLBA of bright methanol (Xu et al. 2006) and water (Hachisuka et al. 2006) masers yielded consistent distance estimates of $1.95 \pm 0.04$ and $2.04 \pm 0.07$ kpc.

Another example that demonstrates the high quality of VLBI parallax measurements is the Orion Nebula. Four independent measurements with two different instruments give, within their joint errors, consistent results. Sandstrom et al. (2007) observed one radio star with the VLBA and obtained a distance of $389^{+24}_{-21}$ pc, while Hirota et al. (2007) used VERA to measure the parallax of water masers in the nebula ($437 \pm 19$ pc). The most accurate measurements of $414 \pm 7$ pc (Menten et al. 2007, VLBA, 4 radio stars) and $418 \pm 6$ pc (Kim et al. 2008, VERA, SiO maser) also agree with each other and with the modeling of the orbit of the binary $\Theta^1$ Ori C from near-infrared interferometry (Kraus et al. 2007, $434 \pm 12$ pc).

# 4 A new model for the Milky Way

## 4.1 Measurements so far

In Reid et al. (2009b), we used 18 sources, with published parallaxes at that time (see Table 1 and Fig. 1), to analyze the spiral structure and the rotation of the Milky Way. We were already able to locate the Local arm and the Perseus spiral arm from Galactic longitudes $l$ = 122°–190°. The measured pitch angle of the Perseus spiral arm of $16° \pm 3°$ favors four rather than two spiral arms for the galaxy. Furthermore, we find that most of the HMSFRs are closer than their kinematic distance. Note that the same data set has been also analyzed by Bovy et al. (2009), McMillan & Binney (2010), and Bobylev & Bajkova (2010).

## 4.2 Solar Motion

Astrometric observations not only yield the distance but also the proper motion of a source. Together with the known position and line-of-sight velocity, this gives the full 6 dimensional phase space information for the observed sources. However, all measurements are heliocentric. Therefore, the peculiar motion of the Sun relative to the LSR is needed to convert the measured heliocentric into Galactocentric motions.

**Table 1:** List of 18 sources which were used to fit a new model of the galaxy in Reid et al. (2009b).

| Source | Parallax [$\mu$as] | Reference |
|---|---|---|
| W3(OH) | $512 \pm 10$ | Xu et al. 2006 |
| Orion Nebula | $2425 \pm 35$ | Menten et al. 2007 |
| S 269 | $189 \pm 16$ | Honma et al. 2007 |
| VY CMa | $876 \pm 76$ | Choi et al. 2008 |
| NGC 281 | $355 \pm 30$ | Sato et al. 2008 |
| S 252 | $476 \pm 6$ | Reid et al. 2009a |
| G232.6+1.0 | $596 \pm 35$ | Reid et al. 2009a |
| Cep A | $1430 \pm 80$ | Moscadelli et al. 2009 |
| NGC 7538 | $378 \pm 17$ | Moscadelli et al. 2009 |
| W51 IRS2 | $195 \pm 71$ | Xu et al. 2009 |
| G59.7+0.1 | $463 \pm 20$ | Xu et al. 2009 |
| G35.2–0.7 | $456 \pm 45$ | Zhang et al. 2009 |
| G35.2–1.7 | $306 \pm 45$ | Zhang et al. 2009 |
| G23.0–0.4 | $218 \pm 17$ | Brunthaler et al. 2009 |
| G23.4–0.2 | $170 \pm 32$ | Brunthaler et al. 2009 |
| G23.6–0.1 | $313 \pm 39$ | Bartkiewicz et al. 2008 |
| WB 89–437 | $167 \pm 6$ | Hachisuka et al. 2009 |
| IRAS00420+5530 | $470 \pm 20$ | Moellenbrock et al. 2009 |

For over a decade, the values of $U_\odot$ = 10.00 $\pm$ 0.36, $V_\odot$ = 5.25 $\pm$ 0.62, and $W_\odot$ = 7.17 $\pm$ 0.38 $\mathrm{km\,s^{-1}}$ for the peculiar motion of the Sun derived from Hipparcos data by Dehnen & Binney (1998) have been widely used. When using this Solar motion, we find that the HMSFRs rotate on average 15 $\mathrm{km\,s^{-1}}$ slower than the Milky Way (Reid et al. 2009b).

This controversial result has sparked new interest in the Solar motion and Binney (2010) argues that this slower rotation is partly induced by an erroneous value of $V_\odot$. Mignard (2000) and Piskunov et al. (2006) already favored a significantly higher value of $V_\odot$, in the range of $\sim$12 $\mathrm{km\,s^{-1}}$. This higher value, which is similar to the pre-Hipparcos value (e.g. Mihalas & Binney 1981), has been also found in a number of recent studies (Coşkunoğlu et al. 2011; Francis & Anderson 2009; Schönrich et al. 2010), while the values of $U_\odot$ and $W_\odot$ have not changed considerably (see also Fuchs et al. 2009). Although the issue of the $V_\odot$ component of the Solar motion is probably still not solved, it seems likely that the value of $V_\odot$ is probably closer to 12 $\mathrm{km\,s^{-1}}$ than to 5 $\mathrm{km\,s^{-1}}$.

## 4.3 Revised values

The analysis in Reid et al. (2009b) was based on the old Hipparcos value of the Solar motion from Dehnen & Binney (1998). Here we give an update using the new values of Schönrich et al. (2010), which are $U_\odot$ = 11.10 $\pm$ 1.2, $V_\odot$ = 12.24 $\pm$ 2.1, and $W_\odot$

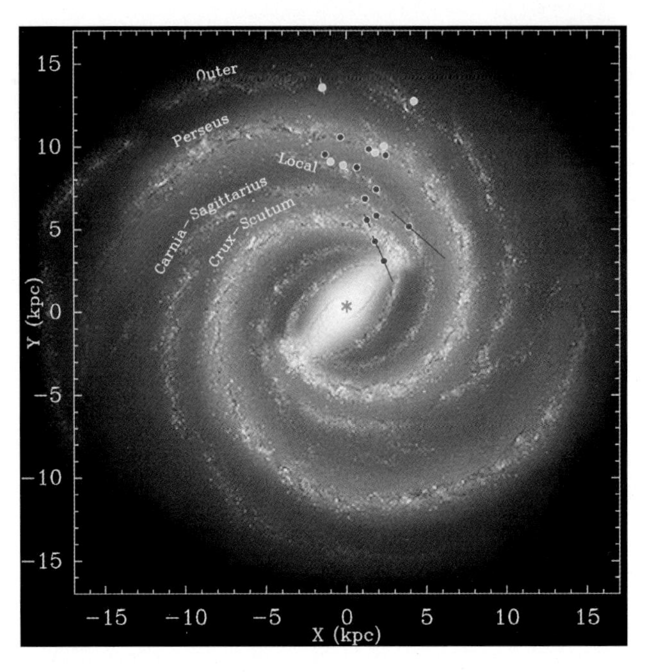


**Figure 1:** (online colour at: www.an-journal.org) Artist conception of the Milky Way (R. Hurt: NASA/JPL-Caltech/SSC) with the positions of the 18 sources from Table 1. Sources in blue are 12.2 GHz methanol masers measured with the VLBA, while the green circles indicate $H_2O$ masers observed with the VLBA or VERA and the Orion Nebula. The red symbols mark the position of the Galactic center and the Sun. Error bars are shown for all sources, but smaller than the symbol size for most sources (taken from Reid et al. 2009b).

= $7.25 \pm 0.6\ \mathrm{km\,s^{-1}}$. Using this value, the HMSFRs rotate $\sim 8 \pm 2\ \mathrm{km\,s^{-1}}$ slower than the Milky Way. By fitting our measurements to a model of the galaxy, we can also estimate the distance to the Galactic center $R_0$ and the circular rotation speed $\Theta_0$. Assuming a flat rotation curve, we get values of $R_0$ = $8.4 \pm 0.6$ kpc and $\Theta_0$ = $247 \pm 16\ \mathrm{km\,s^{-1}}$. A similar analysis allowing for different rotation curves yields results in the range of $R_0$ = 7.9–9 kpc and $\Theta_0$ = 223–280 $\mathrm{km\,s^{-1}}$. However, the ratio of these two values is much better constrained with a value of $\Theta_0/R_0$ = 29.4 $\pm$ 0.9 $\mathrm{km\,s^{-1}\,kpc^{-1}}$. Clearly, the number of sources is currently not large enough to constrain different rotation curves.

The slower average rotation of the HMSFRs together with the higher overall rotation of the Milky Way also explains why most of the kinematic distances are larger than the true distances. A description that takes into account these two effects and gives (in general) more accurate kinematic distances, and a more realistic distance uncertainty estimate is also presented in Reid et al. (2009b). It should be noted, however, that these revised kinematic distances can be still unreliable, because sources may have very large peculiar motions.

## 4.4 Independent measurements of $\vec{R}_0$ and $\Theta_0/\vec{R}_0$

The distance to the Galactic center $R_0$ has been the target of numerous investigations over the last decades (see Reid 1993, for a review). Most measurements in the range between 7.5 and 8.5 kpc and direct geometric distance estimates are rare. The accurate determination of stellar orbits in the Galactic center yields now consistent values from two groups of 8.4 $\pm$ 0.4 kpc (Ghez et al. 2008) and 8.33 $\pm$ 0.35 kpc (Gillessen et al. 2009). A trigonometric parallax to water masers in Sgr B2, a star forming region located within 150 pc from the Galactic center, also yields a consistent value of $R_0$ = $7.9^{+0.8}_{-0.7}$ kpc, but with a larger uncertainty (Reid et al. 2009c). Further observations of the water masers in Sgr B2 will improve the accuracy to a value comparable or even better than from the stellar orbits.

The ratio of $\Theta_0$ and $R_0$ is known with better than 1% accuracy from the proper motion of Sgr A* (Reid & Brunthaler 2004). This motion in the Galactic plane of 6.379 $\pm$ 0.026 mas yr$^{-1}$ is a combination of Galactic rotation and Solar motion, and corresponds to a value for $(\Theta_0 + V_\odot)/R_0$ of 30.24 $\mathrm{km\,s^{-1}\,kpc^{-1}}$. Using the Solar motion from Schönrich et al. (2010), and assuming the distance of 8.4 kpc, yields a value of $\Theta_0/R_0$ = 28.79 $\pm$ 0.26 $\mathrm{km\,s^{-1}\,kpc^{-1}}$, where the uncertainty is dominated by the uncertainty in the Solar motion. This value depends only weakly on the assumed $R_0$ (in the correction of the Solar motion), and does not change by more than 0.7 %, even for extreme values of $R_0$ of 7.5 or 9.5 kpc. This value is perfectly consistent with our results from the independent maser parallaxes but significantly larger ($>11\sigma$) than the IAU value of 25.88 from the combination of $R_0$ = 8.5 kpc and $\Theta_0$ = 220 $\mathrm{km\,s^{-1}}$. Hence, one requires an $R_0$ of less than 7.7 kpc to have a rotation velocity of 220 $\mathrm{km\,s^{-1}}$.

## 4.5 Conclusion and best values for $\vec{R}_0$ and $\Theta_0$

The measured maser parallaxes, and the combination of the stellar orbits with the proper motion of Sgr A* provide independent and consistent evidence for a higher rotation velocity of the Milky Way. Therefore, it seems unavoidable that the IAU recommended values for $R_0$ and $\Theta_0$ need a substantial revision. In particular the combination of 8.5 kpc and 220 $\mathrm{km\,s^{-1}}$ can be ruled out with high confidence.

The weighted average of the four direct measurements for $R_0$ of 8.4 $\pm$ 0.4 kpc (Ghez et al. 2008), 8.33 $\pm$ 0.35 kpc (Gillessen et al. 2009), $7.9^{+0.8}_{-0.7}$ kpc (Reid et al. 2009c), and 8.4 $\pm$ 0.6 kpc Reid et al. (2009b) gives best value of

$$R_0 = 8.3 \pm 0.23\ \mathrm{kpc}.$$

This translates then into a rotation speed of

$$\begin{aligned}\Theta_0 &= \Theta_0/R_0 \times R_0 = 28.79\ \mathrm{km\,s^{-1}\,kpc^{-1}} \times 8.3\ \mathrm{kpc} \\ &= 239 \pm 7\ \mathrm{km\,s^{-1}}.\end{aligned}$$

These values are almost identical to values presented by McMillan (2011) and marginally consistent with the rotation speed of 224 $\pm$ 13 $\mathrm{km\,s^{-1}}$ provided by Koposov et al. (2010) from modelling the stellar stream GD-1. For the old Solar motion of $V_\odot$ = 5.25 $\mathrm{km\,s^{-1}}$, $\Theta_0$ would increase to 246 $\pm$ 7 $\mathrm{km\,s^{-1}}$. The uncertainties are formal uncertainties and are dominated by the uncertainty in $R_0$. The true uncertainties are probably a factor of $\sqrt{2}$ larger, since the two stellar orbit distances may have similar systematic errors (they are independent measurements with different instruments, but of the same source).

A higher value of $\Theta_0$ is also supported by the observation of a retrograde-rotating and metal-poor component in the stellar halo of the Milky Way by Deason et al. (2011) when assuming a value of $\Theta_0$ = 220 $\mathrm{km\,s^{-1}}$. A value of $\Theta_0 \sim 240\ \mathrm{km\,s^{-1}}$ would explain this apparent retrograde rotation.

# 5 The Bar and Spiral Structure Legacy Survey

Motivated by these very encouraging results from only 18 sources, we have started a much larger project, the Bar and Spiral Structure Legacy (BeSSeL)[2] Survey, a VLBA Key science project. The goal of BeSSeL, named in honor of Friedrich Willhelm Bessel who measured the first stellar parallax, is to measure accurate distances and proper motions of up to 400 high-mass star forming regions in the Milky Way between 2010 and 2015. This will result in a catalog of accurate distances to most Galactic high mass star forming regions visible from the northern hemisphere and very accurate measurements of fundamental parameters such as the distance to the Galactic center ($R_0$), the rotation velocity of the Milky Way ($\Theta_0$), and the rotation curve of the Milky Way. Additionally, maps of the maser distribution from the first epoch of each source will be published on the BeSSeL website shortly after the observations.

The BeSSeL Survey will first target 12.2 GHz methanol and 22 GHz water masers. Once the VLBA is equipped with new receivers that also cover the 6.7 GHz methanol maser line (presumably in 2012) these masers will be also observed. In early 2010, preparatory surveys started with the Very Large Array to obtain accurate positions of the target water masers, and with the VLBA to search for extragalactic background sources near the target maser sources (Immer et al. 2011). The first parallax observations started in March 2010, and first results are expected in mid 2011 (see Fig. 2). In parallel, the VERA array will observe additional $H_2O$ and SiO masers

[2]http://www.mpifr-bonn.mpg.de/staff/abrunthaler/BeSSeL/index.shtml

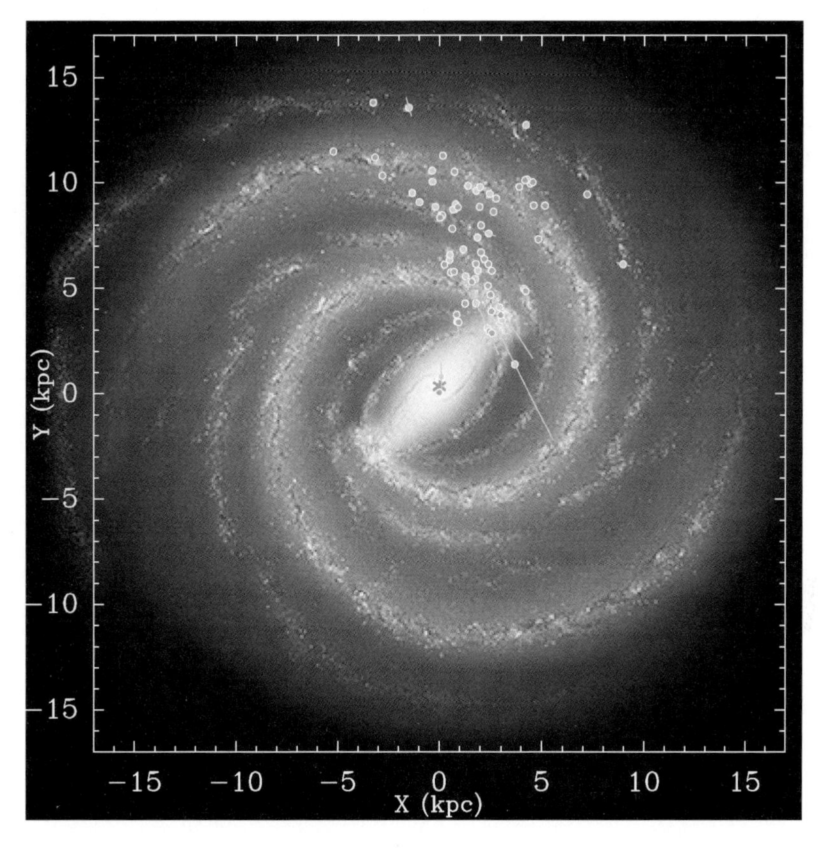


**Figure 2:** (online colour at: www.an-journal.org) Similar to Fig. 1, but showing all sources currently measured (green), including unpublished sources, and all sources observed in the first year of BeSSeL (red), based on their kinematic distances.

throughout the galaxy. Combined with complementary efforts in the southern hemisphere with the Australian Long Baseline Array, this will result into a detailed and accurate map of the spiral structure of the Milky Way. The superior sensitivity and the large field-of-view of the Square Kilometer Array, which will also cover the 6.7 GHz methanol maser line, will reach even higher astrometric accuracies for methanol masers, radio continuum stars, and pulsars. This will result in a very detailed map of the the spiral stucture in the southern hemisphere.

## Acknowledgements

The Very Long Baseline Array (VLBA) is an instrument built and operated by the National Radio Astronomy Observatory, a facility of the National Science Foundation operated under cooperative agreement by Associated Universities, Inc. The European VLBI Network is a joint facility of European, Chinese, South African and other radio astronomy institutes funded by their national research councils.This work was partially funded by the ERC Advanced Investigator Grant GLOSTAR (247078). K.L.J.R. is funded by an ASI fellowship under contract number I/005/07/1.

## References


Bartkiewicz, A., Brunthaler, A., Szymczak, M., van Langevelde, H.J., Reid, M.J.: 2008, A&A490, 787

Benjamin, R.A., Churchwell, E., Babler, B.L., et al.: 2005, ApJ 630, L149

Besla, G., Kallivayalil, N., Hernquist, L., et al.: 2010, ApJ 721, L97

Bessel, F.W.: 1838a, AN 16, 65

Bessel, F.W.: 1838b, MNRAS 4, 152

Binney, J.: 2010, MNRAS 401, 2318

Bobylev, V.V., Bajkova, A.T.: 2010, MNRAS 408, 1788

Bovy, J., Hogg, D.W., Rix, H.: 2009, ApJ 704, 1704

Brunthaler, A., Reid, M.J., Falcke, H., Greenhill, L.J., Henkel, C.: 2005, Sci 307, 1440

Brunthaler, A., Reid, M.J., Falcke, H., Henkel, C., Menten, K.M.: 2007, A&A462, 101

Brunthaler, A., Reid, M.J., Menten, K.M., et al.: 2009, ApJ 693, 424

Burton, W.B.: 1988, in: K.I. Kellermann, G.L. Verschuur (eds.), *The Structure of our Galaxy Derived from Observations of Neutral Hydrogen*, p. 295

Caswell, J.L., Fuller, G.A., Green, J.A., et al.: 2010, MNRAS 404, 1029

Choi, Y.K., Hirota, T., Honma, M., et al.: 2008, PASJ60, 1007

Coşkunoğlu, B., Ak, S., Bilir, S., et al.: 2011, MNRAS 412, 1237

Dame, T.M., Hartmann, D., Thaddeus, P.: 2001, ApJ 547, 792

Deason, A.J., Belokurov, V., Evans, N.W.: 2011, MNRAS 411, 1480

Dehnen, W., Binney, J.J.: 1998, MNRAS 298, 387

Diaz, J., Bekki, K.: 2011, astro-ph/1103.1007

Fomalont, E., Reid, M.: 2004, New A Rev. 48, 1473

Foot, R.: 2010, Phys. Rev. D 82, 095001

Francis, C., Anderson, E.: 2009, Nature 14, 615

Fuchs, B., Dettbarn, C., Rix, H., et al.: 2009, AJ 137, 4149

Georgelin, Y.M., Georgelin, Y.P.: 1976, A&A49, 57

Ghez, A.M., Salim, S., Weinberg, N.N., et al.: 2008, ApJ 689, 1044

Gillessen, S., Eisenhauer, F., Trippe, S., et al.: 2009, ApJ 692, 1075

Green, J.A., Caswell, J.L., Fuller, G.A., et al.: 2010, MNRAS 409, 913

Hachisuka, K., Brunthaler, A., Menten, M.K., et al.: 2006, ApJ 645, 337

Hachisuka, K., Brunthaler, A., Menten, K.M., et al.: 2009, ApJ 696, 1981

Hirota, T., Bushimata, T., Choi, Y.K., et al.: 2007, PASJ59, 897

Honma, M., Bushimata, T., Choi, Y.K., et al.: 2007, PASJ59, 889

Hou, L.G., Han, J.L., Shi, W.B.: 2009, A&A499, 473

Immer, K., Brunthaler, A., Reid, M.J., et al.: 2011, ApJS, subm.

Kim, M.K., Hirota, T., Honma, M., et al.: 2008, PASJ60, 991

Koposov, S.E., Rix, H., Hogg, D.W.: 2010, ApJ 712, 260

Kraus, S., Balega, Y.Y., Berger, J., et al.: 2007, A&A466, 649

Loeb, A., Narayan, R.: 2008, MNRAS 386, 2221

Loinard, L., Torres, R.M., Mioduszewski, A.J., et al.: 2007, ApJ 671, 546

McCabe, C.: 2010, Phys. Rev. D 82, 023530

McMillan, P.J.: 2011, astro-ph/1102.4340

McMillan, P.J., Binney, J.J.: 2010, MNRAS 402, 934

Menten, K.M., Reid, M.J., Forbrich, J., Brunthaler, A.: 2007, A&A474, 515

Mignard, F.: 2000, A&A354, 522

Mihalas, D., Binney, J.: 1981, *Galactic Astronomy: Structure and Kinematics* , 2nd edition, W. H. Freeman and Co., San Francisco, CA

Moellenbrock, G.A., Claussen, M.J., Goss, W.M.: 2009, ApJ 694, 192

Moscadelli, L., Reid, M.J., Menten, K.M., et al.: 2009, ApJ 693, 406

Pandian, J.D., Goldsmith, P.F., Deshpande, A.A.: 2007, ApJ 656, 255

Peebles, P.J.E.: 2010, astro-ph/1009.0496

Perryman, M.A.C., de Boer, K.S., Gilmore, G., et al.: 2001, A&A369, 339

Perryman, M.A.C., Lindegren, L., Kovalevsky, J., et al.: 1997, A&A323, L49

Pestalozzi, M.R., Minier, V., Booth, R.S.: 2005, A&A432, 737

Piskunov, A.E., Kharchenko, N.V., Röser, S., Schilbach, E., Scholz, R.: 2006, A&A445, 545

Reid, M.J.: 1993, ARA&A 31, 345

Reid, M.J., Brunthaler, A.: 2004, ApJ 616, 872

Reid, M.J., Menten, K.M., Brunthaler, A., et al.: 2009a, ApJ 693, 397

Reid, M.J., Menten, K.M., Zheng, X.W., et al.: 2009b, ApJ 700, 137

Reid, M.J., Menten, K.M., Zheng, X.W., Brunthaler, A., Xu, Y.: 2009c, ApJ 705, 1548

Ružička, A., Theis, C., Palouš, J.: 2010, ApJ 725, 369

Rygl, K.L.J., Brunthaler, A., Menten, K.M., et al.: 2010a, astro-ph/1011.5042

Rygl, K.L.J., Brunthaler, A., Reid, M.J., et al.: 2010b, A&A511, A2

Sandstrom, K.M., Peek, J.E.G., Bower, G.C., Bolatto, A.D., Plambeck, R.L.: 2007, ApJ 667, 1161

Sanna, A., Reid, M.J., Moscadelli, L., et al.: 2009, ApJ 706, 464

Sato, M., Hirota, T., Honma, M., et al.: 2008, PASJ60, 975

Schönrich R., Binney J., Dehnen W., 2010, MNRAS , 403, 1829

Shattow, G., Loeb, A.: 2009, MNRAS 392, L21

Torres, R.M., Loinard, L., Mioduszewski, A.J., Rodríguez, L.F.: 2007, ApJ 671, 1813

Ulvestad, J.S., Romney, J.D., Brisken, W.F., et al.: 2010, BAAS 42, 407

Valdettaro, R., Palla, F., Brand, J., et al.: 2001, A&A368, 845

van der Marel, R.P., Guhathakurta, P.: 2008, ApJ 678, 187

van Langevelde, H.J., Vlemmings, W., Diamond, P.J., Baudry, A., Beasley, A.J.: 2000, A&A357, 945

Xu, Y., Reid, M.J., Zheng, X.W., Menten, K.M.: 2006, Sci 311, 54

Xu, Y., Li, J.J., Hachisuka, K., et al.: 2008, A&A485, 729

Xu, Y., Reid, M.J., Menten, K.M., et al.: 2009, ApJ 693, 413

Zhang, B., Zheng, X.W., Reid, M.J., et al.: 2009, ApJ 693, 419

# On the origin of gaseous galaxy halos – Low-column density gas in the Milky Way halo

Nadya Ben Bekhti[1], Benjamin Winkel[1], Philipp Richter[2], Jürgen Kerp[1], and Ulrich Klein[1]

[1]Argelander-Institut für Astronomie (AIfA)
Universität Bonn, Auf dem Hügel 71
53121 Bonn, Germany

nbekhti@astro.uni-bonn.de

[2]Institut für Physik und Astronomie
Universität Potsdam
Karl-Liebknecht-Str. 24/25, 14476 Golm, Germany

**Abstract**

*Recent observations show that spiral galaxies are surrounded by extended gaseous halos as predicted by the hierarchical structure formation scenario. The origin and nature of extraplanar gas is often unclear since the halo is continuously fueled by different circulation processes as part of the on-going formation and evolution of galaxies (e.g., outflows, galaxy merging, and gas accretion from the intergalactic medium). We use the Milky Way as a laboratory to study neutral and mildly ionised gas located in the inner and outer halo. Using spectral line absorption and emission measurements in different wavelength regimes we obtain detailed information on the physical conditions and the distribution of the gas. Such studies are crucial for our understanding of the complex interplay between galaxies and their gaseous environment as part of the formation and evolution of galaxies. Our analysis suggests that the column-density distribution and physical properties of gas in the Milky Way halo are very similar to that around other disk galaxies at low and high redshifts.*

## 1 Introduction

Over the last years great progress has been made in understanding the properties and distribution of extraplanar gas around galaxies. Different gas phases are interacting with each other and influence the evolution of the host galaxy. Therefore, gaseous halos are perfect laboratories to study the aspects of (spiral) galaxy evolution. Using the full spectral regime it is possible to observe the different gaseous phases which helps to obtain a complete view of the galaxy–halo interaction.

*Reviews in Modern Astronomy 23: Zoomimg in: The Cosmos at High Resolution.* First Edition.
Edited by Regina von Berlepsch.

Extraplanar gas has been observed in emission around various spiral galaxies in form of neutral atomic hydrogen (e.g., NGC 253 using the 21-cm line, Boomsma et al., 2005), hot X-ray gas (e.g., NGC 253, Heesen et al., 2009), and diffuse ionised gas (via $H_\alpha$, e.g., NGC 5775, Rossa & Dettmar, 2003). These observations suggest a multiphase medium in the halo of galaxies where warm diffuse and compact cold cloud-like objects are embedded in a hot gaseous galactic corona.

X-ray telescopes are able to detect the hot $T \approx 10^6$ K gas beyond the galactic disk (e.g., Pietz et al., 1998; Nicastro, Mathur & Elvis, 2008). Measurements of O VI absorption lines in the direction of QSOs confirmed the presence of a hot gaseous galactic halo (Sembach et al., 2003). Rossa & Dettmar (2003) showed with their $H_\alpha$ survey that extraplanar diffuse ionised gas is present in all galaxies with a certain star formation rate.

Optical and ultraviolet absorption spectroscopy provides a sensitive tool to analyse different species constituting the halo gas (for a review, see Richter 2006). The observed ionisation stages range from low-ionisation (H I, Na I, Ca II, Mg II) to highly ionised gas traced by O VI Si IV, etc. Especially, Mg II absorption against quasars (QSOs) was extensively used to analyse the absorption characteristics of extragalactic halo structures (e.g., Charlton, Churchill, & Rigby, 2000; Bouché et al., 2006). Absorption spectroscopy has the advantage that it is very sensitive to low-column densities and that it is independent of the distance to the absorbing objects as long as the background continuum source is bright enough.

Spectacular examples for galaxies with extended gaseous H I halos are the nearby spiral galaxies NGC 891 and NGC 262 (e.g., Swaters, Sancisi, & van der Hulst, 1997; Oosterloo, Fraternali, & Sancisi, 2007). NGC 891 is one of the best-studied edge-on galaxies. Observations in different wavelength regimes (Whaley et al., 2009, and references therein) revealed the existence of an extended radio halo, an extended layer of diffuse ionised and hot-ionised gas (e.g., Bregman & Pildis, 1994). Deep H I observations with the Westerbork Radio Sythesis Telescope (WSRT) were performed by Oosterloo, Fraternali, & Sancisi, 2007 and they found that 25% of the total H I mass resides in the halo. The neutral gas extends to large heights of $z \approx 8 \ldots 10\,\mathrm{kpc}$. A long filament extends even up to $z \sim 20\,\mathrm{kpc}$. In the case of NGC 262, a past interaction and gas accretion is probable as suggested by the presence of an enormous H I envelope with 176 kpc in diameter and a large tail-like extension. In the case of NGC 891, NGC 262 and many other spiral galaxies in the Local Volume, the observations indicate that up to 30% of the total H I mass is situated in the halo. The gas in these galaxies is lagging behind the rotation of the host galaxy by up to $20\,\mathrm{km\,s^{-1}}$ and shows a global infall motion.

Our own galaxy, the Milky Way, is surrounded by various gaseous structures as well. The most prominent gaseous objects in the Milky Way halo are the so-called intermediate- and high-velocity clouds (IVCs, HVCs). This population of gas clouds has typically H I column densities of more than $N_{\mathrm{HI}} \approx 10^{19}\,\mathrm{cm^{-2}}$ and radial velocities inconsistent with a model of differential rotation of the galactic disk. Besides the velocity there are other criteria for the distinction between IVCs and HVCs. IVCs are relatively nearby objects with typical distances to the disk of $z \leq 2\,\mathrm{kpc}$ and metallicities of 0.7 to 1.0 whereas HVCs are located at larger distances ($z \leq 50\,\mathrm{kpc}$) and have metallicities of 0.1 to 1.0 (e.g., Wakker, 2001; Wakker, et al., 2008; Richter et al., 2001).

IVCs and HVCs can be detected all over the sky but they are not homogeneously distributed. On one hand, there are extended coherent complexes (like complex A, C, and M) on the other hand one observes large streams (like the Magellanic stream) spanning tens of degrees on the sky. Furthermore, there are numerous isolated and compact clouds (so-called Compact High-Velocity Clouds, CHVCs). A problem of IVC/HVC research is the lack of accurate distance measurements which result in a poor constraint of physical parameters like mass, particle density, and size of the clouds. To set limits, stellar spectroscopy can be used in some cases to find distance brackets (e.g., Thom et al., 2008; Wakker et al. 2007, 2008; Richter et al. 2001) and for a few large features kinematic analyses could be successfully applied (e.g., in case of Smiths cloud; Lockman et al., 2008; or for the Magellanic System; Gould, 2000).

After the discovery of the Milky Way HVC population the question emerged, whether IVCs and HVCs are a common phenomenon around spiral galaxies. Thilker et al. (2004) used the Green Bank Telescope (GBT) to search for this cloud population around M 31. They found 20 discrete features located within 50 kpc of the M 31 disk with line widths in the range of $10 \ldots 70\,\mathrm{km\,s^{-1}}$ and typical H I column densities of $N_{\rm HI} \approx 10^{19}\,\mathrm{cm^{-2}}$ showing the same physical properties as Milky Way IVCs and HVCs. Westmeier, Brüns, & Kerp (2008) mapped a large area around M 31 in 21-cm line emission with the Effelsberg telescope. Their survey extends out to a projected distance of about 140 kpc. The goal was to search for neutral gaseous structures beyond 50 kpc. The nondetection down to an H I column density detection limit of $N_{\rm HI} = 2.2 \cdot 10^{18}\,\mathrm{cm^{-2}}$ (corresponding to $8 \cdot 10^4$ solar masses) suggests that IVCs and HVCs are generally found in the proximity of their host galaxies. A key advantage of studying the gaseous halo objects around M 31 is that their distance can be well-constrained which enables the estimation of distance-dependent parameters and the area filling factor of the gas. In case of M 31 the area filling factor is about 30% where the concentration of the H I gas is decreasing with increasing radius (Richter et al., in prep.).

The properties of the extraplanar gas are manifold. There is a large variety of metal abundances, densities, and ionisation states. This observations make a single origin of the gas unlikely. Today, four major origin scenarios are favored which can be divided into galactic and extragalactic origin. The Galactic fountain model which was proposed by Shapiro & Field (1976) explains the extraplanar gas as a result of supernovae explosions which cause outflows of metal-enriched gas from the disk into the halo. There it cools, condenses and falls back onto the disk. Alternatively, halo gas can be the result of stellar winds of massive stars (e.g., Martin, 2006). Oort, 1966 suggested that the gas is of primordial origin and represents the leftovers of the early galaxy formation.

Finally, major and minor interaction processes between galaxies (and also between galaxies and their environment) expel gas into the halo via ram-pressure interaction or tidal stripping (Gardiner & Noguchi, 1996). Such interactions are probably also the reason for the observed warping of the outer neutral hydrogen gas layers (e.g., Sancisi et al., 1976; Bottema, 1996) and the lopsidedness (e.g., Sancisi et al.,

2008, and references therein) of disk galaxies. Obviously, all these effects substantially influence the evolution of the host galaxy. The complex interplay between the different processes make it often complicated from the observers point of view to relate the various structures seen in the halos of disk galaxies to a certain origin.

The mass and energy exchange between the disk and halo is in many ways fundamental for the galactic life cycle. In order to reach the accretion rates to sustain the observed constant star formation rates for disk galaxies in the Local Volume (Brinchmann et al., 2004), additional infall of large amounts of fresh gas from the Intergalactic Medium (IGM) is necessary.

Another important aspect is that the halo gas is believed to represent the interface between the condensed galactic disk (well observed) and the surrounding IGM (not well observed) (e.g., Fraternali et al., 2007). CDM cosmology predicts that most of the baryonic matter in the local universe is in the IGM (Cen & Ostriker, 2006). Because of the complexity and the many physical processes taking place our knowledge of the IGM is still incomplete. Studying the exchange of material between galaxies and their environment is therefore an efficient way to probe the IGM.

# 2 Motivation of our project

In our project we use the Milky Way as a laboratory to systematically analyse the low-column density halo gas. Although there is a large amount of available 21-cm emission data from large H I surveys like the Leiden-Agentine-Bonn survey (LAB; Kalberla et al., 2005), the Galactic All-Sky survey (GASS; McClure-Griffiths et al.,2009; Kalberla, P.M.W., et al., 2010), and the new Effelsberg-Bonn H I survey (EBHIS; Kerp, 2009; Winkel et al., 2010a) they are all limited to column densities above $N_{@series\mathbf{H}\,\mathrm{I}} \simeq 10^{18}\,\mathrm{cm}^{-2}$. Additionally, the relatively poor angular resolution makes it hard to detect low-column density small-scale structures. The solution to this issue is to use (metal) absorption line spectroscopy against QSOs, which is much more sensitive to low-column density gas (Richter et al., 2009).

Almost all recent absorption studies of IVCs and HVCs were carried out in the UV to study the metal abundances and ionisation conditions of halo clouds. These studies were designed as follow-up absorption observations of known IVCs and HVCs, thus representing an 21-cm emission-selected data set. To statistically compare the absorption characteristics of the extraplanar gas with the properties of intervening metal-absorption systems towards QSOs one requires an absorption-selected data set of IVCs and HVCs.

Our analysis will help us to answer the question whether the known IVCs and HVCs around the Milky Way represent most of the neutral gas mass in the halo or are just the tip of the iceberg of what is observed in 21-cm emission today.

# 3 Data

Our absorption sample consists of 408 archival (Spectral Quasar Absorption Database, SQUAD; PI: M. T. Murphy) high-resolution ($R \approx 40000 \ldots 60000$, corresponding to $6.6\,\mathrm{km\,s^{-1}}$ FWHM) QSO spectra observed with VLT/UVES. We searched for Na I and Ca II (Ca II$\lambda 3934.77, 3969.59$ and Na I$\lambda 5891.58, 5897.56$) absorption of Milky Way halo gas. Ca II and Na I with their low ionisation potentials of 11.9 eV and 5.1 eV are trace species of cold and warm neutral gas.

The absorption measurements were complemented with 21-cm single-dish H I emission line data using the new more sensitive H I surveys EBHIS ($\Theta = 9'$ HPBW, velocity channel separation $\Delta v = 1.3\,\mathrm{km\,s^{-1}}$) and GASS ($\Theta = 14'$ HPBW, velocity channel separation $\Delta v = 0.8\,\mathrm{km\,s^{-1}}$) to search for neutral hydrogen connected with the absorption lines. Both surveys have a column density detection limit of about $N_{\mathrm{HI}} \approx 3 \cdot 10^{18}\,\mathrm{cm^{-2}}$ (calculated for a line width of $20\,\mathrm{km\,s^{-1}}$.) Furthermore, for several sight lines we performed pointed observations with integration times of 15 min with the 100-m telescope Effelsberg. While the latter allow for the detection of lower column densities, the H I survey data provide the opportunity to study the gaseous environment of the detected absorbers.

Having such a large data sample allows for the first time to systematically study the global properties and distribution of the neutral (low-column density) Milky Way halo gas.

To decide whether the detected clouds participate in galactic rotation or not, we used a kinematic model of the Milky Way (Kalberla et al., 2003; Kalberla et al., 2007) and calculated the deviation velocities (Wakker, 1991) for each component. In 126 (75) lines of sight we detect 226 (96) Ca II (Na I) absorption components at intermediate- and high-velocities (Ben Bekhti et al., 2008). Along 100 sight lines (EBHIS/Effelsberg: 38, GASS: 62) the Ca II and/or Na I absorption is connected with H I gas.

Figure 1 shows an all-sky HVC map based on the data of the LAB survey. The different symbols mark the positions of 408 sight lines that were observed with VLT/UVES. Detected absorption components are marked with circles. The boxes indicate sight lines were we find corresponding H I emission lines with EBHIS and/or GASS. The crosses show non-detections. Almost 50% of the intermediate- and high-velocity Ca II and Na I components might be associated to known IVC/HVC complexes considering their spatial position and radial velocity.

Figure 2 displays two example spectra with optical absorption of the Ca II and Na I doublets. Additionally, the H I 21-cm emission line spectra are shown. The solid lines mark the minimal and maximal radial velocities expected for the Galactic disk gas according to the Milky Way model. The dashed lines indicate the location of the absorption and corresponding emission lines. It is remarkable that in many cases one observes distinct absorption lines, but no corresponding 21-cm emission is seen. This suggests that either the H I column densities are below the H I detection limits or that the diameters of the absorbers are very small such that beam-smearing effects render them undetectable.

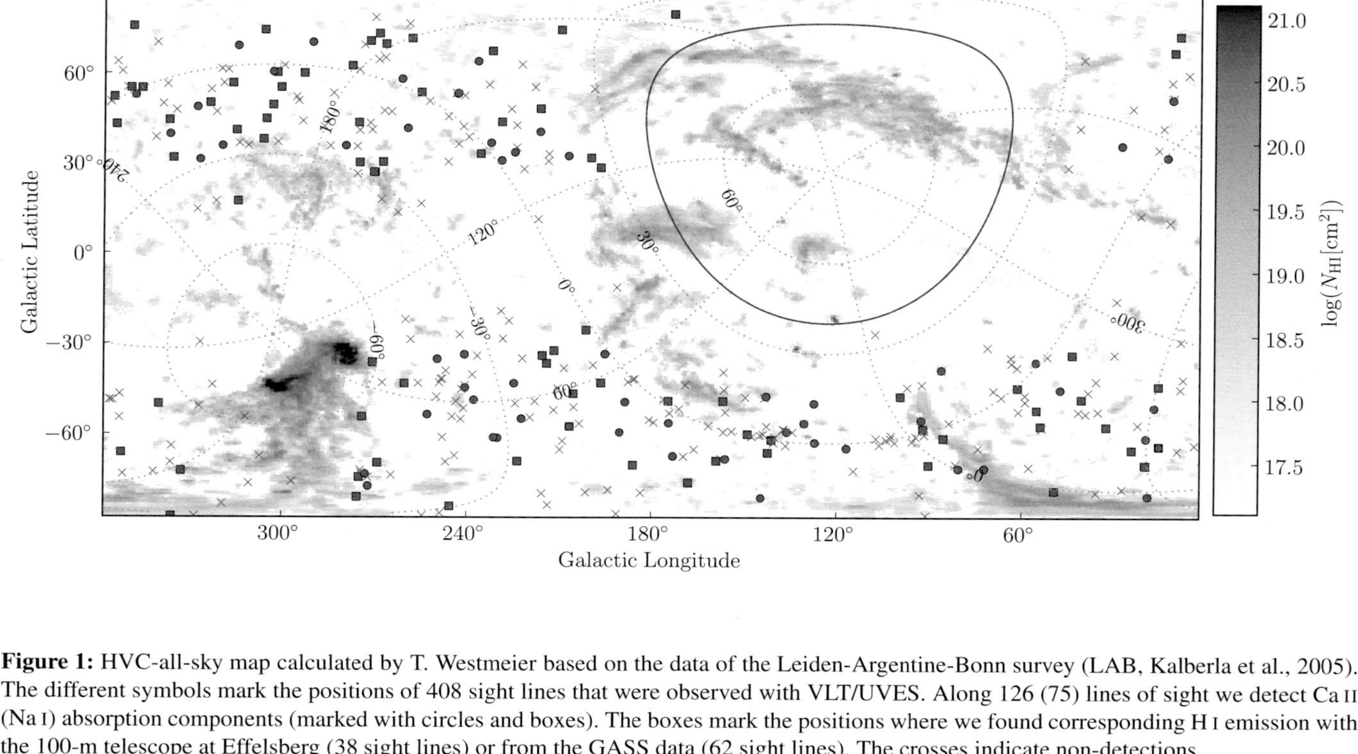


**Figure 1:** HVC-all-sky map calculated by T. Westmeier based on the data of the Leiden-Argentine-Bonn survey (LAB, Kalberla et al., 2005). The different symbols mark the positions of 408 sight lines that were observed with VLT/UVES. Along 126 (75) lines of sight we detect Ca II (Na I) absorption components (marked with circles and boxes). The boxes mark the positions where we found corresponding H I emission with the 100-m telescope at Effelsberg (38 sight lines) or from the GASS data (62 sight lines). The crosses indicate non-detections.

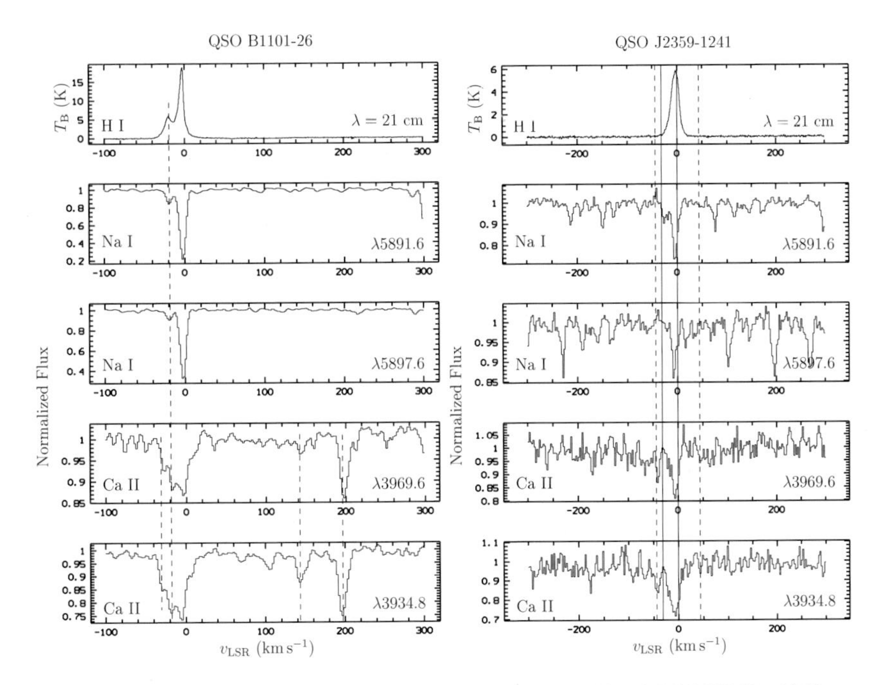


**Figure 2:** Example spectra of the quasars QSO B1126−26 and QSO J2359−1241. The absorption and corresponding emission lines are indicated by dashed lines. The solid lines mark the $v_{\rm lsr}$ velocity range expected for the Milky Way gas in that direction according to a model by Kalberla et al. (2003,2007).

# 4 Results

We performed a detailed statistical analysis using the large data sample (Ben Bekhti et al., 2008; Ben Bekhti et al., in prep.). In the following we will present some of the results.

The absorption systems have Ca II and Na I column densities in the range of $N_{@}series\mathbf{Ca}\,\mathrm{II} \approx 7.8 \cdot 10^{10}\mathrm{cm}^{-2} \ldots 1 \cdot 10^{14}\mathrm{cm}^{-2}$ and $N_{@}series\mathbf{Na}\,\mathrm{I} \approx 3.2 \cdot 10^{9}\mathrm{cm}^{-2} \ldots 1.3 \cdot 10^{13}\mathrm{cm}^{-2}$ and Doppler parameters of $b < 7\,\mathrm{km\,s^{-1}}$ (median value: $b \sim 3\,\mathrm{km\,s^{-1}}$). From the Doppler parameters an upper kinetic temperature limit of $T_{\rm max} \leq 1.2 \cdot 10^5$ K ($T_{\rm max} \leq 2 \cdot 10^4$ K) can be calculated showing that the line widths are likely enhanced due to turbulent effects, as at least Na I usually traces the cold and dense cores of the clouds. From the EBHIS and GASS H I data we get column densities in the range of $N_{@}series\mathbf{H}\,\mathrm{I} \approx 1 \cdot 10^{19}\mathrm{cm}^{-2} \ldots 1 \cdot 10^{20}\mathrm{cm}^{-2}$ and Doppler parameters of $b < 20\,\mathrm{km\,s^{-1}}$ leading to an upper kinetic temperature limit of $T_{\rm max} \leq 2 \cdot 10^4$ K which is typical for warm neutral gas clouds observed in the halo.

Figure 3 shows the column density distribution (CDD) function (Churchill, Vogt & Charlton, 2003) of the Ca II and Na I absorbers derived from the VLT/UVES data. The Ca II (Na I) CDD follows a power law $f(N) = CN^{\beta}$ with $\beta = -1.6 \pm 0.1$ ($\beta = -1.0 \pm 0.1$) for $\log N_{@}series\mathbf{Ca}\,\mathrm{II} > 11.6\,\mathrm{cm}^{-2}$ ($\log(N_{@}series\mathbf{Na}\,\mathrm{I}/\mathrm{cm}^{-2}) >$

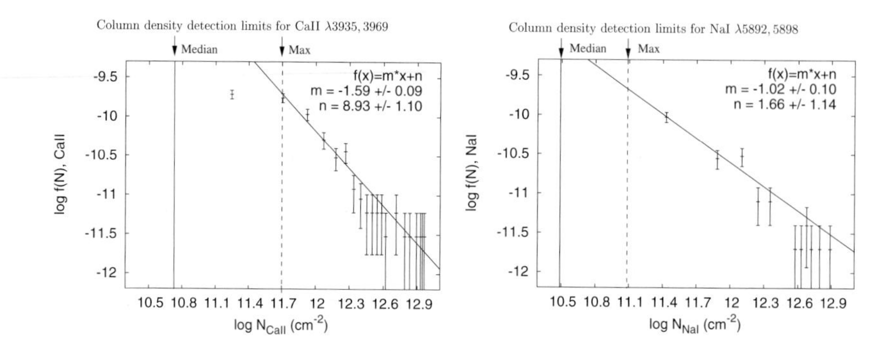


**Figure 3:** The Ca II and Na I column density distributions (CDD), $f(N)$, as derived from the VLT/UVES data, fitted with a power-law $f(N) = CN^{\beta}$. We obtain $\beta_{@}series\mathbf{Ca}\,\mathrm{II} = -1.6 \pm 0.1$, $\log C_{@}series\mathbf{Ca}\,\mathrm{II} = 8.9 \pm 1.1$ and $\beta_{@}series\mathbf{Na}\,\mathrm{I} = -1.0 \pm 0.1$, $\log C_{@}series\mathbf{Na}\,\mathrm{I} = 1.7 \pm 1.1$. The vertical solid lines indicate the $4\sigma$ detection limit $\log(N_{@}series\mathbf{Ca}\,\mathrm{II}, N_{@}series\mathbf{Na}\,\mathrm{I}[\mathrm{cm}^{-2}]) = (10.7, 10.4)$ for the median $S/N_{\mathrm{r}}$ and the dotted lines represent the detection limit $\log(N_{@}series\mathbf{Ca}\,\mathrm{II}^{\max}, N_{@}series\mathbf{Na}\,\mathrm{I}^{\max}[\mathrm{cm}^{-2}]) = (11.7, 11.1)$ for the lowest $S/N_{\mathrm{r}}$ spectra. The latter leads to an incompleteness in the source catalog, hence, the power-law fits were applied to values $\log N_{@}series\mathbf{Ca}\,\mathrm{II} > 11.7\,\mathrm{cm}^{-2}$ and $\log(N_{@}series\mathbf{Na}\,\mathrm{I}/\mathrm{cm}^{-2}) > 11.1$, respectively.

11.1). The vertical solid lines indicate the median detection limit in our sample. The dotted lines represent the spectrum with the highest noise level according to the worst detection limit leading to incompleteness of the sample below the associated column density values. The flattening of the Ca II and Na I distributions towards lower column densities can be attributed to this selection effect.

Churchill, Vogt, & Charlton (2003) derived the CDD function for strong Mg II systems in the vicinity of other galaxies at redshifts $z = 0.4 \ldots 1.2$ and found a slope of $\beta = -1.6 \pm 0.1$ which is in good agreement with our Ca II CDD. Ca II and Mg II have comparable chemical properties and both trace neutral gas in halos. The fact that both slopes agree quite well suggests that the statistical properties of the halo absorption-line systems are similar at low and high redshifts.

Figure 4 shows the number of Ca II and Na I absorbers as a function of deviation velocity. The bulk of absorbing clouds has deviation velocities of $|v_{\mathrm{dev}}| < 50\,\mathrm{km\,s^{-1}}$ with a notable excess towards negative deviation velocities. This fits to previous studies of extraplanar gas indicating an excess of clouds infalling towards the Galactic disk (Oosterloo, Fraternali & Sancisi, 2007, and references therein).

About 35% of the Ca II and 20% of the Na I absorbers show multiple velocity components (e.g., Fig. 2), suggesting the presence of substructure. Richter, Westmeier & Brüns (2005) and Ben Bekhti et al. (2009) observed five sight lines with the Very Large Array (VLA) and the Westerbork Synthesis Radio Telescope (WSRT) to search for such small-scale structures. In all five directions cold ($70 < T_{\mathrm{kin}} < 3700\,\mathrm{K}$, corresponding to a linewidth of $\Delta v = 1.8 \ldots 13\,\mathrm{km\,s^{-1}}$) and compact (down to sub pc scales) clumps were found embedded in a more diffuse environ-

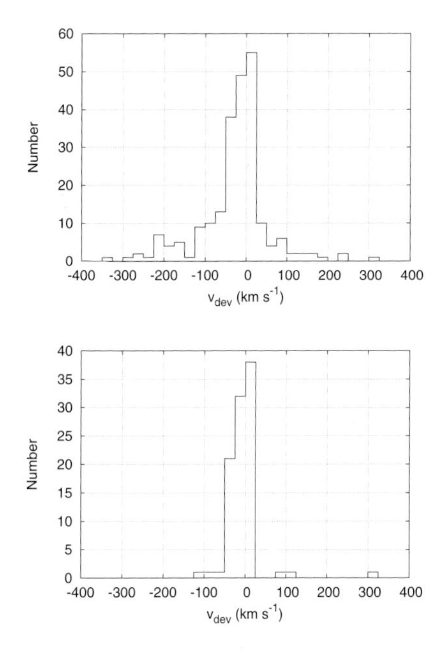


**Figure 4:** Number of Ca II and Na I intermediate- and high-velocity halo absorbers as a function of deviation velocity, $v_{\rm dev}$.

ment; see the white contour maps in Fig. 5 and 6 for two examples in the direction of QSO B1331+170 and QSO J0003+2323.

For many of the sight lines we obtained maps of the H I gas around the absorption systems using the data of EBHIS and GASS, showing both, diffuse and clumpy/filamentary extended structures. In addition to the interferometry data in Fig. 5 and 6 we present the H I column density maps observed with EBHIS and GASS.

QSO B1331+170 shows a two-component structure in Ca II and Na I absorption and H I emission. The lines spread between $v_{\rm LSR} = -40 \ldots 0\,{\rm km\,s^{-1}}$.

In case of QSO J0003+2323 three Ca II absorption components are spread between $v_{\rm LSR} = -120 \ldots -80\,{\rm km\,s^{-1}}$ and one broad 21-cm emission line is found within GASS at $v_{\rm LSR} = -112\,{\rm km\,s^{-1}}$.

Surprisingly, there is no tight spatial correlation between the single-dish and interferometry data. While some structures seen in the high-resolution data are well traced by the EBHIS/GASS (e.g., the southern clump in Fig. 5 bottom panel) other small-scale features are not visibly related to structures in the single-dish maps (e.g., north-western features in Fig. 5 bottom panel). We will study this phenomenon in more detail in a future project (Ben Bekhti et al., in prep.).

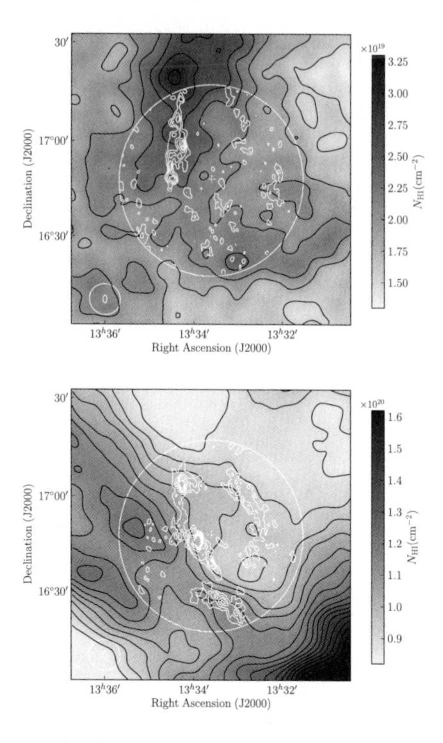


**Figure 5:** H I column density map of the two velocity components (top panel: $v_{\rm lsr} \sim -27\,{\rm km\,s^{-1}}$, bottom panel: $v_{\rm lsr} \sim -9\,{\rm km\,s^{-1}}$) in the direction of QSO B1331+170 as observed with EBHIS. The black contours start at $N_{@series\mathbf{H}\,\rm I} = 1.5 \cdot 10^{19}\,{\rm cm^{-2}}$ and increase in steps of $2.5 \cdot 10^{18}\,{\rm cm^{-2}}$ (top panel) and $N_{@series\mathbf{H}\,\rm I} = 8.5 \cdot 10^{19}\,{\rm cm^{-2}}$ in steps of $5 \cdot 10^{18}\,{\rm cm^{-2}}$ (bottom panel). Overlaid (white) are the contour lines of the high-resolution observation made with the WSRT, starting at $N_{@series\mathbf{H}\,\rm I} = 1 \cdot 10^{18}\,{\rm cm^{-2}}$ in steps of $1 \cdot 10^{18}\,{\rm cm^{-2}}$. In the lower left corner the beam sizes of EBHIS and the WSRT (synthesized beam) are indicated. The line of sight towards QSO B1331+170 is marked with a white cross, the white circle denotes the size of the primary beam of the WSRT ($2 \times$ HPBW).

# 5 Conclusions and outlook

The results presented show how important measurements at different wavelengths and with different resolutions are to get a more complete view of the properties of neutral halo gas. Such observations allow us to study a variety of elements, distinct gas phases, and structure sizes. H I traces the galactic halo with structures on all scales, from tens of AU to kiloparsecs in form of streams, clouds, tiny clumps, and filaments.

In addition to the IVC/HVC clouds and extended complexes, the Milky Way halo contains a large number of low-column density absorbers with with typical H I column densities of $1 \cdot 10^{18}{\rm cm^{-2}} \ldots 1 \cdot 10^{20}\,{\rm cm^{-2}}$. Cold and compact clumps (observed with radio synthesis telescopes) are embedded in a more diffuse envelope

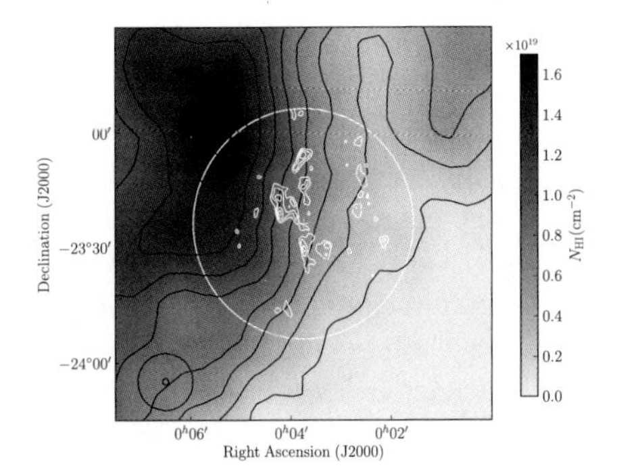


**Figure 6:** H I column density map in the direction of QSO J0003+2323 as observed with GASS, the black contours start at $N_{@}series\mathbf{H}\,\text{I} = 1 \cdot 10^{18}\,\text{cm}^{-2}$ and increase in steps of $2 \cdot 10^{18}\,\text{cm}^{-2}$. Overlaid (white) are the contour lines of the high-resolution observation made with the VLA, starting at $N_{@}series\mathbf{H}\,\text{I} = 1 \cdot 10^{18}\,\text{cm}^{-2}$ in steps of $1 \cdot 10^{18}\,\text{cm}^{-2}$. In the lower left corner the beam sizes of EBHIS and the VLA (synthesized beam) are indicated. The line of sight towards QSO J0003+2323 is marked with a white cross, the white circle denotes the size of the primary beam of the VLA ($2 \times$ HPBW).

(observed with single-dish radio telescopes), which fits perfectly into the picture of a multiphase character of the halo gas.

Assuming that the Milky Way environment is typical for low redshift galaxies, weak Ca II absorption should arise in the neutral disks of galaxies and in their extended neutral gas halo. Richter, et al. (2010) searched for Ca II absorbers at low redshifts ($z < 0.5$). They detect 23 intervening Ca II absorbers out of 304 QSO sightlines with similar physical properties as around the Milky Way. In agreement with H I observations they found that the radial extend of these halo absorbers around their host galaxies is about 55 kpc.

The similar CDD slopes of $\beta = -1.6$ for the Mg II (which traces the halos of other galaxies) and our Ca II absorbers indicate that the statistical properties of halo gas are comparable at low and high redshifts. All these observations lead to the conclusion that around disk galaxies there is a population of gaseous structures with very similar physical properties that likely influence the evolution of the host galaxy substantially.

Although the understanding of halos made great progress in the recent years, there are many open questions. We know that most of the neutral gas seems to reside within 50 kpc of the disks, but the true space distribution is largely unknown. Even for the nearby galaxies it is difficult to measure the neutral hydrogen down to low column densities. While absorption spectroscopy is much more sensitive it has the drawback that it needs suitable background sources, such that only a very incomplete picture of individual galaxies can be obtained. The determination of the mass distribution in the halos, however, is a key to quantify the gas accretion rate of galaxies and connect that to the observed constant star formation rate in disk galaxies of the

Local Volume. The Square Kilometer Array (SKA, e.g., Garrett et al., 2010) and its pathfinders will hopefully allow us to answer this question.

Another issue is the observed lagging halo gas. Is it the result of internal (e.g., galactic fountains, winds) or external (accretion from the IGM) drivers? How do effects like star formation, accretion, and interaction between galaxies contribute to the total mass of extraplanar gas around galaxies? Are warps and lopsidedness observed in many disk galaxies directly related to the material in the halo?

Little is known today when it comes to the smallest scales even for the large HVC complexes. Much of our knowledge today is based on survey data with poor angular resolution like the LAB. Recent studies show, that even with the new single-dish H I surveys much more substructure is found revealing interesting new aspects. One example is the HVC complex GCN where EBHIS/GASS data resolve the previously known "large" clouds into tiny objects mostly even unresolved within the new survey data while no extended diffuse emission is detected (Winkel et al., 2010b; Winkel et al., in prep.). Will the observed fragmentation continue when going to even higher resolution and if so, what are the properties of this scaling?

The best way to confront all these open questions is to combine sensitive, high-resolution multi-wavelength data. Future instruments like SKA, the Atacama Large Millimeter Array (ALMA; e.g., Combes 2010), the Cosmic Origins Spectrograph (COS; Goudfrooij et al., 2010), the Extended Roentgen Survey with an Imaging Telescope Array (eRosita; Predehl, et al. 2010) will shed light on many of the open problems.

From the theoretical point of view, numerical simulations can help to determine the complex dynamics of the multiphase ISM in the halos of galaxies. Especially, magneto-hydrodynamic simulations are the key to understand the interaction between the various gas phases in the ISM and magnetic fields. This will clarify whether the halo structures are stable objects supported by (large-scale) magnetic fields or if they are just transient objects in the turbulently mixed gas phases in the ISM.

#### Acknowledgements

The authors thank the Deutsche Forschungsgemeinschaft (DFG) for financial support under the research grant KE757/7-1 and KE757/9-1. We thank M. T. Murphy for providing the reduced VLT/UVES data which are the basis of our study. We thank P. M. W Kalberla, T. Westmeier, and Gyula Józsa for their support and the fruitful scientific discussions.

## References

Ben Bekhti, N., et al.: 2008, A&A 487, 583

Ben Bekhti, N., et al.: 2009, A&A 503, 483

Bottema, R.: 1996, A&A 306, 345

Boomsma, R. et al.: 2005, ASPC 331, 247

Bouché, N., et al.: 2006, MNRAS, 813

Bregman, J.N., Pildis, R.A.: 1994, ApJ 420, 570
Brinchmann, J., et al.,: 2004, MNRAS 351, 1151
Cen, R., Ostriker, J.P.: 2006 ApJ 650, 560
Charlton, J.C., Churchill, C.W., Rigby, J.R.: 2000, ApJ 544, 702
Churchill, C.W., Vogt, S.S., Charlton, J.C.: 2003, AJ 125, 98
Combes, F.: 2010, AIPC 1294, 9C
Fraternali, F. et al.: 2007, NewAR 51, 95
Gardiner, L.T., Noguchi, M.: 1996, MNRAS 278, 191
Garrett, M. A., et al.: 2010, iska.meetE, 18
Goudfrooij, P. et al.: 2010, cos. rept., 10
Gould, A.: 2000, ApJ 528, 156
Heesen, V., et al.: 2009, A&A 506, 1123
Kalberla, P.M.W., et al.: 2003, ApJ 588, 805
Kalberla, P.M.W., et al.: 2005, A&A 440, 775
Kalberla, P.M.W., et al.: 2007, A&A 469, 511
Kalberla, P.M.W., et al.: 2010, A&A 521A, 17
Kerp, J.: 2009, pra confE, 62
Lockman, F.J., et al.: 2008, ApJ, 679L, 21L
Martin, C.L.: 2006, asup.book 337
McClure-Griffiths, N.M., et al.: 2009, ApJS 181, 398
McPhate, J.B., et al.: 2010, SPIE 7732, 79
Nicastro, F., Mathur, S., Elvis, M.: 2008, Sci 319, 55
Oort, J.H.: 1966, BAN 18, 421
Oosterloo, T., Fraternali, F., Sancisi, R.: 2007, AJ 134, 1019
Pietz, J., et al.: 1998, A&A 332, 55
Predehl, P., et al.: 2010, SPIE 7732, 23
Richter, P.: 2010arXiv1008.2201R
Richter, P.: Reviews in Modern Astronomy 19, 31
Richter, P., et al.: 2001 ApJ 559, 318
Richter, P., Westmeier, T., Brüns, C.: 2005 A&A 442L, 49
Richter, P., et al.: 2009 ApJ 695, 1631
Rossa, J., Dettmar, R.-J.: 2003, A&A 406, 493
Sancisi, R., et al.: 1976, A&A, 53, 159
Sancisi, R., et al.: 2008, A&A, Rv 15, 189
Sembach, K.R., et al.: 2003, ApJS 146, 165
Shapiro, P.R., Field, G.B.: 1976, ApJ 205, 762
Swaters, R. A., Sancisi, R., van der Hulst, J. M.: 1997, ApJ 491, 140

Thilker, D.A., et al.: 2004, ApJ 601L, 39
Thom, C., et al.: 2008, ApJ 684, 364
Wakker, B.P.: 1991, A&A 250, 499
Wakker, B.P.: 2001, ApJS 136, 463
Wakker, B.P., et al.: 2007, ApJ 670, 113
Wakker, B.P., et al.: 2008, ApJ 672, 298
Westmeier, T., Brüns, C., Kerp, J.: 2008, MNRAS 390, 1691
Whaley, C.H., et al.: 2009, MNRAS 395, 97
Winkel, B., et al.: 2010, ApJS 188, 488
Winkel, B., et al.: 2010arXiv1007.3363

# Radio studies of galaxy formation: Dense Gas History of the Universe

Chris L. Carilli[1], Fabian Walter[2], Dominik Riechers[3], Ran Wang[1], Emanuele Daddi[4], Jeff Wagg[5], Frank Bertoldi[6], and Karl Menten[7]

[1]National Radio Astronomy Observatory
PO Box O, Socorro, NM, USA, 87801
ccarilli@nrao.edu

[2]Max-Planck-Institut for Astronomie
Konigstuhl 17, D-69117 Heidelberg, Germany

[3]Department of Astronomy, California Institute of Technology
MC 249-17, 1200 East California Boulevard
Pasadena, CA 91125, USA; Hubble Fellow

[4]Laboratoire AIM, CEA/DSM - CNRS - University Paris Diderot
DAPNIA/Service Astrophysccique, CEA Saclay, Orme des Merisiers
91191 Gif-sur-Yvette, France

[5]European Southern Observatory, Casilla 19001, Santiago, Chile

[6]Argelander Institute for Astronomy, University of Bonn
Auf dem Hügel 71, 53121 Bonn, Germany

[7]Max-Planck Institute for Radio Astronomy
Auf dem Hugel 69,53121, Bonn, Germany

**Abstract**

*Deep optical and near-IR surveys have traced the star formation history of the Universe as a function of environment, stellar mass, and galaxy activity (AGN and star formation), back to cosmic reionization and the first galaxies ($z \sim 6$ to 8). While progress has been truly impressive, optical and near-IR studies of primeval galaxies are fundamentally limited in two ways: (i) obscuration by dust can be substantial for rest-frame UV emission, and (ii) near-IR studies reveal only the stars and ionized gas, thereby missing the evolution of the cool gas in galaxies, the fuel for star formation. Line and continuum studies at centimeter through submillimeter wavelengths address both these issues, by probing deep into the earliest, most active and dust obscured phases of galaxy formation, and by revealing the molecular and cool atomic gas. We summarize the techniques of radio astronomy to perform these studies, then review the progress on radio studies of galaxy formation. The dominant work over the last decade has focused on massive, luminous starburst galaxies (submm galaxies*

*Reviews in Modern Astronomy 23: Zoomimg in: The Cosmos at High Resolution.* First Edition.
Edited by Regina von Berlepsch.

*and AGN host galaxies). The far infrared luminosities are $\sim 10^{13}\ L_\odot$, implying star formation rates, SFR $\geq 10^3\ M_\odot$ year$^{-1}$. Molecular gas reservoirs are found with masses: M(H$_2$)$> 10^{10}(\alpha/0.8)\ M_\odot$. The CO excitation in these luminous systems is much higher than in low redshift spiral galaxies. Imaging of the gas distribution and dynamics suggests strongly interacting and merging galaxies, indicating gravitationally induced, short duration ($\leq 10^7$ year) starbursts. These systems correspond to a major star formation episode in massive galaxies in proto-clusters at intermediate to high redshift. Recently, radio observations have probed the more typical star forming galaxy population (SFR $\sim 10^2\ M_\odot$ year$^{-1}$), during the peak epoch of Universal star formation ($z \sim 1.5$ to 2.5). These observations reveal massive gas reservoirs without hyper-starbursts, and show that active star formation occurs over a wide range in galaxy stellar mass. The conditions in this gas are comparable to those found in the Milky Way disk. A key result is that the peak epoch of star formation in the Universe also corresponds to an epoch when the baryon content of star forming galaxies was dominated by molecular gas, not stars. We consider the possibility of tracing out the dense gas history of the Universe, and perform initial, admittedly gross, calculations. We conclude with a description and status report of the Atacama Large Millimeter Array, and the Expanded Very Large Array. These telescopes represent an order of magnitude, or more, improvement over existing observational capabilities from 1 GHz to 1 THz, promising to revolutionize our understanding of galaxy formation.*

# 1 Introduction

## 1.1 The optical view of galaxy formation

A dramatic advance in the study of galaxy formation over the last decade has been the delineation of the cosmic star formation rate density (the 'star formation history of the Universe'; SFHU) to a look-back time within 0.6 Gyr of the Big Bang (Madau et al. 1996; Bouwens et al. 2010). Three main epochs have been identified (Figure 1): the first is a gradual rise during cosmic reionization at $z \sim 10$ to 6, corresponding to the epoch when light from the first galaxies and quasars reionize the neutral IGM that pervaded the Universe (Fan et al. 2006). Second is the 'epoch of galaxy assembly' at $z \sim 1$ to 3, when the cosmic star formation rate density peaks, and during which about half the stars in the present day Universe form (Marchesini et al. 2009). And third is the order of magnitude decline in the comoving cosmic star formation rate density from $z \sim 1$ to the present, constituting the inexorable demise of galaxy formation with cosmic time, as we run out of cold gas.

These studies have progressed to the next level of detail, namely the SFHU as a function of galaxy environment, stellar mass, and star formation rate (SFR). One interesting result is the observation of 'downsizing' in galaxy formation. This entails the systematic decrease in specific star formation rate (SSFR; star formation rate per unit stellar mass), with increasing stellar mass (Cowie et al. 1997). Downsizing is a manifestation of the general fact that massive galaxies form most of their stars early and quickly, and the more massive, the earlier and quicker. Evidence includes studies of the evolution of the SSFR (Moresco et al. 2010), stellar population synthesis studies of nearby ellipticals (Renzini 2006; Collins et al. 2009), and the direct observation of evolved, passive galaxies at high $z$ (Kurk et al. 2009; Doherty et al. 2009; Andreon & Huertas-Company 2011).

A second interesting result is the shift of the balance of star formation to more actively star forming galaxies with increasing redshift. At $z \sim 0$, the cosmic star formation rate density is dominated by galaxies with star formation rates $\leq 10$ M$_\odot$ year$^{-1}$ (FIR luminosities $\leq 10^{11}$ L$_\odot$). By $z \sim 2$, the dominant contribution shifts to galaxies forming stars at $\geq 100$ M$_\odot$ year$^{-1}$ (Murphy et al. 2011; Magnelli et al. 2011).

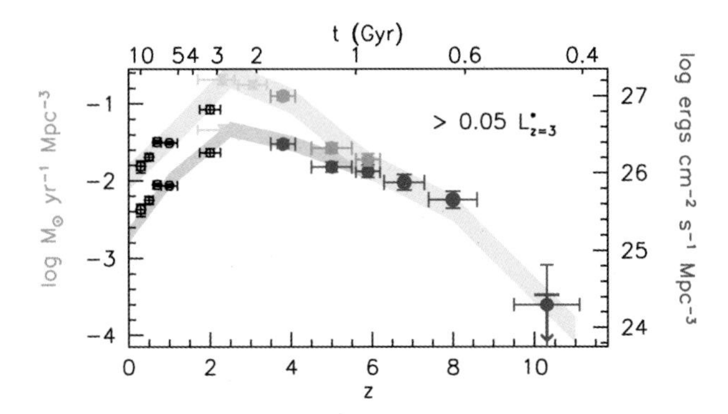


**Figure 1:** The evolution of the comoving cosmic star formation rate density as a function of redshift. The blue curve indicates star formation rates estimated from rest-frame UV measurements with no dust correction. The yellow curve includes the substantial dust correction (from Bouwens et al. 2010).

## 1.2 The role of radio observations

The results above are based, for the most part, on observations at optical through near-IR wavelengths. While truly remarkable in scope, optical through near-IR studies of galaxy formation are limited in two fundamental ways. First, dust obscuration plays a substantial role in determining views of early galaxies, in particular, the most active star forming galaxies. The average correction factor for Lyman Break galaxies (LBGs) when deriving total star formation rates from UV luminosity entails a factor five increase from observed to intrinsic star formation rates. This technique has been refined through the use of UV spectral slopes (Calzetti et al. 1994; Daddi et al. 2004).

And second, optical/near-IR studies reveal the stars, but miss the cold molecular gas, the fuel for star formation in galaxies. There is a well established correlation between the Far-IR and CO luminosity of galaxies (Figure 2). This correlation is not surprising, given that stars form in molecular clouds (Bigiel et al. 2011), and it argues that the rapid evolution of star formation with redshift should be reflected in the evolution of the molecular gas content of galaxies.

Observations at centimeter through submm wavelengths solve both these issues. Radio observations probe deep into the earliest, dusty, most active phases of star

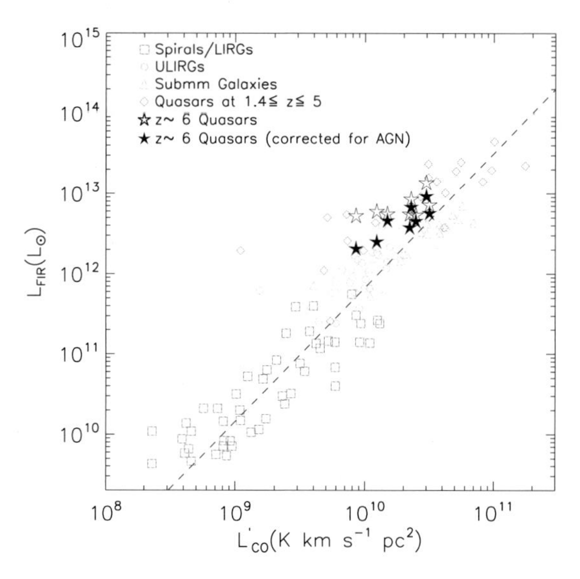


**Figure 2:** The correlation between Far-IR luminosity ($L_{FIR}$ and CO line luminosity ($L'_{CO(1-0)}$ for both low and high redshift star forming galaxies (from Wang et al. 2010).

formation in galaxies. Similarly, cm through submm observations reveal the cool molecular and atomic gas in galaxies.

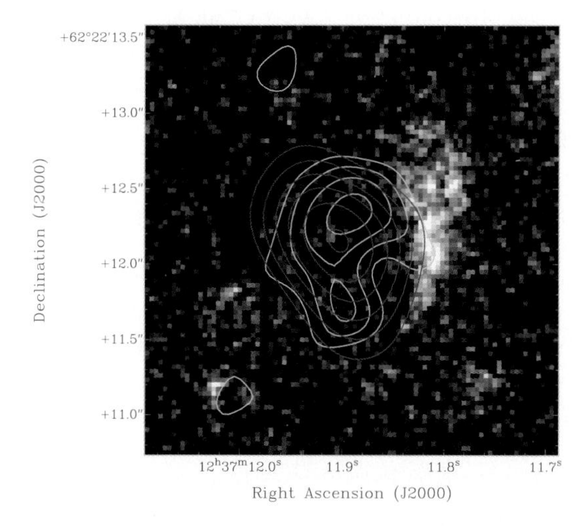


**Figure 3:** Blue contours show the 250GHz continuum emission from the $z = 4.0$ SMG GN20. Green contours show the CO 2-1 emissio. The greyscale shows the HST i-band image (from Carilli et al. 2010).

Figure 3 shows an example of these phenomena in a luminous starburst galaxy at $z \sim 4$. The contours show the thermal emission from warm dust and the CO emission, while the greyscale is the HST i-band image. The dust and gas trace the regions of most active star formation, and these regions are completely obscured in the HST image.

# 2 Tools of radio astronomy

We briefly review some of the observational tools available to study the gas, dust, and star formation in distant galaxies.

## 2.1 Continuum

Figure 4 shows the SED at cm through FIR wavelengths of an active star forming galaxy redshifted to $z = 5$. The cm emission is synchrotron radiation from cosmic ray electrons spiraling in interstellar magnetic fields. These electrons are accelerated in supernova remnant shocks, and hence the radio luminosity will be proportional to the massive star formation rate. The FIR emission is from dust heated by the interstellar radiation field, which in active star forming galaxies is dominated by massive stars.

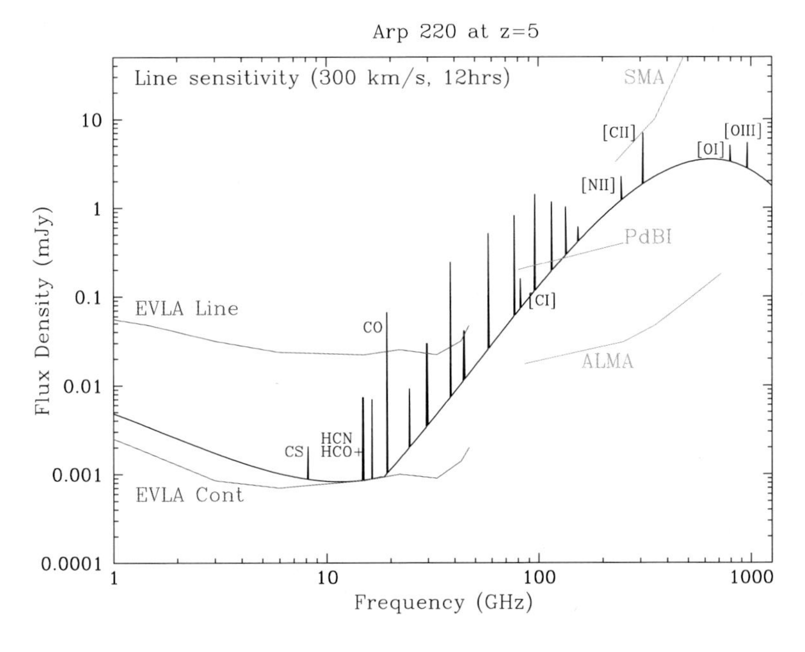


**Figure 4:** Radio through submm spectrum of a star forming galaxy with a star formation rate 100 $M_\odot$ year redshifted to $z = 5$. Also shown are the line and continuum sensitivity for the EVLA in 12 hours, and the line sensitivity for ALMA and the existing (sub)mm interferometers.

Hence, both the FIR and radio emission are a function of the massive star formation rate, resulting in the well quantified correlation between the radio and FIR luminosity from galaxies (Yun et al. 2001). While the tightness and linearity of the relationship over such a large range in luminosity remains puzzling, this correlation is an important tool for studying dust obscured star formation in galaxies, providing two means for estimating massive star formation rates in distant galaxies. Studies of galaxies at least out to $z \sim 2$ show that the correlation remains unchanged (Bourne et al. 2011), although at the highest redshifts it may be inevitable that inverse Compton scattering losses off the CMB by the relativistic electrons leads to a departure from the low $z$ relationship (Carilli et al. 2008).

An important point is that FIR emission peaks around 100$\mu$m. For high redshift galaxies, this emission shifts into the submm bands. A related point is the well studied 'inverse-K' correction in the submm: the rapid rise in luminosity density on the Rayleigh-Jeans side of the modified grey body emission curve offsets distance losses, leading to a roughly constant observed flux density with redshift for a fixed observing frequency and intrinsic luminosity (Blain et al 2002). In essence, submm observations provide a distance independent method with which to study star forming galaxies.

For reference, for a typical dust temperature of $\sim$ 45K, the relationships between observed 350 GHz flux density ($S_{350GHz}$, in mJy), dust mass, integrated far-IR luminosity, and total star formation rate (SFR; Chabrier IMF) are approximately:

$$L_{IR} = 1.5 \times 10^{12} S_{350\mathrm{GHz}}\ \mathrm{L}_\odot$$

$$M_{dust} = 7.6 \times 10^{-5}\ L_{IR}\ \mathrm{M}_\odot$$

$$\mathrm{SFR} = 1.0 \times 10^{-10} L_{IR}\ \mathrm{M}_\odot\ \mathrm{year}^{-1}$$

## 2.2 Molecular rotational transitions

A rich spectrum of rotational transitions of common molecules redshifts into the cm and mm bands for distant galaxies (Figure 4). Most prominent are the emission lines from CO. CO has long been used as a tracer for the total molecular gas mass (dominated by $H_2$):

$$\mathrm{M(H_2)} = \alpha L'_{CO1-0}\ \mathrm{M}_\odot$$

where $\alpha$ is the CO luminosity to $H_2$ mass conversion factor. The units for $L'_{CO1-0}$ are K km s$^{-1}$ pc$^2$. These units were originally designed for spatially resolving observations of CO in the Galaxy, where brightness temperature was paramount.

Values of the CO luminosity to $H_2$ mass conversion factor $\alpha$, range from 0.8 $\mathrm{M}_\odot$/[K km s$^{-1}$ pc$^2$] for luminous starburst nuclei in nearby galaxies, to 3.6 $\mathrm{M}_\odot$/[K km s$^{-1}$ pc$^2$] for Milky Way-type spiral disks. Since the CO emission is optically thick on small scales (ie. molecular cloud cores), $\alpha$ is calibrated essentially based on dynamical measures of masses, from virialized GMCs to gas dominated rotating disks in starburst nuclei (Downes & Solomon 1998). There is evidence for both values in the different populations of high $z$ galaxies, again separated according to compact starbursts and star forming disk galaxies (Daddi et al 2010b; Tacconi et al. 2010; Genzel et al. 2010; Naraynan et al. 2011).

Solomon & vanden Bout (2005) derive the following relationships between CO luminosity and observed flux density and line width:

$$L'_{CO} = 3.3 \times 10^{13} S\Delta V D_L^2 \nu_o^{-2} (1+z)^{-3} \quad \mathrm{K\ km\ s^{-1}\ pc^2}$$

where $\nu_o$ is the observing frequency in GHz, the luminosity distance, $D_L$, is in Gpc, the flux density, $S$, is in Jy, and velocity width, $\Delta V$, is in km s$^{-1}$. For CO luminosity in solar units the relationship is:

$$L_{CO} = 1.0 \times 10^{3} S\Delta V (1+z)^{-1} \nu_r D_L^2 \quad \mathrm{L_\odot}$$

where $\nu_r$ is the rest frequency in GHz. Solving for $S\Delta V$ in (4) and (5), and equating (Carilli 2011), yields:

$$L_{CO} = 3 \times 10^{-11} \nu_r^3 L'_{CO} \quad \mathrm{L_\odot}$$

Other critical contributions of molecular line observations include:

- Gas velocities determine the dynamical masses of high$z$ galaxies, and internal gas dynamics in star forming regions.
- Multi-transition studies provide the gas excitation, which gives a rough estimate of gas density and/or temperature.
- Observations of high dipole moment molecules, such as HCN and HCO+, provide an estimate of the dense gas content of galaxies ($\mathrm{n(H_2)} > 10^4\ \mathrm{cm^{-3}}$).

### 2.3 Atomic fine structure transitions

The atomic fine structure lines are emitted predominantly in the rest frame FIR, and hence redshift into the submm range for distant galaxies (Figure 4). Being metastable transitions, and hence typically optically thin, these lines, and in particular the [CII] 158$\mu$m line, are the principle coolant of interstellar gas (Spitzer 1998). The [CII] can carry up to 1% of the total IR luminosity from galaxies, and is typically the brightest line from IR through meter wavelengths (Malhotra et al. 2001; Bennett et al. 1994). The [CII] line traces the CNM and photon-dominated regions associated with star formation (Cormier et al. 2010). Fine structure line ratios can be used as an AGN versus star formation diagnostic (Genzel & Cesarsky 2000).

Herschel is providing a revolutionary view of these lines in nearby galaxies (Cormier et al. 2010). Submm telescopes are beginning to make serious in-roads into the study of these lines in distant galaxies (Stacey et al. 2010).

## 3 Molecular gas at high redshift

Figure 5 shows a histogram of the number of detections of CO emission from high redshift galaxies versus year. To date, there are 115 detections of CO at $z > 1$. The number of detections has almost doubled in the last two years due to two factors. First is improved instrumentation, in particular continued improvements at the Plateau de Bure Interferometer (PdBI), and the coming on-line of the EVLA and the GBT Zpectrometer.

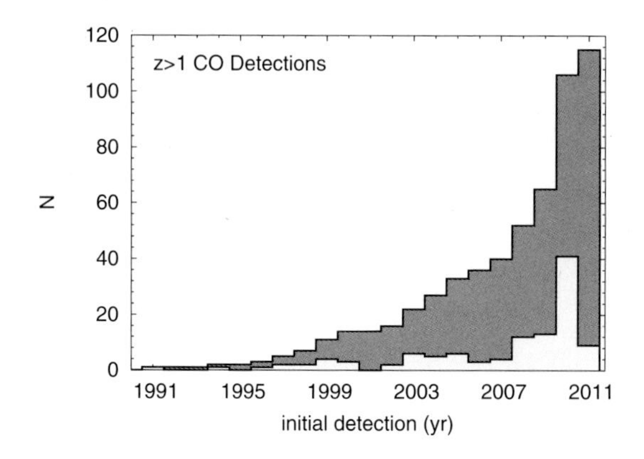


**Figure 5:** The number of new CO detected galaxies at $z > 1$ per years (yellow). Red is the cumulative curve.

More important is the change in the type of galaxies that are being discovered in CO emission in that last two years. Prior to 2009, the only galaxies that were detected in CO emission at high $z$ were extreme starburst galaxies selected in wide field submm surveys (the submm galaxies, or SMGs), as well as the host galaxies of some very luminous quasars and radio galaxies. The only exceptions were a few highly gravitationally magnified Lyman Break Galaxies (LBG; Baker et al. 2004; Riechers et al. 2010b; Coppin et al. 2007). However, in the last two years a new class of galaxy has been detected in CO emission: more typical star forming disk galaxies at $z \sim 1$ to 3 selected via standard color-color techniques in optical and near-IR deep fields. These are a hundred times more numerous than the SMGs, and yet are often as luminous in CO emission.

# 4 Extreme starbursts: massive galaxy formation in the early Universe

## 4.1 General properties

Molecular gas detections of high redshift galaxies have focused primarily on extreme starburst galaxies at $z > 1$. These include highly dust obscured galaxies identified in wide field submm surveys (SMGs), as well as the host galaxies of optically selected luminous quasars, and some radio galaxies (Miley & de Breuck 2008). Typical submm surveys at 350 GHz detect galaxies at the few mJy level, implying $L_{FIR} \geq 10^{13}$ $L_\odot$ (ie. 'Hyperluminous infrared galaxies', HyLIRGs), dust masses $\sim 10^9$ $M_\odot$, and star formation rates $\geq 10^3$ $M_\odot$ year$^{-1}$. In parallel, about 1/3 of optically selected quasars from eg. SDSS, are detected with similar FIR luminosities.

Solomon & Vanden Bout (2005) and Blain et al. (2002) present extensive reviews of these 'Extragalactic Molecular Galaxies,' and we update some of the information herein.

The areal density of SMGs at $S_{250} \geq$ 3mJy is about 0.05 sources arcmin$^{-2}$ (Bertoldi et al. 2010; Blain et al. 2002). The redshift distribution has been determined for about 50% of the SMGs selected using 1.4GHz continuum observations to determine arcsecond positions. The median redshift is $z \sim 2.3$, with most of the radio detected sub-sample within $z \sim 1$ to 3 (Chapman et al. 2003). However, recently it has become clear that there is a tail of SMGs extending to high redshift, with possibly 20% of the sources extending to $z \sim 5$ (Riechers et al. 2010a; Daddi et al. 2009a; 2009b; Schinnerer et al. 2008; Coppin et al. 2009).

The mean space density of the SMGs is $\sim 10^{-5}$ Mpc$^{-3}$ at $\sim 2.$ (comoving; Blain et al. 2002; Chapman et al. 2003). This space density is about 1000 times larger than for galaxies of similar FIR luminosity at $z = 0$, demonstrating the dramatic evolution in the number density of FIR luminous galaxies with redshift. Daddi et al. (2009a) conclude, based on SMG space densities and duty cycles, that there are likely enough SMGs at $z > 3.5$ to account for the known populations of old massive galaxies at $z \sim 2$ to 3. Study of the clustering properties of SMGs suggests a minimum halo mass of $3 \times 10^{11}$ M$_{\odot}$ (Amblard et al. 2011). However, this calculation is somewhat problematic due to the low space density, the broad redshift selection function, and the likely low duty cycle of SMGs (Chapman et al. 2009).

Michalowski et al. (2010) present a detailed study of of the radio through UV SED of SMGs. They find a median stellar mass of $3.7 \times 10^{11}$ M$_{\odot}$. The sources follow the radio-FIR correlation for star forming galaxies, except perhaps at the highest luminosities, where a low luminosity radio AGN may contribute. The dust temperatures span a broad range (10 K to 100 K), with a typical value $\sim$ 40K. They calculate that SMGs contribute about 20% of the cosmic star formation rate density at $z \sim 2$ to 4.

## 4.2 Molecular gas

Extensive observations have been performed of the molecular gas in SMGs and high redshift AGN host galaxies. Typical gas masses derived from observations of low order CO transitions range from $10^{10}$ M$_{\odot}$ to $10^{11} \times (\alpha/0.8)$ M$_{\odot}$ (Hainline et al. 2006;; Ivison et al. 2011; Riechers et al. 2010a, 2011c; Carilli et al. 2010; Wang et al. 2010).

Figure 6 shows the spectral energy distribution of the CO emission lines from a number of sources (Weiss et al. 2007). The excitation is uniformly high, significantly higher than is seen for the CO in the inner disk of the Milky Way. The excitation is comparable to what is found in the nuclear starburst regions on 100pc scales in M82 and NGC 253. Radiative transfer model fitting to the mean excitation indicates warm ($\geq$ 50 K), dense ($\geq 10^4$ cm$^{-3}$) molecular gas dominates the integrated CO emission from these galaxies. Such conditions are only found in the star forming cores of Giant Molecular Clouds in the Milky Way on parsec scales. The SMGs show systematically lower excitation than the QSO host galaxies, and there is mounting evidence for a lower excitation, more spatially extended molecular gas distribution

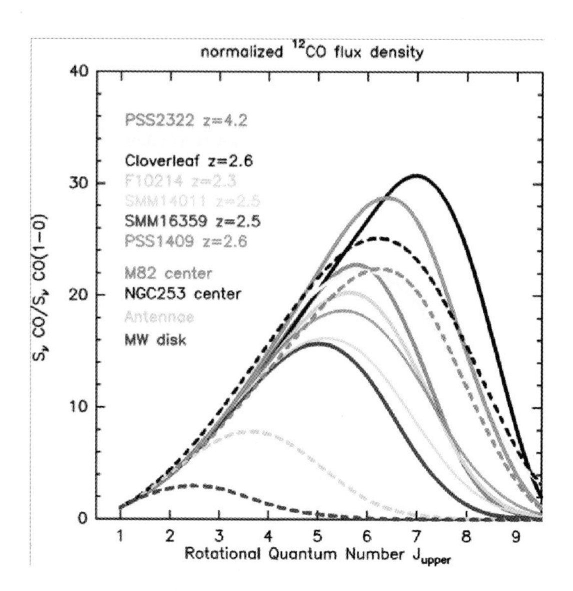


**Figure 6:** The CO excitation ladder for the integrated emission from high $z$ SMGs and AGN hosts, plus the dashed lines show the inner disk of the Milky Way and selected nearby galaxies (from Weiss et al. 2007). The X-axis is the upper rotational level of the CO transition, while the Y-axis is the normalized flux from the CO line.

in SMGs (Ivison et al. 2011, Harris et al. 2010; Riechers et al. 2010a; 2011c, Carilli et al. 2010, Papadopoulos et al. 2002; Scott et al. 2011).

High resolution CO imaging has been performed on a number of SMGs and AGN host galaxies. Figure 7 shows an example of the CO intensity and velocity field for the $z = 4.4$ quasar host galaxy, BRI1335-0417 (Riechers et al. 2008). In most cases, the CO emission appears to be complex, with two or more compact knots of emission separated by a few kpc (Tacconi et al. 2008; Carilli et al. 2002; Riechers 2008; 2011c). The velocity fields often appear chaotic, with little indication of rotation, however, this may not be universal for the low order emission (Carilli et al. 2011).

Figure 2 shows the integrated star formation law for both low redshift, lower luminosity star forming galaxies, and low and high redshift HyLIRG. There is a non-linear correlation between these two quantities over a broad range in luminosity. The relation is consistent with a powerlaw of index 1.5 between $L'_{CO}$ and $L_{FIR}$. These two quantities are linearly related to total molecular gas mass and star formation rate (Section 2.1). Extensive analysis has gone into understanding this relationship physically, both for the integrated correlation, and the spatially resolved correlation in galaxies (Leroy et al. 2008; Bigiel et al. 2011; Kennicutt 1998; Krumholz et al. 2009; Narayanan et al. 2011). Regardless of the physical interpretation and $\alpha$, the empirical implication of Figure 2 is that the FIR luminosity increases super-linearly relative to the CO luminosity of galaxies. A review of the physics of the star formation law is well beyond the scope of this review. Herein, we make a few simple, empirical points (see also Section 5.4).

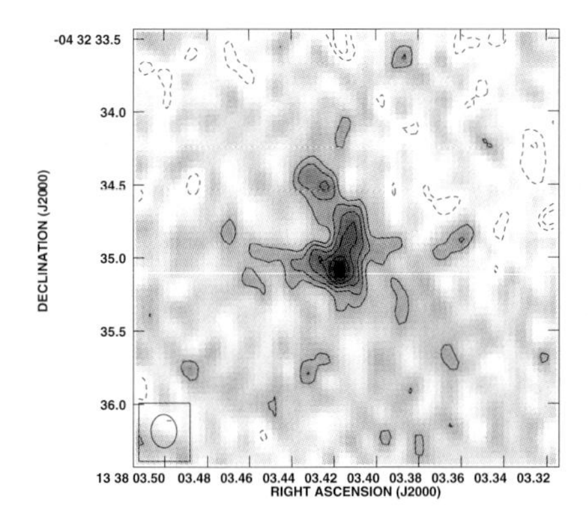


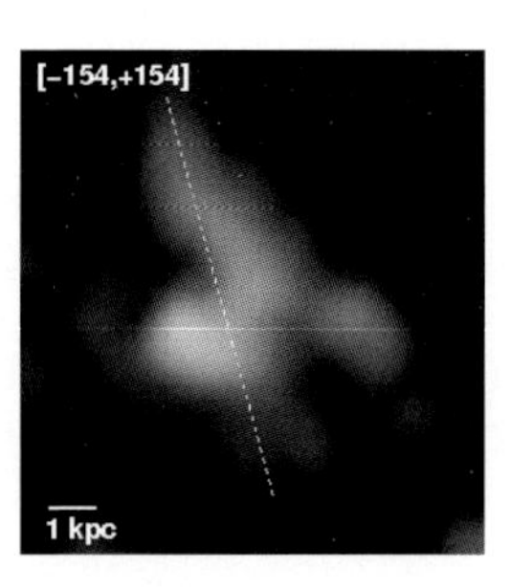


**Figure 7:** The CO emission from the $z = 4.4$ quasar host galaxy BRI1335-0417 (from Riechers et al. 2008). Left is the velocity integrated intensity and right the mean CO velocity.

First is that even the AGN sources follow this relationship, which is circumstantial evidence for star formation in the host galaxies as the dominant heating source for the warm dust (see section 4.3). And second is that the non-linearity of the relationship implies shorter gas consumption timescales ($\equiv$ $M_{gas}$/SFR) with increasing luminosities. For spiral galaxies like the Milky Way, with $L_{FIR} \sim 10^{10}$ $L_\odot$, the gas consumption timescale is a few $\times 10^8$ years, while for HyLIRGs this decreases to $\leq 10^7$ years, although this depends strongly on the assumed value of $\alpha$ (see section 5.3).

Progress has also been made on detecting the dense gas tracers, such as HCO+ and HCN, from high redshift galaxies. These molecules are much less abundant than CO, but have much higher dipole moments, and hence stronger rotational transitions. This implies that the radiative lifetimes are much shorter, and hence maintenance of a Boltzmann distribution via collisions requires high densities, $\mathrm{n(H_2)} > 10^4$ cm$^{-3}$, even for the lower states, and considerably higher for the higher order states. Hence, the high order states can be significantly sub-thermally populated, and emission from these molecules only comes from the densest molecular gas in galaxies. Interestingly, $L_{FIR}$ and $L'_{HCN}$ form a linear correlation (as opposed to the non-linear correlation with $L'_{CO}$; Gao & Solomon 2004). The simplest interpretation is that observations of these high density tracers simply 'count' star forming clouds in galaxies, and that the properties of the dense clouds are relatively universal (Wu et al. 2005).

The emission lines from the dense gas tracers are typically an order of magnitude weaker than CO, although the ratio varies dramatically between galaxies (Gao & Solomon 2004). A few galaxies have been detected in the dense gas tracers at high redshift, predominantly strongly gravitationally lensed systems (Riechers et al. 2011a,b; Wagg et al. 2005, Carilli et al. 2005; Solomon et al. 2003). The high redshift galaxies generally follow the low redshift correlations. Interestingly, in some cases even the high order transitions are excited, indicating either extremely dense gas, or a contribution to the excitation by the AGN (Riechers et al. 2011a,b; Wagg et al. 2005).

## 4.3 Topics on quasar hosts

### 4.3.1 Dust and molecular gas in the most distant galaxies

Detection of molecular line emission at the very highest redshifts ($z > 6$) has thus far been limited to quasar host galaxies. These galaxies generally follow the trends discussed above for SMGs and lower redshift quasar hosts in terms of their warm dust and molecular gas properties, and in particular, the 1/3 fraction of submm detections of quasar host galaxies at $S_{250GHz} \geq$ 2mJy remains constant to the highest redshifts, implying HyLIRG host galaxies.

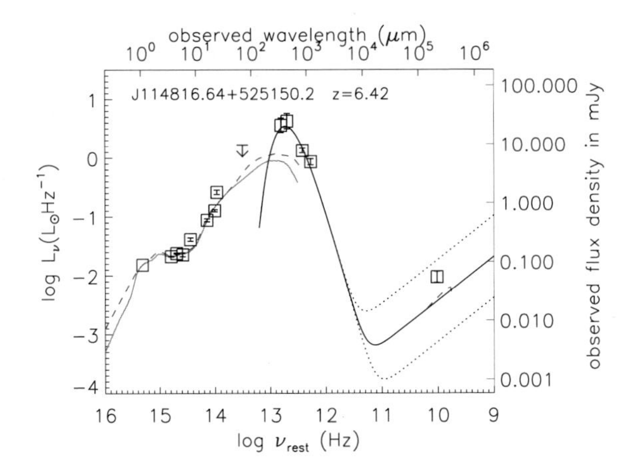


**Figure 8:** The UV through radio SED of the $z = 6.42$ quasar J1148+5258 (from Wang et al. 2008). The curves in the UV through mid-IR are local quasar templates from Elvis et al. (1994) and Richards et al. (2006). The curves in the Far-IR through radio are for a 50K dust model that obeys the radio-FIR correlation for star forming galaxies.

Figure 8 shows the SED of the most distant SDSS quasar J1148+5258 from UV to radio wavelengths. This SED is typical of the submm detected high $z$ quasars (Wang et al. 2009, 2010; Leipski et al. 2010). From the UV through mid-IR ($\sim 30\mu$m rest frame), the SED is consistent with that of lower redshift SDSS quasars, including a hot dust (1000K) seen in the mid-IR, heated by the AGN. However, the submm detected sources show a substantial excess over the lower $z$ SDSS quasar SED. This excess is well fit by a warm dust component ($\sim$ 50K). Extrapolating this component to the radio also shows that most of the sources follow the radio-FIR correlation for star forming galaxies. These results argue that the warm dust component is heated by star formation in the host galaxy. The star formation rates are $\sim 10^3$ $M_\odot$ year$^{-1}$, implying a major starburst coeval with the AGN in the host galaxy.

CO has been detected in every $z \sim 6$ quasar host galaxy that was selected via a previous submm detection of the dust. To date, 11 quasar host galaxies have been detected in CO emission between $z = 5.7$ and 6.4 (Wang et al. 2010), with gas masses $\geq 10^{10}$ $M_\odot$.

The detection of large dust masses within 1Gyr of the Big Bang immediately raises an interesting question: how does so much dust form so early? One standard dust formation mechanism in the ISM involves coagulation in the cool winds from low mass stars, which, naively would take too long. The large dust masses have led to a number of theoretical studies of early dust formation, with models involving: dust formation associated with massive star formation in eg. supernova remnants (Dwek et al. 2007; Venkatesen et al 2006), dust formation in outflows from the broad line regions of quasars (Elitzur in prep; Elvis et al. 2002), and dust formation in the gas phase ISM (Draine 2003). Michalowski et al. (2010) consider this problem in detail, and show that AGB stars are insufficient, and even SNe require a very top-heavy IMF and unrealistic dust yields.

Recent observations of the UV-extinction curves in a few $z \sim 6$ quasars and GRBs suggest a different dust composition at $z > 6$ relative to the Milky Way or the SMC, as well as relative to quasars at $z < 4$. The extinction can be modeled by larger silicate and amorphous carbon grains (vs. eg. graphite), as might be expected from dust formed in supernova remnants (Stratta et al. 2007; Perley et al. 2010). The formation of dust in the early Universe remains an interesting open question.

### 4.3.2 Fine structure lines: [CII] 158$\mu$m

At high redshift the FIR atomic fine structure lines are observed in the submm band, and hence can be studied with existing ground-based telescopes. In particular, substantial progress has been in the study of [CII] in distant galaxies.

We have started a systematic search for [CII] emission from $z > 4$ quasars quasar host galaxies (Maiolino et al. 2005; Wagg et al. 2010). At the highest redshifts, we now have three detections of the [CII] line from $z > 6.2$ quasar host galaxies (Bertoldi et al. in prep).

Figure 9 shows the [CII] images of the $z = 6.42$ quasar J1148+5258, as well as the mm continuum and VLA CO 3-2 images, at $0.25''$ resolution from the PdBI (Walter et al. 2009). The [CII] emission is extended over about 1.5 kpc, while the CO is even more extended. If [CII] traces star formation, the implied star formation rate per unit area $\sim 10^3$ $M_\odot$ year$^{-1}$ kpc$^{-2}$. This value corresponds to the predicted upper limit for a 'maximal starburst disk' by Thompson et al. (2005), ie. a self-gravitating gas disk that is supported by radiation pressure on dust grains. Such a high star formation rate areal density has been seen on pc-scales in Galactic GMCs, as well as on 100 pc scales in the nuclei of nearby ULIRGs. For J1148+5251 the scale for the disk is yet another order of magnitude larger.

One potential difficulty with using the [CII] line as a star formation diagnostic is the very broad range in the ratio of [CII] to Far-IR luminosity, in particular for galaxies with $L_{FIR} > 10^{11}$ $L_\odot$. Stacey et al. (2011) present an analysis of this ratio versus $L_{FIR}$ for low and high redshift galaxies. At high luminosity, the distribution is essentially a scatter plot, with the ratio ranging by 3 orders of magnitude, although there appears to be less scatter if AGN are removed. Figure 10 shows this ratio versus dust temperature. Malhotra et al. (2001) consider temperature to be the more fundamental quantity, due to inefficiency of photoelectric heating of charged dust grains in high radiation environments. Figure 10 shows a possible correlation of the [CII] to

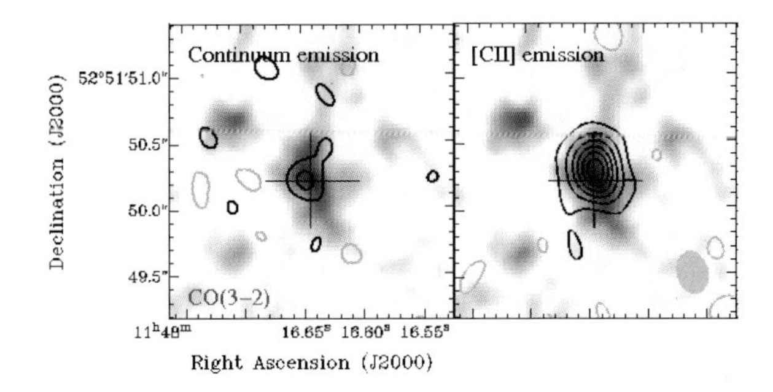


**Figure 9:** The images of the $z = 6.42$ quasar J1148+5251. Images from the PdBI show the dust and [CII] emission (contours, left and right, respectively) at $0.25''$ resolution, plus the VLA CO 3-2 in color (from Walter et al. 2009; 2004).

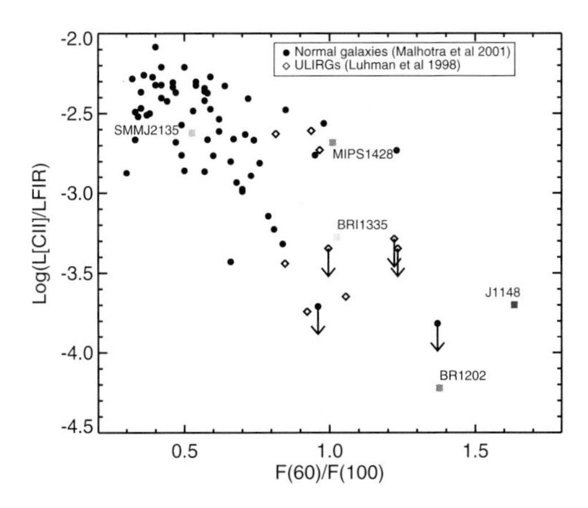


**Figure 10:** The [CII]/$L_{FIR}$ ratio versus dust temperature for low and high redshift galaxies (Wagg et al. in prep.).

$L_{FIR}$ ratio with dust temperature, but again, the scatter is very large. Papadopoulos et al. (2010) also point out that dust opacity might play a role in decrease the [CII] luminousity from extreme starburst galaxies.

### 4.3.3 The Black Hole – Bulge mass relation within 1 Gyr of the Big Bang

There is a well studied correlation between the masses of black holes at the centers of galaxies, and the velocity dispersion of the host galaxies. This $M_{BH}$ - $\sigma_V$ relation implies a roughly linear correlation between black hole and spheroidal galaxy mass, with a proportionality constant of: $\mathrm{M_{BH}} = 0.002\,\mathrm{M_{bulge}}$. This correlation has been used to argue for a 'causal connection between the formation of supermassive black holes and their host spheroidal galaxies' (Gebhardt et al. 2000; Kormendy & Bender 2011; Häring & Rix 2004; Gultekin et al. 2009). While AGN feedback via winds or jets has been invoked to explain the effect, the details remain obscure.

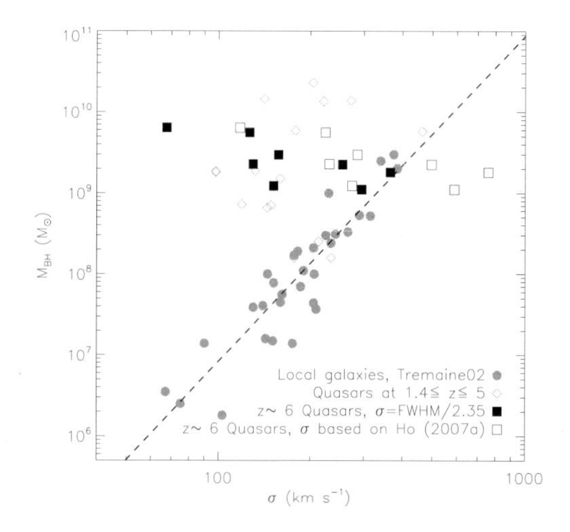


**Figure 11:** The relationship between the black hole mass and the host spheroidal galaxy velocity dispersion for low and high redshift quasars (from Wang et al. 2010a). For the high $z$ sources the velocity dispersion is estimated from the CO line imaging.

High resolution imaging of gas dynamics allows for study of the evolution of the black hole – bulge mass relation to high redshift redshift. Imaging of a few $z \geq 4$ quasars shows a systematic departure from the low $z$ relationship (eg. Walter et al. 2004; Riechers et al. 2008). Figure 11 shows a compilation by Wang et al. (2010). They find that, assuming random inclination angles for the molecular gas, the $z \sim 6$ quasars are, on average, a factor 15 away from the black hole – bulge mass relation, in the sense of over-massive black holes. Alternatively, all of the $z \sim 6$ quasars could be close to face-on, with inclination angles relative to the sky plane all $< 20^o$. High resolution imaging of the CO emission from these systems is required to address the interesting possibility that the black holes form before the host spheroids.

## 4.4 Massive galaxy formation at high redshift

Overall, the observations of the molecular gas and dust in extreme starburst galaxies at high redshift (SMGs, AGN hosts) indicate a major star formation episode during the formation of massive galaxies in group or cluster environments.

A key question is: what drives the prolific star formation? Tacconi et al. (2006; 2008) argue, based on imaging of higher-order CO emission from a sample of $z \sim 2$ SMGs, that SMGs are predominantly nuclear starbursts, with median sizes $< 0.5''$ ($< 4$ kpc), 'representing extreme, short-lived, maximum star forming events in highly dissipative mergers of gas rich galaxies.' This conclusion is supported by VLBI imaging of the 1.4 GHz emission from star forming regions in two SMGs (Momjian et al. 2005; 2010).

However, recent EVLA imaging of the lower order CO emission in SMGs (Ivison et al. 2011; Carilli et al. 2010; Riechers et al. 2011c), suggests that the lower-excitation molecular gas reservoirs can be significantly more extended. Riechers et al. (2011c) suggest a sequence in which the SMG phase is an early stage of a major gas rich merger, with the quasar phase arising later in the evolution (Sanders et al. 1988).

# 5 Secular galaxy formation during the epoch of galaxy assembly

## 5.1 sBzK and other typical star forming galaxies at $z \sim 1$ to 3

Optical through near-IR color selection techniques have become remarkably efficient at finding both star forming and passive galaxies during the peak epoch of galaxy assembly ($z \sim 1$ to 3). These include (rest frame) UV-selected samples (BX/BM; Steidel et al. 2004) and near-IR selected samples (sBzK; Daddi et al. 2005). Grazian et al. (2003) analyze the substantial overlap between the populations. The critical aspect for these samples is that they are not rare, pathological galaxies, such as luminous quasar hosts or the hyper-starburst submm galaxies. These galaxy samples are generally representative of the broad distribution of star forming galaxies at these epochs (Section 1), with areal densities of a few arcmin$^{-2}$, or volume densities $\geq 10^4$ Mpc$^{-3}$.

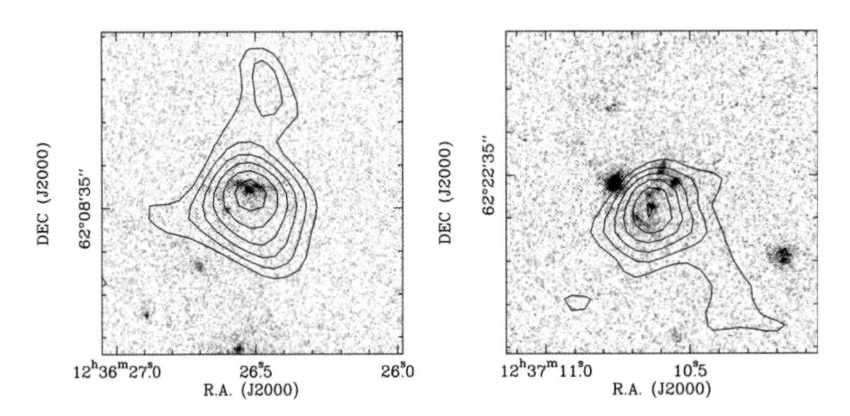


**Figure 12:** The CO emission from two $z \sim 1.5$ sBzK galaxies, with two CO velocity intervals shown as contours, and the greyscale is the HST i-band image (from Daddi et al 2010a).

HST imaging of these galaxies show clumpy but predominantly disk-like galaxies with sizes of order 10 kpc (Figure 12; Daddi et al. 2010a), with stellar masses $\geq 10^{10}$ $M_{\odot}$. Imaging of the H$\alpha$ and CO emission, reveals turbulent, but systematically rotating gas disks, and giant star forming clumps up to 1kpc in size (Genzel et al. 2008, 2011; Förster-Schreiber et al. 2011; Tacconi et al. 2010).

## 5.2 COSMOS radio stacking: dawn of downsizing

Pannella et al. (2010) have selected a sample of 30,000 sBzK galaxies from the COSMOS survey to determine the mean dust-unbiased star formation rates using stacking of 1.4 GHz observations. The number of sources, and the sensitivity of the radio observations, allow for substantial binning of the galaxies as a function of stellar mass, star formation rate, color, and blue magnitude. There is no correlation between median star formation rate and blue magnitude. This lack of correlation of SFR with blue magnitude occurs because there is a strong correlation of star formation rate with B-z color, ie. the extinction increases with star formation rate. There is also a positive correlation of star formation rate with stellar mass.

Combining these data, Figure 13 shows the specific star formation rate (SFR/$M_{stars}$) versus stellar mass for the COSMOS sBzK sample. For comparison, the relation at $z = 0.3$ from Zheng et al. (2007) is also shown. The red line shows the inverse Hubble time at $z = 1.8$. Galaxies above this line have SSFR that are sufficient to form the observed stars in the galaxy over their Hubble time. Galaxies below this line required a substantial increase in SFR rate in the past to form the stars that are seen. Lastly, the open points showing the SSFR based on dust-uncorrect star formation rates from the UV measurements.

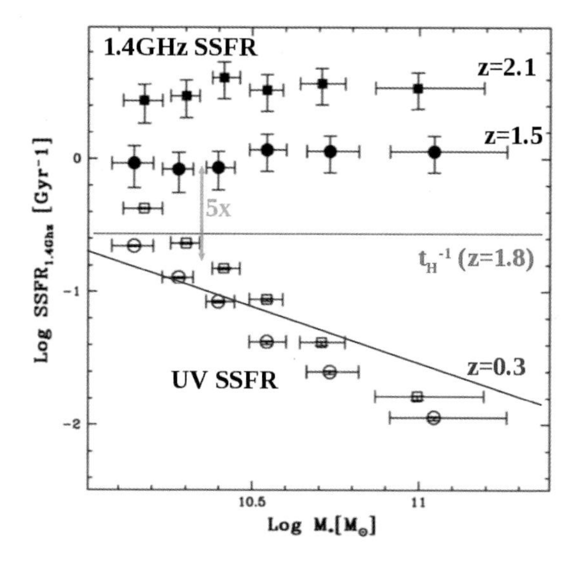


**Figure 13:** The specific star formation rate for $z \sim 2$ COSMOS sBzK galaxies (Pannella et al. 2009). Black filled symbols are SFR results from stacking of 1.4 GHz emission. Open symbols are the UV derived SFR without an extinction correction. The blue line indicates the SSFR vs. stellar mass for $z \sim 0.3$ galaxies in the Zheng et al. (2007) sample.

From Figure 13, the SSFR increases with redshift, even between $z = 1.5$ and 2.1. The high redshift galaxies are all above the 'red and dead' line indicated by the inverse Hubble time. Pannella et al. find that the substantial negative slope with stellar mass seen at $z = 0.3$ (ie. the decrease in SSFR with increasing stellar mass) becomes essentially flat at $z \geq 1.5$. Hence, $z \sim 2$ corresponds to an epoch when even fairly massive but common galaxies are actively forming stars. Lastly, it is clear that the dust extinction increases with increasing star formation rate. Interestingly, the standard factor 5 UV extinction correction for LBGs occurs at a stellar mass of $\sim 2 \times 10^{10}$ $M_\odot$, which is typical of LBG samples (Shapley et al. 2003).

## 5.3 Molecular gas: gas-dominated galaxies

Perhaps the most interesting result from the study of the sBzK and BX/BM galaxy samples comes from the searches for molecular gas. These samples show a remarkably high detection rate ($> 50\%$), in CO emission (Daddi et al. 2010a; Tacconi et al. 2010). The line strengths are comparable to those seen in SMGs and quasar hosts, but the star formation rates are an order of magnitude smaller. Figure 12 shows some examples. High resolution imaging shows that the CO is extended on the same scale as the optical disks, ($\sim 10$ kpc), with large condensations of size $\sim 1$ kpc, and masses $> 10^9$ $M_\odot$ (Tacconi et al. 2010; Aravena et al. 2010; Daddi etal. 2009a).

Some of these galaxies have been observed in the CO 1-0 transition with the VLA, as well as higher order transitions with the PdBI (Figure 14; Dannerbauer et al. 2009; Aravena et al. 2010). The excitation up to CO 3-2 appears to be sub-thermal, and substantially lower than is seen in either quasar hosts and SMGs (Figure 6). The excitation up to CO 3-2 is comparable to the Milky Way disk. Likewise, Figure 15 shows that the ratio of CO luminosity to FIR luminosity in these galaxies is similar to the Milky Way, and not to the SMG population, or compact nuclear starbursts.

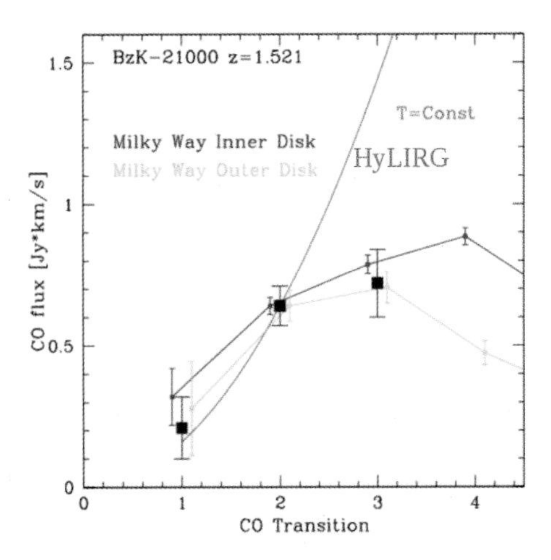


**Figure 14:** The CO ladder for sBzK galaxies (CO flux versus rotational quantum number for the upper state. The redline indicates constant brightness temperature (flux $\propto \nu^2$), typical for SMGs and AGN hosts in this J range (Dannerbauer et al. 2009; Aravena et al. 2010).

Daddi et al. (2010a) have done a detailed analysis of the possible CO luminosity to $H_2$ mass conversion for the sBzK sample. They employ dynamical models of forming disk galaxies including dark matter, as well as extensive observations of the stellar content and the CO dynamics. They conclude that the CO conversion factor is likely similar to the Milky Way value, rather than the nuclear starburst value employed for SMGs and quasar hosts. This conclusion is consistent with the Milky Way-like excitation and $L'CO/L_{FIR}$ ratio, and the large scale for the CO disks. Tacconi et al. (2010) reach a similar conclusion on $\alpha$ for the BX/BM samples.

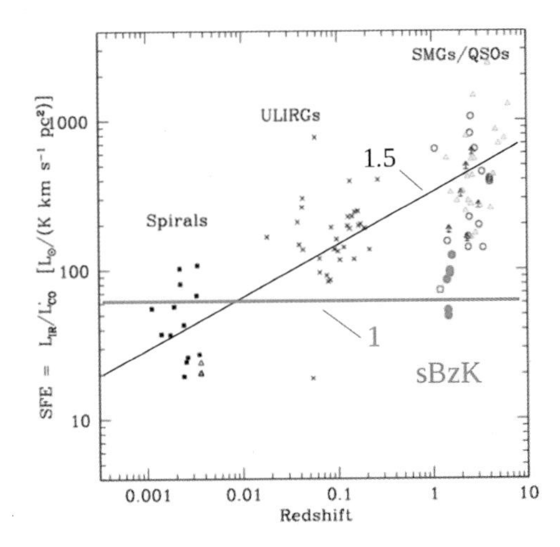


**Figure 15:** The ratio of FIR to CO luminosity for sBzK galaxies (red), and other galaxy samples (from Daddi et al. 2010a). The lines indicate powerlaw relationships of different indices.

The implied $H_2$ masses are then of order $10^{11}$ $M_\odot$. The gas masses are comparable to, or larger than stellar masses in the sBzK and BX/BM samples (Daddi et al. 2010a; Tacconi et al. 2010). This is very different with respect to low redshift disk galaxies, where the baryon content is dominated by stars (Figure 16). Hence, the peak epoch of cosmic star formation also corresponds to an an epoch when the dominant baryon component in star forming galaxies is gas, not stars.

## 5.4 Normal galaxy formation in the gas rich-era

The implied gas consumption timescales for the sBzK and BX/BM galaxies is $\sim$ few $\times$ $10^8$ years. This is an order of magnitude longer than the gas consumption timescale for the hyperstarbursts in SMGs and quasar host galaxies. Genzel et al. (2010) consider this point in detail, and conclude that the most likely explanation relates to different global dynamical effects in disks versus compact starbursts. They also emphasize that the gas consumption timescales are much shorter than the Hubble time in either case, and hence star formation must be a balance between gas accretion and feedback (see also Bauermeister et al. 2010).

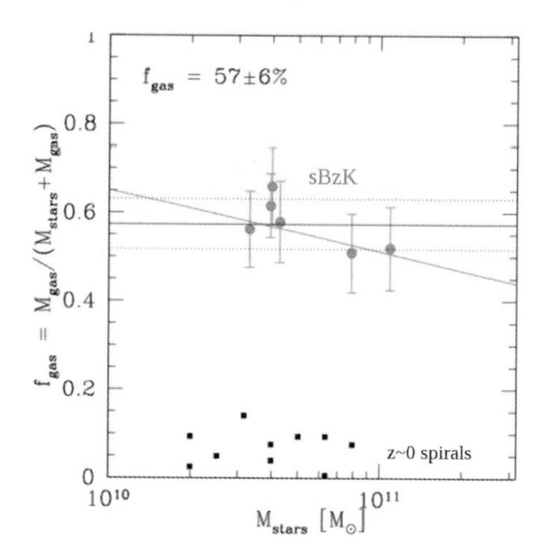


**Figure 16:** The molecular gas baryon fraction for sBzK galaxies at $z \sim 1.5$ (red circles), and low redshift spirals (black squares; from Daddi et al. 2010a)).

Theoretical and numerical studies have recently suggested that the dominant mode of star formation in the Universe, again peaking around $z \sim 1$ to 3, is not related to major gas rich merger events, but is driven by the slower process of 'cold mode accretion' (Dekel et al. 2009; Keres et al. 2009). In the CMA model, gas flows into galaxies from the IGM along cool, dense filaments. The flow never shock-heats due to the rapid cooling time, but continuously streams onto the galaxy at close to the free-fall time. This gas forms a thick, turbulent, rotating disk which efficiently forms stars across the disk, punctuated by giant clouds of enhanced star formation on scales $\sim$ few kpc. Genzel et al. (2008; 2011) show that the process is consistent with marginally Toomre-unstable gaseous disks. These star forming regions then migrate to the galaxy center via dynamical friction and viscosity, forming compact stellar bulges (Genzel et al. 2008; Bournaud et al. 2009; Elmegreen et al. 2009). The process is regulated by feedback, both within the giant star forming clouds themselves, and possibly from an active nucleus (Dave et al. 2011; Genzel et al. 2010).

The CMA process leads to relatively steady and active ($\sim$ 100 $M_\odot$ $yr^{-1}$) star formation in galaxies over timescales approaching 1 Gyr. The process slows down dramatically as gas supply decreases, and the halo mass increases, generating a virial shock in the accreting gas. Subsequent dry mergers at lower redshift then lead to continued total mass build up, and morphological evolution, but little subsequent star formation.

The H$\alpha$ and CO dynamical analyses of these samples are generally consistent with the morphologies expected from CMA, including turbulent but systematically rotating, gas disks, and giant star forming clumps. Shapiro et al. (2008) show that such a study of disk kinemetry enables an 'empirical differentiation between merging and non-merging systems'. Of course, this remains just a consistency check, and not direct observation of CMA.

Interestingly, observations of even more luminous star forming galaxies at even high redshift ($z > 4$) suggest that CMA may scale up to SMG luminosities at the highest redshifts (Carilli et al. 2010).

# 6 Dense gas history of the Universe

While admittedly compressing much information, the star formation history of the Universe has been a dominant tool in the study of galaxy formation over the last decade (Figure 1). However, the relationship between star formation and the molecular gas content of galaxies (Figure 2), implies that the SFHU should be reflected in the evolution of molecular gas.

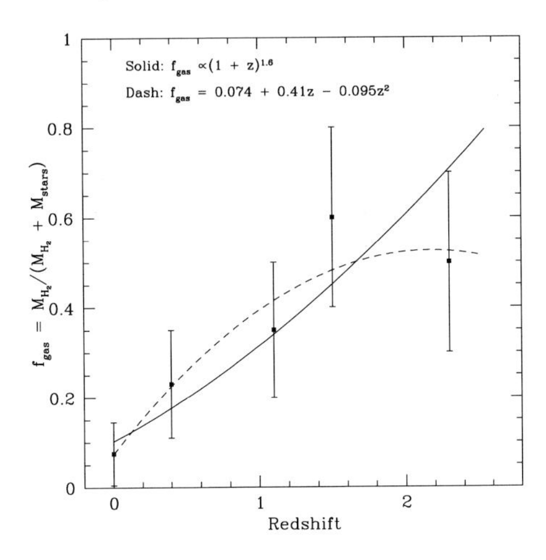


**Figure 17:** The molecular gas baryon fraction for star forming galaxies with $M_{stars} \geq 10^{10}$ $M_\odot$ (adapted and revised from Daddi et al. 2010a; Geach et al. 2011; Bauermeister et al. 2010).

Section 5.3 discusses how the average molecular gas content of star forming galaxies rises substantially with redshift. This is quantified in Figure 17, which shows the mean molecular gas baryon fraction ($\equiv \mathrm{M_{gas}}/[\mathrm{M_{gas}} + \mathrm{M_{stars}}]$), for star forming galaxies with $\mathrm{M_{stars}} > 10^{10}$ $M_\odot$ (adapted from Daddi et al. 2010b; Geach et al. 2011; Bauermeister et al. 2010). Admittedly there are many selection effects and assumptions that enter this calculation, but the current data support the idea of a substantial increase in the molecular gas content of galaxies at the peak epoch of cosmic star formation.

The next level of abstraction is to sum the gas mass to obtain the evolution of the cosmic density of molecular gas. Figure 18 shows an initial, very gross attempt at such a calculation, based on the 115 CO detections at $z > 1$ to date. The gas mass calculation entails a simple scaling from stellar mass densities from Grazian et al. (2007) using the mean gas baryon fraction for the different samples (Daddi et al. 2009b; Tacconi et al. 2010; Riechers et al. 2010b). We include the $z = 0$ value from

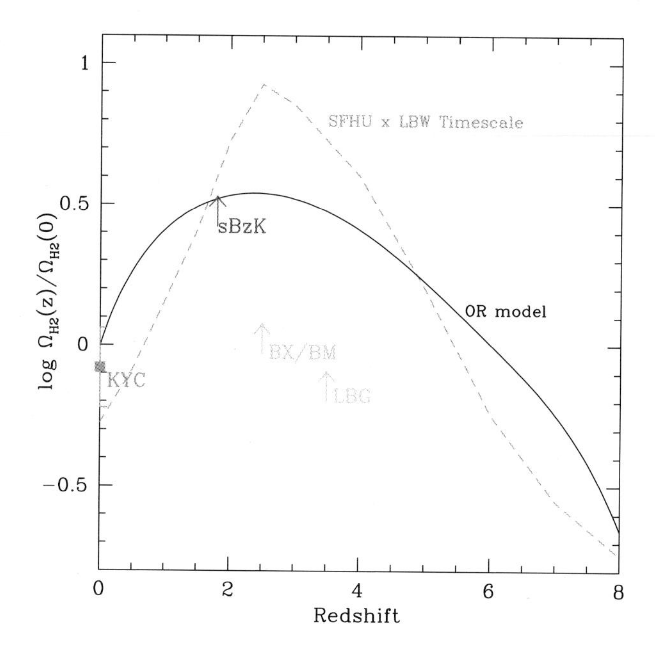


**Figure 18:** Gross estimates of the cosmic density (comoving) of molecular gas versus redshift based on the samples of star forming galaxies observed at $z > 1$. These include the near-IR selected sBzK galaxies (Daddi et al. 2010a), optically selected (BX/BM) galaxies (Tacconi et al. 2010; Genzel et al. 2010), and LBGs (Riechers et al. 2010b). Also plotted are two basic models: one boot-strapped from the SFHU plot assuming a simple timescale for gas conversion to stars in galaxies (Leroy et al. 2008). The second is based on the work of Obreschkow and Rawlings (2009) and the Millenium simulations. The zero redshift point is from the CO survey of Keres et al. (2003).

Keres et al. (2003). This figure includes two basic models: one boot-strapped from the SFHU plot assuming a constant timescale for gas conversion to stars in galaxies (Bigiel et al. 2011; Leroy et al. 2008; Bauermeister et al. 2010). The second is based on the work of Obreschkow and Rawlings (2009), who predict the molecular gas content of galaxies based on the Millenium simulations.

We emphasize this plot remains highly speculative due to the many assumptions involved, in particular, the conversion factor of CO luminosity to molecular gas mass, the open question of the evolution of the star formation law, and very limited samples.

# 7 ALMA and EVLA

The last decade has seen the opening of the high $z$ Universe to radio observations of the dust, gas, and star formation in the first galaxies. These results emphasize the critical need for pan-chromatic studies of galaxy formation to reveal all the key elements of the complex process. Fortunately, the immediate promise for a dramatic improvement in these studies is being realized with the science commissioning of the Expanded Very Large Array and the Atacama Large Millimeter Array.

The EVLA is a complete reinvention of the VLA, building on existing infrastructure (telescopes, rail track), but completely replacing the entire (1970's) electronic systems, from receivers through LO/IF to the correlator, to establish an essentially completely new telescope for the coming decade (Perley et al. 2011). The EVLA has complete frequency coverage from 1 GHz to 50GHz, and the bandwidth has increased by a factor 80, to 8GHz with thousands of spectral channels, greatly improving capabilities for spectral line searches and studies. The continuum sensitivity has increased by up to an order of magnitude. The array still provides compact configurations to image larger structures, and extended configurations that provide resolutions down to 40mas at 40GHz.

The EVLA construction project is close to completion, and the full array is operating at $\geq 18$ GHz with up to 2 GHz of bandwidth. Early science observations have been proceeding since March 2010, and indeed, some of the results discussed above are based on these observations.

The ALMA telescope will consist of 54 12m antennas for sensitive observations, plus 12 7m antennas for wide field imaging (Wootten & Thompson 2009), located at one of the best submm observing sites in the world, at 5000m elevation in Chile. The array will work from 80GHz to 750 GHz, initially in four bands, with an instantaneous bandwidth of 8GHz. The array configurations can be adjusted for wide field imaging as well as high resolution imaging to 20mas resolution at 350 GHz. Figure 4 shows the sensitivity of ALMA compared to existing (sub)mm arrays. The 3 orders of magnitude improvement in sensitivity in the submm is particularly dramatic.

The ALMA is also well into construction, with about half the antennas either operating at the high site, or under construction at the observational support facility. Demonstration science observations have already shown the power of even a limited set of ALMA antennas to perform ground-breaking observations of the dust and gas in galaxies, and early science observations will start at the end of 2011.

Taken together, ALMA and the EVLA represent an order of magnitude, or more, improvement in observational capabilities from 1 GHz up to 1 THz. Such a jump in capabilities is essentially unprecedented in ground-based astronomy. Following are a few examples of the potential of these instruments.

Figure 19 shows the sensitivity of ALMA, and other telescopes, to the [CII] line emission versus redshift. Included are the expected signals of star forming galaxies of different luminosity. ALMA will detect the [CII] line from typical LBGs well into cosmic reionization. Indeed, the 8GHz bandwidth and sensitivity present the interesting potential for determining the redshifts for these first galaxies. Determining optical spectroscopic redshifts is particularly hard as the Ly$\alpha$ line shifts into the near-IR, and the presence of a partially neutral IGM can attenuate the Ly$\alpha$ emission as well.

Figure 20 shows what an 8GHz spectrum of a luminous starburst, like J1148+5258 at $z = 6.42$, might look like with ALMA. Numerous interesting transitions of important diagnostic molecules will be detected, including dense gas tracers, isotopes, and isomers. ALMA opens up full astrochemical studies of the first galaxies.

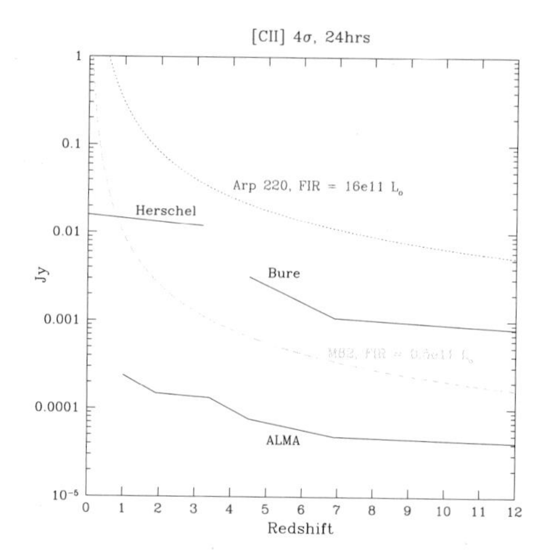


**Figure 19:** The expected [CII] line peak flux density versus redshift for active and dwarf star forming galaxies, plus the sensitivity of ALMA and other instruments.

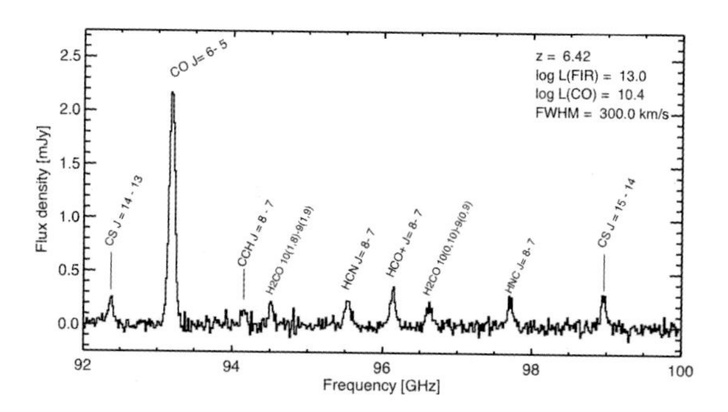


**Figure 20:** A simulated spectrum of the HyLIRG J1148+5258 at $z = 6.42$ with 8GHz using ALMA in 24 hours.

The EVLA has a 25% fractional bandwidth at 30 GHz. Hence, every observation with the EVLA at high frequency will include a blind search for molecular line emitting galaxies at high redshift. For instance, an EVLA observation at 19 to 27 GHz covers CO 1-0 at $z = 3.2$ to 5.0 instantaneously, thereby obviating the need for optical spectroscopic redshifts. Figure 21 shows a recent example from EVLA early science of this potential. Three molecule rich galaxies have been observed in a single pointing and 256MHz bandpass with the EVLA. Essentially every long observation of the EVLA over 20GHz will detect molecular line emission from distant galaxies, whether intended or not.

## Acknowledgements

CC thanks the Astronomische Gesellschaft organizers for their hospitality. We thank A. Weiss and R. Bouwens for figures.

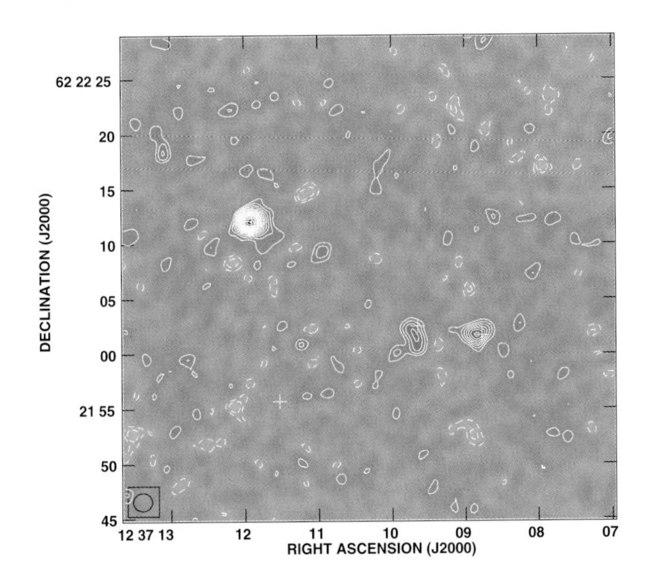


**Figure 21:** The GN20 field seen with the EVLA with a 256 MHz bandwidth at 45 GHz. Three galaxies are detected in CO 2-1 emission at $z = 4.0$ (from Carilli et al. 2011).

# References

Amblard, A. et al. 2011, Nature, 470, 510

Andreon, S. & Huertas-Company, M. 2011, A& A, 526, 11

Aravena, M., Bertoldi, F., Carilli, C. et al. 2010, ApJL, 708, L36

Baker, A. et al. 2004, ApJ, 613, L113

Bauermeister, A. et al. 2010, 717, 323

Blain, A. et al. 2002, PhR 369 111

Bennett, C. et al. 1994, ApJ, 434, 587

Bertoldi, F. et al. 2010, COSMOS mm

Bigiel, F. et al. 2011, 730, L13

Bournaud, F. et al. 2009, ApJ, 707, L1

Bourne, N. et al. 2011, MNRAS, 410, 1155

Bouwens, R. et al. 2010, ApJ, in press, arXiv 1006.4360

Calzetti, D. et al. 1994, ApJ, 429, 582

Carilli, C.L. et al. 2008, ApJ, 689, 883

Carilli C.L. et al. 2010, 714, 1407

Carilli C.L. et al. 2002, AJ, 123, 1838

Carilli C.L. 2011, ApJ, 730, L30

Carilli C.L. et al. 2011, ApJL, in press

Carilli, C.L. et al. 2005, ApJ, 618, 586
Chapman, S., Blain, A., Ivison, R., Smail, I. 2003, Nature 422 695
Chapman, S., Blain, A., Ibata, R. et al. 2009, ApJ, 691, 560
Collins, D., Stott, J.P., Hilton, M. et al. 2009, Nature, 458, 603
Coppin, K, et al. 2010, MNRAS, 407, L103
Coppin, K, et al. 2010, ApJ, 665, 936
Cormier, D. et al. 2010, A& A, 518, L57
Cowie, L. et al. 1997, 481, L9
Daddi, E. et al. 2004, ApJ, 617, 746
Daddi, E., Dickinson, M., Chary, R. et al 2005, ApJ 631 L13
Daddi, E., Dannerbauer, H., Stern, D. et al. 2009a, ApJ, 694, 1517
Daddi, E., Dannerbauer, H., Krips, M. et al. 2009b, ApJ, 695, L176
Daddi, E. et al. 2010a, ApJ, 713, 686
Daddi, E. et al. 2010b, ApJ, 714, L118
Dannerbauer, H., Walter, F., Morrison, G. 2008, ApJ, 673, L127
Dave, R. et al. 2011, MNRAS, in press ArXiv1103.3528
Dekel, A. et al. 2009, ApJ, 703, 785
Downes & Solomon 1998, ApJ, 507, 615
Doherty, M., Tanaka, M., de Breuck, C. et al. 2009, A& A, 509, 83
Draine, B. 2003, ARAA 41 , 241
Dwek, E. et al. 2007, ApJ 662 , 927
Elvis, M. et al. 1994, ApJS, 95, 1
Elvis, M. et al. 2002, ApJ 567 , L107
Elmegreen, B. et al. 2009, 692, 12
Fan, X., Carilli, C. Keating, B. 2006, ARAA, 44, 415
Förster-Schreiber, N. et al. 2011, ApJ, in press, arXiv1104.0248
Geach, J. et al. 2011, ApJ, 730, L19
Gao, Y. & Solomon, P. 2004, ApJ 606 , 271
Gebhardt, K. et al. 2000, ApJ, 543, L5
Genzel, R. & Cesarsky, C. 2000, ARAA, 38, 761
Genzel, R. et al. 2011, ApJ, in press ArXiv 1011.5360
Genzel, R. et al. 2010, MNRAS, 407, 2091
Genzel, R. et al. 2008, ApJ, 687, 59
Grazian, A. et al. 2007, A& A, 465, 393
Gultekin, K. et al. 2009, 706, 404
Hainline, :. et al. 2006, ApJ, 650, 614
Häring, N. & Rix, W. 2004, ApJ 604 , L89

Harris, A. et al. 2010, ApJ, 723, 1139
Ivison, R. et al. 2011, MNRAS, 412, 1913
Kennicutt, R. 1998, ARAA, 36, 189
Keres, D., Yun, M.S., Young, J. 2003, ApJ, 582, 659
Keres, D. et al. 2009, MNRAS, 395, 160
Kormendy, J. & Bender, R. 2011, Nature, 469, 377
Krumholz et al. 2009, ApJ, 699, 850
Kurk, J., Cimatti, A., Zamorani, G. et al. 2009, A& A, 504, 331
Leroy, A. et al. 2008, 136, 2782
Leipski, C. et al. 2010, A& A, 518, L34
Madau, P. et al. 1996, MNRAS, 283, 1388
Maiolino, R. et al. 2005, A& A 40 , L51
Malhotra, S. et al. 2001, ApJ, 561, 766
Marchesini, D. et al. 2009, ApJ, 701, 1765
Magnelli, B. et al. 2011, A& A, 528, 35
Michalowski, M. et al. 2010, A& A, 522, 15
Michalowski, M. et al. 2010, A& A, 515, 67
Miley, G. & de Breuck, C. 2008, A& ARv, 15, 67
Momjian, E., Carilli, C. & Petric, A. 2005, AJ, 129, 1809
Momjian, E. et al. 2010, AJ, 139, 1622
Moresco, M. et al. 2010, A& A, 524, 67
Murphy, E. et al. 2011, ApJ, in press arXiv1102.3920
Narayanan, D. et al. 2011, ApJ, in press arXiv 1104.4118
Obreschkow, D. & Rawlings, S. 2009, ApJ, 696, L129
Pannella, M. et al. 2009, ApJ, 698, 116
Papadopoulos, P. et al. 2002, ApJ, 564, L9
Papadopoulos, P. et al. 2010, ApJ, 711, 757
Perley, R. et al. 2011, ApJL, in press
Perley, D. et al. ApJ in press (2010)
Renzini, A. 2006, ARAA, 44, 141
Richards, G.T. et al. 2006, ApJS, 166, 470
Riechers, D. et al. 2008, ApJ, 686, L9
Riechers, D. et al. 2010a, ApJL, 720, L131
Riechers, D. et al. 2010b, ApJ, 724, L153
Riechers, D. et al. 2011a, ApJ, 726, 50
Riechers, D. et al. 2011b, ApJ, 725, 1032
Riechers, D. et al. 2011b, ApJ, in press arXiv1104.4348

Sanders, D. et al. 1988, ApJ, ApJ 325, 74
Schinnerer, E. et al. 2008, ApJ, 689, L5
Scott, K. et al. 2011, ApJ, in press, arXiv1104.4115
Shapiro, K. et al. 2008, ApJ, 682, 231
Shapley, A. et al. 2003, ApJ, 651, 688
Spitzer, L. 1998, *Physical Processes in the Interstellar Medium,* (Wiley)
Solomon, P. & Vanden Bout, P. 2005, ARAA, 43, 677
Solomon, P. et al. 2003, Nature, 426, 636
Stacey, G. et al. 2010, ApJ, 724, 957
Stratta, G. et al. 2007, ApJ 661 , L9
Steidel, C. et al. 2004, ApJ, 604, 534
Tacconi, L., Neri, R., Chapman, S. et al. 2006, ApJ, 640, 228
Tacconi, L., Genzel, R., Smail, I., et al. 2008, ApJ, 680, 246
Tacconi, L. et al. 2010, Nature, 464, 781
Thompson, T. et al. 2005, ApJ 630 , 167
Venkatesan, A. et al., ApJ 640 , 31 (2006)
Wagg, J. et al. 2010, A& A, 519, L1
Wagg, J. et al. 2005, ApJ, 634, L13
Walter, F. et al. 2009, Nature 457 , 699
Walter, F et al. 2004, ApJ 615 , L17
Walter, F. et al. 2003, Nature 424 , 406
Wang, R. et al. ApJ 714 , 699 (2010b)
Wang, R. et al. ApJ 687 , 848 (2008)
Weiss, A. et al. 2007, ASPC, 375, 25
Wootten, A. & Thompson, A. 2009, IEEEP, 97, 1463
Wu, J. et al. 2005, ApJ, 635, L173
Yun, M.S. et al. 2001, ApJ 554 , 803
Zheng, X. et al. 2007, ApJ, 661, L41

*Highlight talk Astronomische Gesellschaft 2010*

# Water in star-forming regions with Herschel[1,2]

Lars E. Kristensen[1] and Ewine F. van Dishoeck[1,2]

[1]Leiden Observatory, Leiden University
PO Box 9513, 2300 RA Leiden, The Netherlands
ewine@strw.leidenuniv.nl

[2]Max Planck Institut für Extraterrestrische Physik
Giessenbachstrasse 1, 85748 Garching, Germany

**Abstract**

*The Herschel Space Observatory is well suited to address several important questions in star- and planet formation, as is evident from its first year of operation. This paper focuses on observations of water, a key molecule in the physics and chemistry of star-formation. In the WISH Key Program, a comprehensive set of water lines is being obtained with the HIFI and PACS instruments toward a large sample of well-characterized protostars, covering a wide range of luminosities and evolutionary stages. Lines of $H_2O$, CO and their isotopologues, as well as chemically related hydrides, [O I] and [C II] are observed. Together, the data determine the abundance of water in cold and warm gas, reveal the entire CO ladder up to 4000 K above ground, elucidate the physical processes responsible for the warm gas (passive heating, UV or X-ray-heating, shocks), quantify the main cooling agents, and probe dynamical processes associated with forming stars and planets.*

## 1 Introduction

Water is a key molecule in the formation of stars and planetary systems, both as a cornerstone molecule in the oxygen chemistry and as a diagnostic of the physical structure of young stellar objects (YSOs). It is thus important to follow the abundance and excitation of water from the earliest stages of the collapsing cloud to the end stages of planet-forming disks (see, e.g., reviews by Cernicharo & Crovisier (2005) and Melnick (2009)). The Herschel Key Program "Water in star-forming regions with Herschel" (WISH; van Dishoeck et al. (2011)) uses high spectral and spatial

[1]Herschel is an ESA space observatory with science instruments provided by European-led Principal Investigator consortia and with important participation from NASA.

[2]This article has already appeared in Astron. Nachr./AN 332, no. 5 (2011).

*Reviews in Modern Astronomy 23: Zoomimg in: The Cosmos at High Resolution.* First Edition.
Edited by Regina von Berlepsch.

resolution HIFI and PACS observations of water through all evolutionary stages of star formation to follow the "water trail" and to use water as a probe of physical and dynamical processes taking place.

Previous data have suggested that the abundance of water is very low, $10^{-9}$–$10^{-8}$, in the cold quiescent parts of the molecular envelope, where the bulk of water is in the form of ice on dust grains. As the temperature increases closer to the protostar, all water is expected to evaporate and the abundance to increase by several orders of magnitude, up to $\sim 10^{-5}$–$10^{-4}$ with respect to $H_2$. Water is also frozen out onto dust grains in shocks, but sputtering of the grain mantles effectively releases all water into the gas phase. At the same time, atomic oxygen may be pushed into water through neutral-neutral reactions in the warm post-shock gas. Both processes lead to a jump in the abundance by up to four orders of magnitude, to $\sim 10^{-4}$, and this variation makes water such an excellent diagnostic of energetic phenomena. These processes have been previously inferred by a combination of SWAS and ISO data (e.g., Boonman et al. (2003)). SWAS was able to spectrally resolve the $H_2O$ $1_{10}$–$1_{01}$ transition at 557 GHz (Bergin et al. (2003), Franklin et al. (2008)), but in a much larger beam and at a lower sensitivity than what is currently possible with the Heterodyne Instrument for the Far-Infrared on Herschel. These factors make it possible to detect water in a larger number of protostars at any evolutionary stage, as well as detecting higher-excited lines and the rare isotopologue $H_2^{18}O$.

# 2 Observational strategy

The Heterodyne Instrument for the Far-Infrared (HIFI; de Graauw et al. (2010)) is used as the instrument of choice for observing water. Because of its high spectral resolution and sensitivity, it is possible to resolve the line profiles down to a velocity resolution of $<0.1\,\mathrm{km\,s^{-1}}$. Depending on the type of object, up to ten $H_2O$ and isotopologue transitions are targeted with HIFI, the transitions having upper-level energies of 50–500 K above the ground-state. For the faintest objects (pre-stellar cores and protoplanetary disks), the integration time is long enough (up to 10 hours or more) that an rms of $\sim$1 mK is achieved in the $H_2O$ $1_{10}$–$1_{01}$ and $1_{11}$–$0_{00}$ ground state lines, thus representing the longest single-pointing integrations done with HIFI. The bulk of the observations are carried out as single-pointing observations with beam-sizes ranging from $12''$ at 1834 GHz to $39''$ at 557 GHz. $H_2^{18}O$ and $H_2^{17}O$ lines are targeted to constrain the optical depth of emission lines across the entire mass spectrum and in protostellar outflows. High-$J$ CO and isotopologue lines are also targeted up to $J$ = 10–9 ($E_{\mathrm{up}}$ = 300 K).

HIFI observations are complemented by data from the Photodetector Array Camera and Spectrometer (PACS; Poglitsch et al. (2010)). PACS has a resolving power of only 1500-4000 and does not generally resolve the line profiles. However, its 5×5 array receiver with $9\farcs4$ pixels provides spatial information. Full spectral scans are carried out for a small number of low- and high-mass YSOs, whereas line scans of a specific sub-set of lines are done for the entire mass spectrum of sources. The lines targeted include high-$J$ CO lines, OH, [O I] at 63 and 145 $\mu$m, and, of course, higher

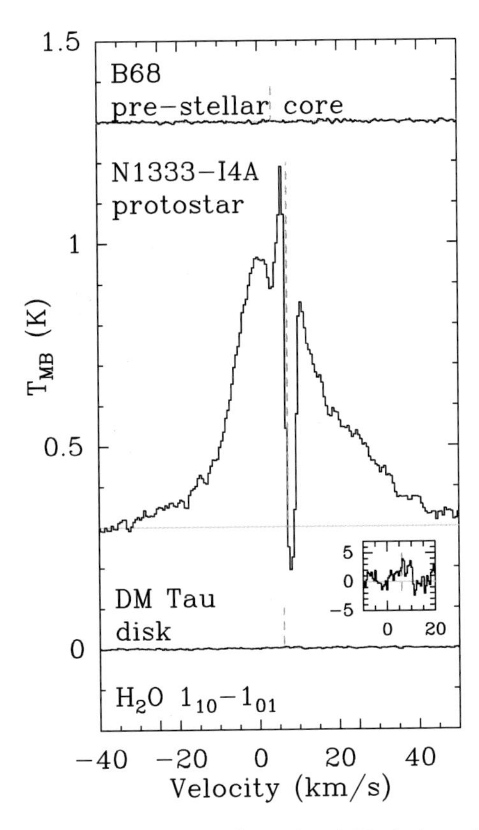


**Figure 1:** (online colour at: www.an-journal.org) Evolution of the $H_2O$ $1_{10}$–$1_{01}$ line at 557 GHz for the pre-stellar core B68, the low-mass embedded YSO NGC1333-IRAS4A and the protoplanetary disk DM Tau. The red dashed line indicates the systemic velocity, and the inset shows the tentative detection of $H_2O$ in a disk. Water emission is only bright in the embedded phase of star formation.

excited $H_2O$ lines (up to $E_{\rm up} \sim 1000$ K). Fully sampled line maps are obtained for three protostellar outflows in the 179.5 $\mu$m $H_2O$ line ($2_{12}$–$1_{01}$).

WISH targets about 80 sources, most of them in the embedded phase of star formation. The luminosity of the YSOs ranges from $<1\,L_\odot$ to $>10^5\,L_\odot$, covering low-, intermediate- and high-mass star formation. A few pre-stellar cores and protoplanetary disks are included in the sample to probe the different evolutionary stages from the pre-collapse stage of the cloud to the phase when the envelope has been dispersed and only a young star with a disk is left. The full source list and associated line list can be found in van Dishoeck et al. (2011).

## 3 Results

Figure 1 shows the evolution of the $H_2O$ $1_{10}$–$1_{01}$ line at 557 GHz from the pre-stellar core B68 (Caselli et al. (2010)), to the low-mass embedded source NGC 1333-

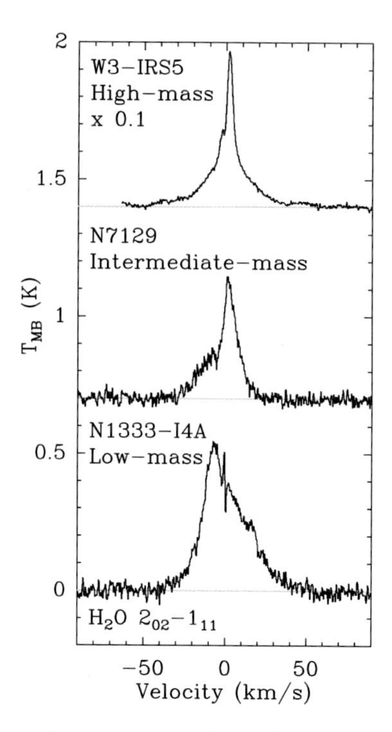


**Figure 2:** (online colour at: www.an-journal.org) Evolution of the $H_2O$ $2_{02}$–$1_{11}$ line at 988 GHz for the high-mass YSO W3-IRS5, the intermediate-mass YSO NGC7129-IRS1 and the low-mass YSO NGC 1333-IRAS4A. The line centers have been shifted to $0\,\mathrm{km\,s^{-1}}$. Note the presence of broad (FWHM $\geq 25\,\mathrm{km\,s^{-1}}$ and medium-broad (FWHM $\approx$ 5–10 $\mathrm{km\,s^{-1}}$) components in all sources irrespective of luminosity, as well as an inverse P-Cygni profile on top of the broad line in NGC 1333-IRAS4A.

IRAS4A (Kristensen et al. (2010)) ending with the protoplanetary disk DM Tau (Bergin et al. (2010)). This temporal evolutionary sequence highlights that during the quiescent phases of star formation, the water abundance is very low. The embedded stage of star formation, however, shows very broad and complex line profiles consisting of many dynamical components. An example is emission from the low-mass source NGC1333-IRAS4A, which shows a broad underlying component centered at the source velocity with an inverse P-Cygni profile on top indicative of large-scale infall. In disks, again most of the water is frozen out onto grains.

The mass-evolutionary track is very different from the temporal track. Comparing typical low-, intermediate-, and high-mass YSOs (Fig. 2) shows that absolute intensity aside, the profiles are remarkably similar. They are all broad, complex and consist of multiple components, clearly indicating that the water emission is primarily generated in shocks. In the following, the initial results from each sub-program of WISH are summarized.

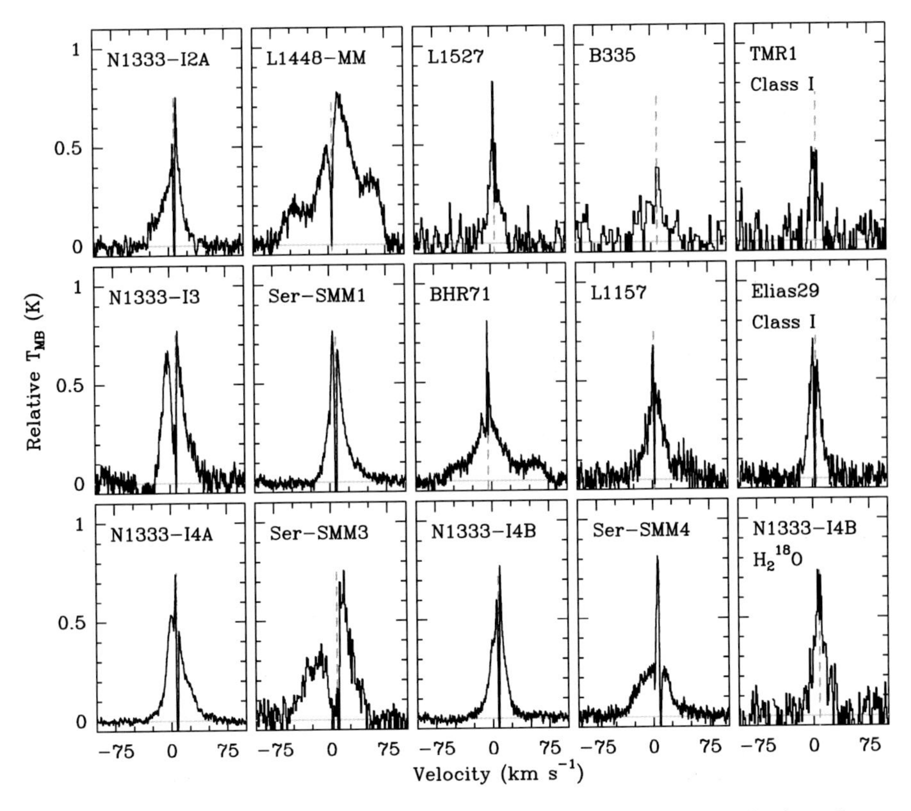


**Figure 3:** (online colour at: www.an-journal.org) HIFI spectra of a selection of low-mass Class 0 and I sources in the $H_2O$ $1_{10}$–$1_{01}$ transition at 557 GHz. The spectra have been scaled to a peak intensity of 0.8 K. The *bottom right panel* shows the detection of the $H_2^{18}O$ $1_{10}$–$1_{01}$ transition towards the embedded class 0 source NGC1333-IRAS4B. All spectra are continuum-subtracted and the red dashed line indicates the source velocity.

## 3.1 Pre-stellar cores

Caselli et al. (2010) present early results for the two cores L1544 and B68, with the B68 o-$H_2O$ $1_{10}$–$1_{01}$ spectrum included in Fig. 1. These spectra took ~4 h integration (on + off) each. No emission is detected in B68 but interestingly, the L1544 spectrum shows a tentative *absorption* feature against the submillimeter continuum at velocities where the CO 1–0 line is seen in emission. The B68 limit down to 2.0 mK rms in 0.6 km s$^{-1}$ bins is more than an order of magnitude lower than the previous upper limit obtained by SWAS for this cloud (Bergin & Snell (2002)) as well as most pre-Herschel model predictions. The $3\sigma$ upper limit corresponds to an o-$H_2O$ column density $< 2.5 \times 10^{13}$ cm$^{-2}$ and a mean line-of-sight abundance $< 1.3 \times 10^{-9}$. If the water absorption in L1544 is confirmed by deeper integrations currently planned within WISH, it provides a very powerful tool to determine the water abundance profile along the line of sight.

## 3.2 Low-mass YSOs

The embedded low-mass sources are often dominated by outflows (see, e.g., Arce et al. (2007) for a recent review). $H_2^{16}O$ emission is detected towards all sources, both Class 0 and the more evolved Class I (Fig. 3; Kristensen et al., in prep.). These results constitute the first detection of cold $H_2O$ in a Class I object as well as the detection of cold $H_2^{18}O$ in a low-mass protostar. The line profiles appear to be unique to each source, with very few common traits. The main characteristic of all detected $H_2O$ lines, including the $H_2^{18}O$ isotopologue lines, is the width of the line profiles. It is typically $>20\,\mathrm{km\,s^{-1}}$, but may go up to $>50\,\mathrm{km\,s^{-1}}$. The profiles show many dynamical components, including inverse P-Cygni-type profiles, emission from molecular "bullets" at high velocities, and deep absorptions going below the continuum level. The $H_2^{18}O$ transition is also self-absorbed, indicating that the envelope is optically thick even in $H_2^{18}O$.

The bulk of $H_2O$ emission arises in shocks associated with the molecular outflows (Kristensen et al. (2010)), and the water abundance is found to increase with velocity up to $H_2O/CO$ abundance ratios of $\sim$1. The only clear signature of the envelope is the deep absorption by gas-phase water in the outer, cold envelope, where the abundance is constrained to $\sim 10^{-8}$. The lack of any envelope emission signature, in particular the absence of excited $H_2^{18}O$ emission, is used to constrain the inner envelope abundance to $<10^{-5}$ (Visser et al., in prep.). This upper limit is significantly lower than expected if all of the water ice on grains evaporates in the inner envelope.

The next step consists of detailed 2D modeling for each protostar. Such a model has already been developed and tested for the low-mass protostar HH46 (van Kempen et al. (2010); Visser et al., in prep.). The model consists of three separate physical components: the molecular envelope heated by the accretion luminosity of the protostar, UV-heating of the outflow cavity walls and shocks along the cavity walls (see Fig. 4). The 'passively-heated' model uses the DUSTY radiative transfer code to compute the temperature structure through the envelope for a given power-law density structure, and assumes that the gas temperature is equal to the dust temperature. The UV model adds a photon-heated gas layer along the outflow cavity walls, either using a parametrized temperature grid based on the PDR models of Kaufman et al. (1999) or by computing the gas temperature explicitly in a 2D code such as presented by Bruderer et al. (2010b).

The line emission from the combined passively heated and photon-heated models is then computed using the radiative transfer code LIME (Brinch et al. (2010)). Shocks are added using the model results from Kaufman & Neufeld (1996). With this model it is possible to reproduce the CO ladder from $J$ = 2–1 up to $J$ = 40–39 with only two free parameters: the UV luminosity from the source and the shock velocity, as shown in Fig. 5. The passively heated envelope reproduces the low-$J$ lines ($\leq$6), the UV-heated layer the intermediate-$J$ lines ($J \approx 10$–20), and the shocks the high-$J$ lines ($J \geq 20$). Although there is some degeneracy in the models resulting in uncertainties in the inferred values of these two parameters, the conclusion that both components are needed is robust. Work is currently underway to do the same for $H_2O$.

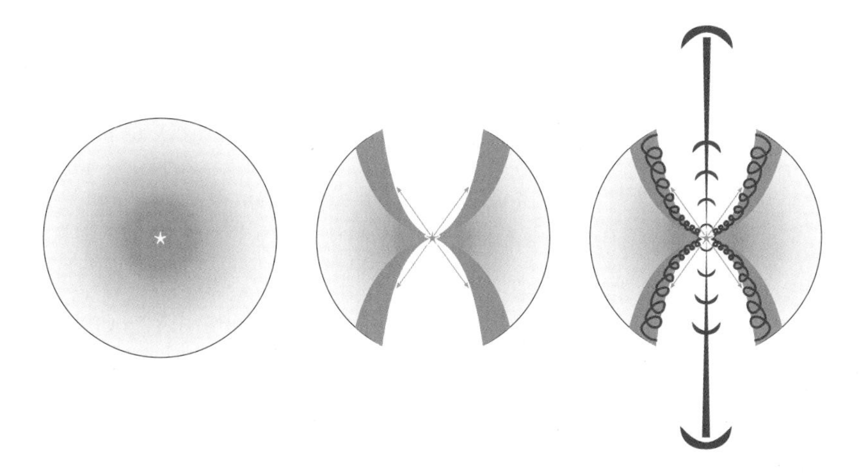

**Figure 4:** (online colour at: www.an-journal.org) Cartoons showing the different physical components of a protostar on scales of a few thousand AU. *Left*: the molecular envelope heated by the accretion luminosity of the protostar; *center*: outflow cavity walls heated by UV photons from the protostar; *right*: shell shocks running along the outflow cavity walls and molecular jet with internal working surfaces (Visser, priv. comm.).

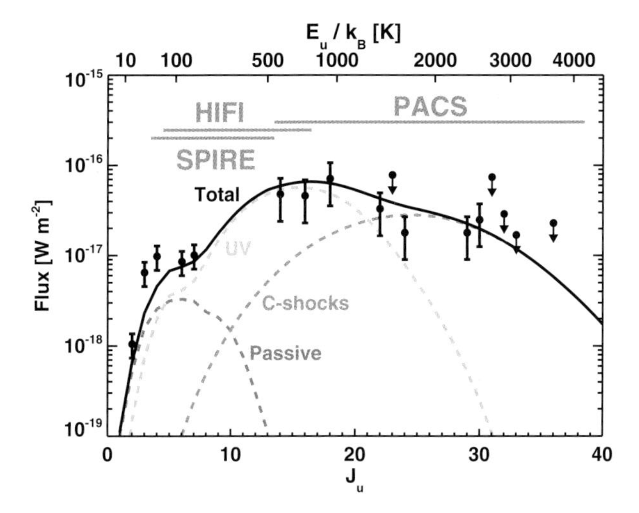


**Figure 5:** (online colour at: www.an-journal.org) CO ladder for the embedded low-mass protostar HH46. The ladder is decomposed into three components arising from the passively heated envelope, UV heating of the outflow cavity walls, and shell shocks running along the cavity walls. The wavelength ranges of the three instruments on-board Herschel are shown.

### 3.3 Intermediate-mass YSOs

Observations of various water lines toward NGC 7129 FIRS2 (430 $L_\odot$, 1260 pc) are presented in Fich et al. (2010) and Johnstone et al. (2010) and one of the excited p-$H_2O$ lines detected with HIFI is included in Fig. 2. The line profile is similar to that of the low-mass YSO NGC 1333 IRAS4A, showing both a broad component (FWHM $\sim$ 25 km s$^{-1}$) due to shocks along the outflow cavity as well as a medium-broad component (FWHM $\sim$ 6 km s$^{-1}$) associated with small-scale shocks in the inner dense envelope. Quantitative analysis of all the $H_2O$ and isotopologue lines observed for this object result in an outer envelope abundance $X_D$ of order $\sim 10^{-7}$, an order of magnitude higher than for the low- and high-mass YSOs.

Alternatively, both the high-$J$ CO and $H_2O$ lines seen with HIFI and PACS can be analyzed in a single slab model with a temperature of $\sim$1000 K and density $\sim 10^7$–$10^8$ cm$^{-3}$, representative of a shock in the inner dense envelope (Fich et al. (2010)). This leads to a typical $H_2O$/CO abundance ratio of 0.2–0.5.

### 3.4 High-mass YSOs

Figure 2 includes the excited p-$H_2O$ line for the high-mass YSO W3 IRS5 (Chavarría et al. (2010)). The profile reveals the same broad and medium-broad components as seen for their low- and intermediate-mass counterparts. In addition to self-absorption and absorption of the cold envelope against the continuum, absorptions due to foreground clouds lying along the line of sight are seen at various velocities. In particular, the high S/N absorption data of the ground-state p-$H_2O$ line show cold clouds surrounding the protostellar envelope at nearby velocities, revealing the detailed structure of the protostellar environment (Marseille et al. (2010)). Broad line profiles due to outflows associated with the high-mass protostars are commonly seen, testifying to the importance of shocks. Moreover, the presence of blue-shifted absorption suggests expansion of the outer envelope rather than infall.

The different physical components composing the high-mass protostellar environment can be disentangled to first order with simple models, as demonstrated by initial performance verification data on DR21 (OH) (van der Tak et al. (2010)). Using a slab model and assuming an o/p ratio of 3, the broad component due to the outflow is found to have an $H_2O$ abundance of a few $\times 10^{-6}$. Water is much less abundant in the foreground cloud where p-$H_2O$ has an abundance of $4\times10^{-9}$. The outer envelope abundance is even lower, a few $\times 10^{-10}$.

### 3.5 Outflows

Figure 6 shows the Herschel-PACS image of $H_2O$ in the 179 $\mu$m line toward the Stage 0 source L1157 obtained during the science demonstration phase of Herschel (Nisini et al. (2010)). The map clearly reveals where the shocks interact with the molecular cloud, lighting up the water emission along the outflow. The $H_2O$ emission is spatially correlated with that of pure rotational lines of $H_2$ observed with Spitzer (Neufeld et al. (2009)), and corresponds well with the peaks of other shock-produced molecules such as SiO and $NH_3$ along the outflow. In contrast with these species,

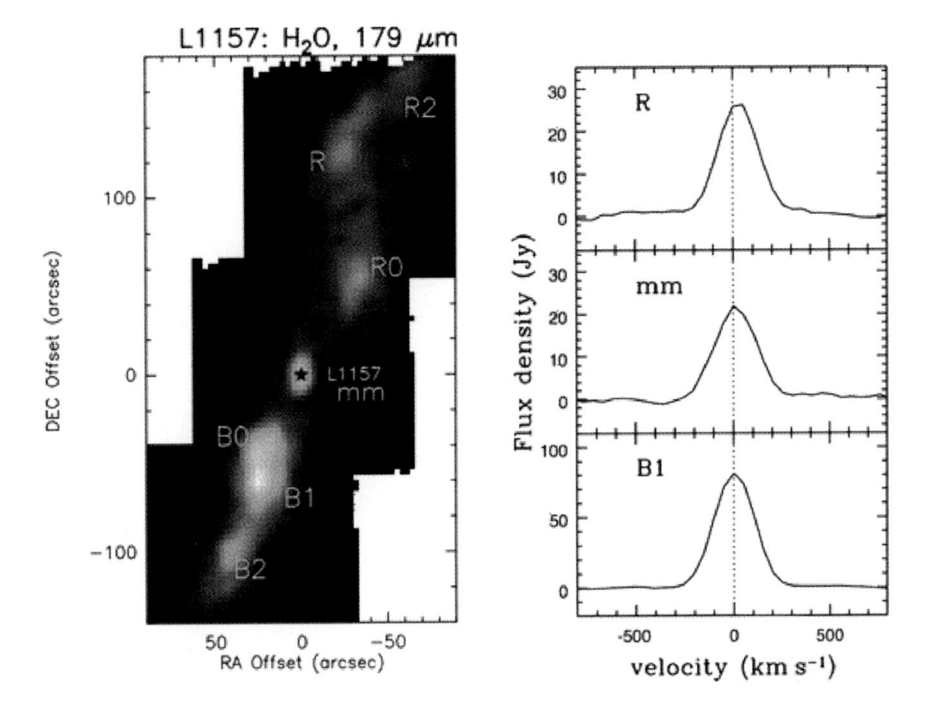


**Figure 6:** (online colour at: www.an-journal.org) Map of the L1157 outflow in the $H_2O$ $2_{12}$–$1_{01}$ transition at 179.5 $\mu$m obtained with PACS (*left*). Three representative spectra are shown to the *right* from the red outflow lobe (R), the source position (mm) and the blue outflow lobe (B1). The lines are unresolved.

$H_2O$ is also strong at the source position itself. The analysis of the $H_2O$ 179 $\mu$m emission, combined with existing Odin and SWAS data, shows that water originates in warm compact shocked clumps of a few arcsec in size, where the water abundance is of the order of $10^{-4}$, i.e., close to that expected from high-temperature chemistry. The total $H_2O$ cooling amounts to 23 % of the total far-IR energy released in shocks estimated from the ISO-LWS data (Giannini et al. (2001)).

The PACS spectra contained in Fig. 6 are spectrally unresolved. HIFI spectra of $H_2O$ and CO at the B1 position in the blue outflow lobe (see Fig. 6) have been presented and analyzed by Lefloch et al. (2010) as part of the CHESS key program, showing that the $H_2O$ abundance increases with shock velocity. They propose that this reflects the two mechanisms for producing $H_2O$: ice evaporation at the lower velocities and high temperature chemistry at the higher velocities where the O + $H_2$ and OH + $H_2$ reactions become significant. The presence of $NH_3$ (another grain surface product) in only the lower velocity gas is consistent with this picture (Codella et al. (2010)), but other explanations are not excluded.

### 3.6 Radiation diagnostics

An early surprise from all Herschel-HIFI key programs is the detection of widespread $H_2O^+$ absorption in a variety of galactic sources including diffuse clouds, the envelopes of massive YSOs and outflows (see, e.g., Gerin et al. (2010)). $H_2O^+$ and $OH^+$ lines are even seen in emission in SPIRE-FTS spectra of the AGN Mkr231 (van der Werf et al. (2010)).

Within WISH, several hydride molecules have been targeted explicitly as diagnostics of the presence of UV radiation and/or X-rays deep inside the envelope from which the radiation cannot be observed directly due to the large extinction. Pre-Herschel models had predicted the importance of species like $OH^+$, $H_2O^+$, $SH^+$ and $CH^+$ in probing this radiation (e.g., Stäuber et al. (2004, 2005, 2007)). Subsequent models have also highlighted the importance of the 2D geometry and UV-illuminated cavity walls in producing these species (Bruderer et al. (2010b)). Deep HIFI integrations on two high-mass YSOs reveal indeed most of these molecules, a confirmation of the model results. CH, NH, $OH^+$ and $H_2O^+$ are detected toward both W3 IRS5 and AFGL 2591 (Benz et al. (2010); Bruderer et al. (2010a)). In addition $SH^+$ and $H_3O^+$ are seen toward W3 and $CH^+$ toward AFGL 2591. Only SH and $NH^+$ are not detected. $H_2O^+$, $OH^+$ and $SH^+$ are new molecules, with $OH^+$ and $SH^+$ only recently detected for the first time in ground-based APEX observations by Wyrowski et al. (2010) and Menten et al. (2011), respectively.

The observed range in line profiles and excitation conditions (absorption versus emission) immediately indicates that the species originate in different parts of the YSO environment. The $OH^+/H_2O^+$ abundance ratio of $>1$ for the blue-shifted gas seen in absorption in both sources indicates low density and high UV fields (Gerin et al. (2010), as expected along the cavity walls at large distances from the source. Low densities ($<10^4$ cm$^{-3}$) and a low molecular fraction (i.e., a $H_2/H$ ratio of a few %) are needed to prevent $H_2O^+$ from reacting rapidly with $H_2$ to form $H_3O^+$.

The CH, $CH^+$ and $H_3O^+$ emission lines must arise closer to the protostar, at densities $>10^6$ cm$^{-3}$ (Bruderer et al. (2010a)). Their abundances are roughly consistent with the 2D photochemical models, in which high temperature chemistry and high UV fluxes boost their abundances along the outflow walls (Bruderer et al. (2009, 2010b)). The UV field is enhanced by at least four orders of magnitude compared with the general interstellar radiation field in these regions.

Because the $H_2O^+$ ground state line occurs close to that of p-$H_2O$, observations of both species are available for a large sample of high-mass protostars within WISH. Using also $H_2^{18}O$ data, Wyrowski et al. (2010) provide column densities for both species for all components along 10 lines of sight. $H_2O^+$ is always in absorption, even when $H_2O$ is seen in emission. Overall, $H_2O^+/H_2O$ ratios range from 0.01 to $>1$, with the lower values found in the dense protostellar envelopes and the larger values in diffuse foreground clouds and outflows, as expected from the above mentioned models.

### 3.7 Disks

Deep integrations on the DM Tau disk show a tentative detection of the o-$H_2O$ line with a peak temperature of $T_{\rm MB}$ = 2.7 mK and a width of 5.6 km s$^{-1}$ (Bergin et al. (2010)). Whether detected or not, the inferred column density is a factor of 20–130 weaker than predicted by a simple model in which gaseous water is produced by photodesorption of icy grains in the UV illuminated regions of the disk (Dominik et al. (2005)). The most plausible implication of this lack of water vapor is that more than 95–99 % of the water ice is locked up in coagulated grains that are so large that they are not cycled back up to the higher disk layers but remain settled in the midplane.

## 4 Conclusions

The initial results of the WISH program show that water and related species can indeed be used as physical and chemical probes of star-forming regions. Water shows large abundance variations between cold and warm gas, with abundances of gaseous water in pre-stellar cores and disks even lower than most pre-Herschel predictions. In contrast, shocks are seen brightly in broad water emission lines, obscuring any narrower emission from quiescent hot cores, even in isotopologue lines. Ions and hydrides involved in the water chemistry schemes are surprisingly easy to detect and are found to be widely distributed throughout the interstellar medium. The CO ladder, with detections of lines as high as $J$ = 44–43, forms a powerful tool to unravel the physical components of the protostellar envelope. Observations of all major coolants allow the total far-infrared cooling to be quantified. So far, water has not yet been found to be the major coolant in any of these regions.

Quantitative analysis of Herschel data will require the further development of multidimensional models of protostellar envelopes, outflows and disks. Together with the results from related Herschel key programs, they will greatly enhance our understanding of water in the galactic interstellar medium and solar system, and provide a true legacy for decades to come. Eventually, they will allow the determination of the formation and destruction of water from the most diffuse gas to planet-forming disks, and eventually comets and planets in our own solar system.

#### 4.0.1 Acknowledgements

This paper is presented on behalf of the entire WISH team and is made possible thanks to the HIFI guaranteed time program. HIFI has been designed and built by a consortium of institutes and university departments from across Europe, Canada and the US under the leadership of SRON Netherlands Institute for Space Research, Groningen, The Netherlands with major contributions from Germany, France and the US. Consortium members are: Canada: CSA, U.Waterloo; France: CESR, LAB, LERMA, IRAM; Germany: KOSMA, MPIfR, MPS; Ireland, NUI Maynooth; Italy: ASI, IFSI-INAF, Arcetri-INAF; Netherlands: SRON, TUD; Poland: CAMK, CBK; Spain: Observatorio Astronomico Nacional (IGN), Centro de Astrobiología (CSIC-INTA); Sweden: Chalmers University of Technology – MC2, RSS & GARD, Onsala Space Observatory, Swedish National Space Board, Stockholm University – Stockholm Observatory; Switzerland: ETH Zürich, FHNW; USA: Caltech, JPL, NHSC. We thank many funding agencies for financial support, in particular the Netherland Organization for Scientific Research (NWO) and the Netherlands Research School for Astronomy (NOVA).

## References

Aikawa, Y., Herbst, E., Roberts, H., Caselli, P.: 2005, ApJ 620, 330

Arce, H.G., Shepherd, D., Gueth, F., et al.: 2007, Protostars and Planets V, 245

Benz, A.O., et al.: 2010, A&A 521, L35

Bergin, E.A., Snell, R.L.: 2002, ApJ 581, L105

Bergin, E.A., Kaufman, M.J., Melnick, G.J., et al.: 2003, ApJ 582, 830

Bergin, E.A., Hogerheijde, M.R., Brinch, C., et al.: 2010, A&A 521, L33

Boonman, A.M.S., Doty, S.D., van Dishoeck, E.F., et al.: 2003, A&A 406, 937

Brinch, C., Hogerheijde, M.R.: 2010, A&A 523, A25

Bruderer, S., Benz, A.O., Doty, S.D., van Dishoeck, E.F., Bourke, T.L.: 2009, ApJ 700, 872

Bruderer, S., et al.: 2010a, A&A 521, L44

Bruderer, S., Benz, A.O., Stäuber, P., Doty, S.D.: 2010b, ApJ 720, 1432

Caselli, P., Keto, E., Pagani, L., et al.: 2010, A&A 521, L29

Cernicharo, J., Crovisier, J.: 2005, Space Sci. Rev. 119, 29

Chavarría, L., et al.: 2010, A&A 521, L37

Codella, C., et al.: 2010, A&A 518, L112

van Dishoeck, E.F., Kristensen, L.E., Benz, A.O., et al.: 2011, PASP 123, 138

Dominik, C., Ceccarelli, C., Hollenbach, D., Kaufman, M.: 2005, ApJ 635, L85

Fich, M., et al.: 2010, A&A 518, L86

Franklin, J., Snell, R.L., Kaufman, M.J., et al.: 2008, ApJ 674, 1015

Gerin, M., et al.: 2010, A&A 518, L110

Giannini, T., Nisini, B., Lorenzetti, D.: 2001, ApJ 555, 40

de Graauw, T., Helmich, F.P., Phillips, T.G., et al.: 2010, A&A 518, L6

Johnstone, D., Fich, M., McCoey, C., et al.: 2010, A&A 521, L41

Lefloch, B., et al.: 2010, A&A 518, L113

Keto, E., Caselli, P.: 2010, MNRAS 402, 1625

Kaufman, M.J., Neufeld, D.A.: 1996, ApJ 456, 611

Kaufman, M.J., Wolfire, M.G., Hollenbach, D.J., Luhman, M.L.: 1999, ApJ 527, 795

Kristensen, L.E., Visser, R., van Dishoeck, E.F., et al.: 2010, A&A 521, L30

Kristensen, L.E., et al.: 2011, in prep.

Marseille, M.G., et al.: 2010, A&A 521, L32

Melnick, G.J.: 2009, in: D.C. Lis et al. (eds.), *Submillimeter Astrophysics and Technology: a Symposium Honoring Thomas G. Phillips*, ASPC 417, p. 59

Menten, K.M., Wyrowski, F., Belloche, A., Güsten, R., Dedes, L., Müller, H.S.P.: 2011, A&A 525, A77

Neufeld, D.A., et al.: 2009, ApJ 706, 170

Nisini, B., et al.: 2010, A&A 518, L120

Poglitsch, A., Waelkens, C., Geis, N., et al.: 2010, A&A 518, L2

Roberts, H., Millar, T.J.: 2007, A&A 471, 849

Stäuber, P., Doty, S.D., van Dishoeck, E.F., Jørgensen, J.K., Benz, A.O.: 2004, A&A 425, 577

Stäuber, P., Doty, S.D., van Dishoeck, E.F., Benz, A.O.: 2005, A&A 440, 949

Stäuber, P., Benz, A.O., Jørgensen, J.K., van Dishoeck, E.F., Doty, S.D., van der Tak, F.F.S.: 2007, A&A 466, 977

van der Tak, F.F.S., et al.: 2010, A&A 518, L107

van Kempen, T.A., Kristensen, L.E., Herczeg, G.J., et al.: 2010, A&A 518, L121

van der Werf, P.P., et al.: 2010, A&A 518, L42

Visser, R., priv. comm.

Visser, R., et al.: 2011, in prep.

Wyrowski, F., et al.: 2010, A&A 521, L34

Wyrowski, F., Menten, K.M., Güsten, R., Belloche, A.: 2010, A&A 518, A26

*Highlight talk Astronomische Gesellschaft 2010*

# Light-element abundance variations in globular clusters[1]

Sarah L. Martell

Zentrum für Astronomie der Universität Heidelberg
Mönchhofstraße 12–14, 69120 Heidelberg, Germany
martell@ari.uni-heidelberg.de

**Abstract**

*Star-to-star variations in abundances of the light elements carbon, nitrogen, oxygen, and sodium have been observed in stars of all evolutionary phases in all Galactic globular clusters that have been thoroughly studied. The data available for studying this phenomenon, and the hypotheses as to its origin, have both co-evolved with observing technology; once high-resolution spectra were available even for main-sequence stars in globular clusters, scenarios involving multiple closely spaced stellar generations enriched by feedback from moderate- and high-mass stars began to gain traction in the literature. This paper briefly reviews the observational history of globular cluster abundance inhomogeneities, discusses the presently favored models of their origin, and considers several aspects of this problem that require further study.*

## 1 Introduction

Light-element abundance inhomogeneities in globular clusters have been a subject of active study for the past thirty years. The phenomenon was first noticed as anticorrelated variations in the broad CN and CH absorption features in red giants in a few clusters, and with the development of larger-aperture telescopes and higher-resolution spectrographs, the data set has expanded in several directions. Star-to-star abundance variations have been found in stars from the main sequence to the tip of the red giant branch, the set of abundances involved has expanded to include carbon, nitrogen, oxygen, sodium, magnesium and aluminium, and individual studies often survey tens of clusters rather than one or two.

Intriguingly, stars with these unusual abundances are found universally in globular clusters, but are apparently only formed in that environment. The overall picture that has developed is that stars at all evolutionary phases in all Galactic globular

[1] This article has already appeared in Astron. Nachr./AN 332, no. 5 (2011).

*Reviews in Modern Astronomy 23: Zoomimg in: The Cosmos at High Resolution.* First Edition.
Edited by Regina von Berlepsch.

clusters occupy a wide range in light-element abundance that is not observed in any other Galactic stellar population, or in the fields of Local Group dwarf galaxies. This abundance range is typically observed in anticorrelated abundance pairs, C vs. N or O vs. Na, because roughly half of cluster stars have abundance patterns like Population II field stars while the other half are relatively depleted in carbon, oxygen and magnesium and enhanced in nitrogen, sodium and aluminium, with no variations in any other elemental abundances.

There are a few globular clusters known to exhibit metallicity variations along with light-element variations, such as $\omega$ Cen and M54, and they are unusual in other ways as well, with large masses and likely extragalactic origins (e.g., Carretta et al. 2010). However, they are an exception, while clusters with only light-element variations appear to be the rule. Comparative studies of light-element abundances in globular cluster stars and halo field stars (e.g., Langer, Suntzeff & Kraft 1992) find that the field star population shows very little or no light-element abundance variation. The field-star studies of Pilachowski et al. (1996a) and Gratton et al. (2000) found no stars with cluster-like abundance variations in the field, and concluded that the cluster environment must play a vital role in creating or permitting light-element abundance variations.

More recently, Martell & Grebel (2010) searched a sample of field giants from the SEGUE survey (Yanny et al. 2009) and found that 2.5 % of those stars have strong CN and weak CH features relative to the majority of the field at the same metallicity and luminosity, suggesting that they may have the full cluster-like light-element abundance pattern. Those authors claim that the CN-strong field stars formed in globular clusters and later migrated into the halo as a result of cluster mass-loss and dissolution processes. A similar paucity of stars with cluster-like light-element abundances has been reported in nearby dwarf galaxies. McWilliam & Smecker-Hane (2005) and Sbordone et al. (2007) both report quite unusual abundance patterns in the Sagittarius dwarf galaxy, but no stars with cluster-like C-N or O-Na anticorrelations. Shetrone et al. (2003) surveyed stars in Sculptor, Fornax, Carina and Leo I and found none with cluster-like light-element abundances, and Letarte et al. (2010) confirm that result with more stars in Fornax.

Studies of Galactic globular cluster abundances are experiencing something of a revival at present, driven by the discovery that the (presently) most massive clusters have complex color-magnitude diagrams (CMDs). There is a variety of unexpected behavior found in carefully constructed, highly accurate CMDs: multiple main sequences in M54 (e.g., Siegel et al. 2007; Carretta et al. 2010), $\omega$ Centauri (e.g., Bedin et al. 2004; Sollima et al. 2007; Bellini et al. 2010) and NGC 2808 (Piotto et al. 2007), multiple subgiant branches in M54, $\omega$ Cen, 47 Tucanae (e.g., Anderson et al. 2009), NGC 1851 (e.g., Cassisi et al. 2008; Milone et al. 2008; Yong & Grundahl 2008), NGC 6388 (Piotto 2009) and M22 (Piotto 2009), and broadened red giant branches (RGBs) in M4 (Marino et al. 2008) and NGC 6752 (Milone et al. 2010).

There are several tantalizing potential connections between the photometric multiplicity in clusters and light-element abundance variations. Following the suggestion in Cassisi et al. (2008) that the two distinct subgiant branches discovered in NGC 1851 by Milone et al. (2008) have very similar ages but total [C+N+O/Fe] abundances that differ by a factor of two, Yong et al. (2009) measured abundances of carbon, nitrogen and oxygen in four bright red giants in NGC 1851. The total [C+N+O/Fe] abundance in their data set does vary fairly widely, which is consistent with a model in which NGC 1851 contains two populations in total light-element abundance. The case of NGC 1851 also provides a useful illustration of the importance of a careful choice of photometric systems: while the division of the subgiant branch is visible in the Milone et al. (2008) (F336W, F814W) and (F606W, F814W) color-magnitude diagrams, and in the Johnson-Cousins ($U, I$) color-magnitude diagram presented in Han et al. (2009b), it is not visible in the Han et al. (2009b) ($V, I$) color-magnitude diagram or the Strömgren ($vby$) color-magnitude diagram presented in Yong et al. (2009).

In M4, there is also RGB multiplicity that is found or not depending on the filter set used to construct a color-magnitude diagram: Marino et al. (2008) report a split of the red giant branch in the Johnson-Cousins ($U, B$) color-magnitude diagram, which had not been observed in previous color-magnitude diagrams based on redder photometric passbands. However, in M4, the two giant branches appear to correlate with typical globular cluster variations in [O/Fe] and [Na/Fe] abundances rather than variations in total [C+N+O/Fe] abundance.

It is of course not surprising that abundance variations with significant spectral effects should have noticeable effects on photometry, particularly in UV-blue photometric bands where CN and NH molecular bands can be dominant. Photometric determinations of stellar parameters and iron abundance were the motivation for defining medium-band filter systems like Strömgren (Strömgren 1963), DDO (McClure 1973) and Washington (Canterna 1976). It is, however, unexpected that the oxygen-sodium anticorrelation would affect photometry, particularly in the very broad Johnson/Cousins system, and it is unclear whether variations in oxygen and sodium abundance, which are observed in all Galactic globular clusters (e.g., Carretta et al. 2009), correlate with variations in $U - B$ color.

The complex phenomenology of photometrically multiple globular clusters, and the excitement about them, serve to underline the importance of the rhetorical framework used in discussing an observed phenomenon: as an anomaly, star-to-star abundance variations are a curiosity to be catalogued, but as a result of light-element self-enrichment and multiple star-formation events they are powerful markers of the conditions in early Galactic history. The present challenge is to understand the connections between the new photometric observations and the well-studied abundance variations. This requires that we develop a comprehensive model for the origin of chemical and photometric complexity in globular clusters, and further that we ground that model in the larger cosmological environment to enable studies of the effects of formation time and environment on the ability of a star cluster to self-enrich and to form a second stellar generation.

## 2 Development of the observational data set

Photographic color-magnitude diagrams constructed for globular clusters (e.g., Arp & Johnson 1955; Sandage & Wallerstein 1960) revealed simple stellar populations: unlike the wide variety found in surveys of the Solar neighborhood, stars in globular and open clusters all apparently shared a single age and metallicity. The clusters were quickly recognized as ideal laboratories for testing theories of stellar structure and evolution (e.g., Sandage 1958, Preston 1961), and are used to the present day as anchors for metallicity scales (e.g., Kraft & Ivans 2003; Carretta et al. 2009).

## 2.1 Carbon and nitrogen in bright red giants

It was therefore surprising when these orderly, predictable stellar systems turned out to host a number of stars with wide variations in the strength of molecular features. Unusually weak absorption in the CH G band (the phenomenon of "weak-G-band stars") was noted among giants in M92 (Zinn 1973; Butler, Carbon & Kraft 1975), in NGC 6397 (Mallia 1975), in M13 and M15 (Norris & Zinn 1977), in $\omega$ Cen (Dickens & Bell 1976), in M5 (Zinn 1977), and in 47 Tuc (Norris 1978). It was quickly shown (Norris & Zinn 1977; Zinn 1977) that most of the weak-G-band stars were on the asymptotic giant branch, implying the existence of some process that dramatically reshapes the surface abundances of stars between the RGB and the AGB, later named the "third dredge-up" by Iben (1975).

It was also noted (by, e.g., Norris & Cottrell 1979; Norris et al. 1981; Hesser et al. 1982; Norris, Freeman & Da Costa 1984) that those RGB stars with unusually weak CH bands also had relatively strong CN absorption at 3883 Å and 4215 Å. Since molecular abundance is controlled by the abundance of the minority species, CH traces carbon abundance, while CN reflects nitrogen abundance. This general association of bandstrength and abundance was confirmed by spectral-synthesis studies such as Bell & Dickens (1980), which implies that the CN-strong, CH-weak stars found only in globular clusters have atmospheres that are depleted in carbon and enriched in nitrogen.

Since the CNO cycle, operating in equilibrium, tends to convert both carbon and oxygen into nitrogen, stars with strong CN and weak CH bands were interpreted as having some amount of CNO-processed material in their atmospheres. Several theories were put forward to explain this extra CNO-cycle processing: McClure (1979) surveyed the data available at the time and concluded that internal mixing, specifically the meridional circulation described by Sweigart & Mengel (1979), could be responsible for "some or all" of the surface abundance variations. The connection between angular momentum and the efficacy of meridional circulation prompted Suntzeff (1981) to propose that different rotational velocities might explain the different levels of carbon depletion seen in giants in M3 and M13, an idea further explored in the Norris (1987) study of the relation between CN anomalies and overall globular cluster ellipticity. Langer (1985) suggested that a uniform mixing efficiency, in combination with star-to-star variations in CNO-cycle fusion rates, could produce the observed ranges in surface carbon and nitrogen abundance. Cohen (1978) proposed that star-to-star scatter in [Na/Fe] and [Ca/Fe] in M3 giants required a non-homogeneous initial gas cloud, a scenario that implies a primordial origin for [C/Fe] and [N/Fe] variations as well. In addition, D'Antona et al. (1983) proposed that light-element abundance variations were merely surface pollution, a consequence of mass loss from evolved stars and the high density of globular clusters.

## 2.2 Other elemental abundances

Hoping to learn more about the source of apparently CNO-processed material in the atmospheres of some globular cluster giants, researchers began obtaining higher-resolution spectra for cluster giants. These were difficult observations to make with the 4 m-class telescopes available at the time, and as a result the data sets were typically small and limited to stars brighter than $V \simeq 14$. Despite these challenges, it was quickly discovered that stars depleted in carbon and enhanced in nitrogen were also depleted in oxygen and magnesium, and enhanced in sodium and aluminium. CN-strong giants in M5 were found by Sneden et al. (1992) to have systematically lower oxygen abundances and higher sodium abundances than their CN-normal counterparts. Cottrell & Da Costa (1981) found positive correlations between CN bandstrength and both sodium and aluminium abundances in NGC 6752. A correlation between aluminium and sodium abundances, and anticorrelations between aluminium and both oxygen and magnesium, were found by Shetrone (1996) in M92, M13, M5 and M71, clusters that span the range of halo globular cluster metallicity.

Oxygen depletion was consistent with an evolutionary explanation, as the hydrogen-burning shell of a red giant is hot enough to host the CNO-cycle reactions $^{14}$N+2p $\rightarrow$ $^{12}$C+$\alpha$. However, changes in the abundances of heavier elements were unexpected from fusion reactions occurring within 0.8 $M_{\odot}$ red giants: the NeNa and MgAl cycles operate similarly to the CNO cycle, with the nuclei acting as catalysts to convert hydrogen into helium, but both require significantly higher temperatures. Some authors (e.g., Pilachowski et al. 1996b) interpreted the extension of the light-element abundance variations to sodium, magnesium and aluminium as a sign that the hydrogen-burning shells in red giants must have the ability to operate the hotter hydrogen-fusion cycles, while others (e.g., Peterson 1980) considered it a sign that the initial gas cloud from which globular clusters formed must have been inhomogeneous.

## 2.3 Main-sequence and turnoff stars

With the construction of 8-meter-class telescopes came access to fainter stars within globular clusters. Harbeck et al. (2003) observed around 100 stars at or below the main-sequence turnoff in 47 Tuc, and found clear bimodality in the distribution of CN bandstrength in those stars, implying variations in C and N abundance as large as those already known in red giants in the cluster. Main-sequence and turnoff stars in M13 were observed by Briley et al. (2004), and significant, anticorrelated ranges in C and N abundance ($\Delta$[N/Fe] $\simeq$ 1.0, $\Delta$[C/Fe] $\simeq$ 0.5) were also found among those stars.

These discoveries had a major impact on theoretical explanations for light-element abundance variations in globular clusters, and prompted a serious evaluation of the possibility that they are not simple stellar populations. While evolutionary explanations could conceivably be stretched to include modifications of surface aluminium abundance, they could not accommodate abundance variations in low-mass main-sequence stars, which are not capable of either high-temperature hydrogen fusion or mixing between the surface and the core. Additionally, the fact that the abundance ranges are as large above the "bump" in the RGB luminosity function as below it indicates that the abundance variations cannot be mere surface pollution, because such a signal would be greatly diminished at the first dredge-up (Iben 1965), when the surface convective zone briefly deepens well into the interior of the star.

# 3 Current models for globular cluster formation

The presently favored explanation for the presence of primordial light-element abundance variations in globular clusters is that the CN-strong, N- and Na-rich, C- and O-poor stars are a second generation formed from material processed by intermediate- or high-mass stars in the first generation. There have been several types of stars proposed as the source of this feedback material, each with its own strengths and weaknesses. AGB stars with masses between 4 and 8 $M_\odot$ (e.g., Parmentier et al. 1999) are popular because they are relatively common, they are known to have slow, massive winds, they are a site of hot hydrogen burning, and they evolve on timescales of $\sim 10^8$ years, quite fast compared to the lifetime of a globular cluster. In addition to AGB stars, rapidly rotating massive stars (Decressin et al. 2007b) and massive binary stars undergoing mass transfer (de Mink et al. 2009) have been proposed as alternative sources of feedback material, and both could deliver more feedback mass in a shorter amount of time than AGB stars, though the stars themselves are less common. As is pointed out by Sills & Glebbeek (2010), while it may be observationally determined that one particular feedback source is dominant, they all certainly contribute to the cluster ISM at some level.

In a present-day globular cluster with a mass of $5 \times 10^5$ $M_\odot$ and a $1:1$ ratio of first- to second-generation stars, assuming a 30 % star formation efficiency, the second generation of stars must have formed from $\simeq 8 \times 10^5$ $M_\odot$ of gas. The first stellar generation, with a mass of $2.5 \times 10^5$ $M_\odot$, clearly cannot have produced enough feedback material to form the second generation (indeed, typical values for AGB mass loss are $\leq 10$ %). There have been four solutions proposed for this "mass budget problem": a top-heavy first-generation mass function (e.g., Decressin et al. 2007b), a second-generation mass function that is truncated above 0.8 $M_\odot$ (e.g., D'Ercole et al. 2008), a first generation that is initially 10–20 times as massive as at present (e.g., D'Ercole et al. 2010), and infall of pristine gas (e.g., Carretta et al. 2010b; Conroy & Spergel 2010).

A top-heavy first generation would alleviate the mass budget problem by placing more 5–15 $M_\odot$ feedback sources in the first generation. A truncated (bottom-heavy) second generation would require the first generation to produce less feedback material: assuming a Kroupa IMF (Kroupa et al. 1993) and a mass range from 0.1 to 100 $M_\odot$, half of the mass is in stars with $M \leq 0.8$ $M_\odot$. However, it is unclear what physical process would cause overproduction of massive stars in the first generation or underproduction in the second, and neither effect is significant enough to solve the mass budget problem. Additionally, observations of young star clusters (e.g., Boudreault & Caballero 2010; Gennaro et al. 2010) do not find either of these effects, and present-day globular clusters do not contain unusually large populations of neutron stars or other compact remnants of massive first-generation stars relative to the number of low-mass first-generation stars still on the main sequence (e.g., Bogdanov et al. 2010; Lorimer 2010).

Massive first generations and significant gas infall, in contrast, are central to the leading theoretical models of globular cluster formation. In the model of D'Ercole et al. (2008), the first stellar generation has a mass 10 to 20 times its present mass, and produces all the material needed for the formation of the second generation from AGB winds. The authors assume that Type Ia supernovae begin occurring 40 Myr after the formation of the first generation. The supernovae conclusively end star formation in the cluster, meaning that all second-generation stars must have formed within that time. The second generation stars form near the cluster center, and as a result, when the cluster expands in response to the supernovae, the stars that become dissociated from the cluster are mostly or exclusively first-generation stars. The authors also consider a model with infall of pristine gas from near the globular cluster, and find that it prolongs star formation significantly beyond the onset of Ia supernovae.

The globular cluster formation model of Conroy & Spergel (2011) also relies on AGB stars to provide the chemical inhomogeneities between the first and second stellar generations, but requires significantly more gas accretion to provide the mass for second-generation stars. The authors justify this by speculating about the cosmological environment of proto-globular clusters, namely self-gravitating, gas-dominated proto-galactic systems with gas masses between $10^8$ $M_\odot$ and $10^{10}$ $M_\odot$. They find that globular clusters orbiting in such systems can accrete significant amounts of gas over $10^8$ years, through a combination of Bondi accretion and "sweeping up" of material in the cluster's path, with little vulnerability to ram-pressure stripping for clusters above $10^4$ $M_\odot$. In this model, Lyman-Werner flux ($912\,\text{Å} \leq \lambda \leq 1100\,\text{Å}$) from massive first-generation stars prevents the gathering gas from forming stars by dissociating $H_2$ molecules, creating a gap of roughly $10^8$ years between the two generations and allowing time for the mass in accreted gas to become large enough to create a second generation as massive as the first.

Any globular cluster formation model that requires significant gas accretion implicitly assumes that the accreted gas must have a metallicity very similar to that of the first stellar generation, since systematic [Fe/H] differences are not observed between first- and second-generation cluster stars. There are a few models that are explicit about the importance of this coincidence, such as Carretta et al. (2010a) and Smith (2010). Both of these models incorporate supernova material into the second generation but require that it mix well with accreted lower-metallicity gas before the formation of the second generation in order to make the metallicity of the second stellar generation be the same as the first. While these conditions are certainly conceivable for certain proto-globular clusters, they are unlikely to hold for all globular clusters in the Milky Way.

# 4 Recent observational progress

There have recently been several large-scale light-element abundance studies, which have the distinct advantage over smaller-scale studies that the abundance behavior of multiple clusters can be directly compared within homogeneous observations, data reduction and analysis. The low-resolution study of Kayser et al. (2008) measured CN and CH bandstrengths in stars from the main sequence to the red giant branch in 8 Southern globular clusters. Those authors confirmed the presence of RGB CN bandstrength variations and CN-CH anticorrelations, as found in previous single- and few-cluster studies. They also demonstrated that CN bandstrength variations can be found on the main sequence in the clusters NGC 288, NGC 362, M22 and M55, clusters that had not previously been observed to contain main-sequence abundance variations. To explore the question of the source of light-element abundance variations, they also evaluate possible correlations between the ratio of CN-strong to CN-weak stars and several cluster parameters, and find mild positive correlations to cluster luminosity and tidal radius. These trends are interpreted as signs that globular clusters with larger masses or outer-halo orbits would be more efficient at producing second-generation stars.

The comprehensive study of Carretta et al. (2009b) reported homogeneous oxygen and sodium abundances for 1958 stars of all evolutionary phases in 19 Southern globular clusters. This study made a significant statement about the universality of light-element abundance variations in globular clusters, and also explicitly adopted the language of self-enrichment and multiple stellar generations. By identifying groups of stars as "primordial", "intermediate" or "extreme" depending on oxygen and sodium abundances, this study made the claim that the degree of abundance variation can differ between clusters, and may be a function of environment and feedback source. Main-sequence stars in many of the same clusters were observed by Pancino et al. (2010), who measured CN and CH bandstrength variations from low-resolution spectra. These authors found that the fraction of CN-strong (second-generation) stars was $\sim$30 %, distinctly lower than the 70 % reported in Carretta et al. (2009b). This discrepancy is curious, and Pancino et al. (2010) suggest that it may indicate that C-N abundance variations are contributed, at least in part, by a different feedback source from the O-Na abundance variations studied by Carretta et al. (2009b).

From recent large-scale studies it appears that light-element abundance variations are universal in Galactic globular clusters, but the question of the dominant first-generation feedback source remains unsolved. Future studies will need to measure the C-N and O-Na variations simultaneously in order to address the mismatch in frequency of second-generation stars found in low- versus high-resolution spectroscopic studies, and will need to measure other specific elemental abundances to evaluate various aspects of cluster formation scenarios. For example, abundances of s-process elements like Ba would be useful for placing limits on AGB contributions (e.g., Smith 2008; Yong et al. 2008), and the abundance of Li carries a great deal of information about the importance of infalling pristine gas (e.g., D'Orazi & Marino 2010; Shen et al. 2010).

The APOGEE survey, one of four components of the SDSS-III project, will obtain high-resolution near-infrared spectra for 100 000 stars in all components of the Galaxy, including red giants in globular clusters. The data reduction pipeline will automatically determine 14 elemental abundances, including overall [Fe/H] metallicity, $\alpha$ elements, and most of the light elements that vary in globular clusters. It will provide a database for studying cluster light-element variations that is unparalleled in sample size, amount of abundance information per star, and start-to-finish homogeneity, and ought to shed significant light on many aspects of cluster light-element abundance variations.

# 5 Evolution of the Galactic globular cluster system

Although efforts have been made to understand the presence or degree of light-element abundance variations as a function of present-day globular cluster properties, correlations with present-day mass, concentration or ellipticity are loose at best. While we expect the total mass or central density of a cluster during the formation of the first and second generations to have an influence on its ability to self-enrich, those properties have clearly evolved significantly over each cluster's lifetime.

## 5.1 Self-enrichment and escape velocity

One of the more perplexing elements of the question of globular cluster light-element self-enrichment is the fact that most of the Galactic globular cluster population is unable to retain AGB or massive-star winds at the present day. This is observable both in the lack of intracluster material in globular clusters (Evans et al. 2003; van Loon et al. 2006; Boyer et al. 2006), and in low present-day escape velocities. The census of Galactic globular clusters conducted by McLaughlin & van der Marel (2005) includes values for $v_{\rm esc}$, and roughly half of the clusters have $v_{\rm esc}$ below 20 km s$^{-1}$. Since these clusters all have light-element abundance variations, and since all proposed sources of feedback material have wind speeds $\geq$20 km s$^{-1}$, these clusters must have had higher escape velocities in the past. The massive first generation in the D'Ercole et al. (2008) model provides one natural solution to this problem, as do suggestions (e.g., Palouš et al. 2009; Sills & Glebbeek 2010) that collisions between winds from multiple stars should result in a lower bulk wind velocity, trapping wind material that otherwise would escape in the dense inner regions of proto-globular clusters. It seems clear that the Galactic globular cluster population has evolved strongly since its formation, both in terms of the overall cluster mass function (e.g., Parmentier & Gilmore 2005) and in the structural properties of individual clusters (e.g., de Marchi et al. 2010).

## 5.2 The initial cluster mass function

It is curious that light-element abundance variations are apparently universal among present-day Galactic globular clusters, considering that the initial cluster mass function included many low-mass clusters that should not have been able to self-enrich according to current globular cluster formation models. There are two possible explanations for this coincidence that are quite simple: that self-enrichment in globular clusters is very common, and occurs at lower cluster masses than we expect, or that cluster dissolution was extremely effective early in the lifetime of the Milky Way, with only a small percentage of the highest-mass clusters surviving to the present day. Globular cluster formation scenarios that rely on significant gas infall tend to promote the first explanation, allowing clusters with lower-mass first generations to form a second stellar generation. The numerical study of Marks & Kroupa (2010) found that the expulsion of residual gas following star formation is very effective at destroying globular clusters with low initial masses and concentrations. This result both supports the second explanation and implies that globular clusters have contributed significant numbers of stars with first-generation abundances to the construction of the stellar halo of the Milky Way, as is also suggested by the result of Martell & Grebel (2010).

If it is simply coincidental that the minimum mass for a globular cluster to survive to the present day in the Milky Way is larger than the minimum mass for a globular cluster forming in the Milky Way to host two stellar generations, then it is instructive to consider environments where those conditions are not met. In galactic environments that are more hospitable to long-lived low-mass globular clusters, the present-day cluster populations ought to include Milky Way-like, high-mass, two-population clusters along with lower-mass, chemically homogeneous globular clusters. In galactic environments in which it is more difficult for clusters to self-enrich, there would be some fraction of high-mass globular clusters with homogeneous light-element abundances. Regarding the first possibility, the theoretical study of Conroy & Spergel (2011) suggests that intermediate-aged clusters in the Large Magellanic Cloud, with masses between $10^4$ $M_\odot$ and $10^5$ $M_\odot$ should be able to retain first-generation winds and self-enrich because of the relatively low ram pressure they experience. This claim is bolstered by the observational study of Milone et al. (2009), in which clearly broad and/or bifurcated main-sequence turnoffs were found to be common in intermediate-aged LMC clusters.

# 6 Future challenges

In order to correctly interpret the photometric complexities observed in some globular clusters (e.g., Marino et al. 2008; Milone et al. 2008; Han et al. 2009; Lardo et al. 2011), we must understand the photometric shifts caused by changes in light-element abundances and helium, in addition to those caused by age, overall metallicity and [$\alpha$/Fe]. Current theoretical isochrones (e.g., Bertelli et al. 2008; Dotter et al. 2008; Han et al. 2009) are built from stellar models that allow variations in age, overall metallicity, and sometimes the abundances of $\alpha$-elements and helium. Considering the correlations between light-element abundances and $U$-band photometry reported by, e.g., Marino et al. (2008), it seems prudent to expand the theoretical grid of stellar models to test for photometric sensitivity to light-element abundance variations. The study of Dotter et al. (2007) considered exactly this question, constructing isochrones with enhancements in one of C, N, O, Ne, Mg, Si, S, Ca, Ti, or Fe while maintaining a constant overall heavy-element abundance Z in order to explore the effects of individual-element abundance variations on stellar structure. They find that enhancement in C, N, or O abundance caused the isochrones to shift to the blue and reduced main-sequence lifetimes by as much as 15 %, while an enhanced Mg abundance caused isochrones to be redder but had a minimal positive effect on main-sequence lifetimes. They did not calculate isochrones for the anticorrelated light-element abundance pattern found in globular clusters, but such an exercise would be extremely helpful to our understanding of photometric complexity in globular clusters.

It will also be important to understand whether photometric variations are a generic result of light-element abundance variations, or if not, which globular cluster properties permit or prohibit them from being observed. As an example, large variations in CN and CH bandstrength are almost certainly responsible for $U$-band variations among red giants in relatively high-metallicity ([Fe/H $\geq -1.5$) globular clusters, but not all relatively high-metallicity clusters are known to have complex $U - B, B$ CMDs. Additionally, multiplicities in different regions of the CMD do not always correspond. For instance, the cluster $\omega$ Cen has three main sequences and five distinct subgiant branches (Villanova et al. 2007), making it unclear how many distinct populations it contains. A search by Piotto (2009) of archival HST/ACS photometry uncovered multiple turnoffs in several clusters, and further searches for UV-blue photometry of globular cluster stars in public databases (as done in SDSS by Lardo et al. 2011) or observatory archives could be a quick and profitable way to confirm or deny the presence of photometric complexity in a large number of Galactic globular clusters.

Developing tools for interpreting integrated spectra of extragalactic globular clusters will dramatically expand our ability to study the effects of cosmological environment on globular cluster formation and self-enrichment. Methods for deriving ages and mean elemental abundances from low-resolution spectra have been adapted from galactic stellar populations studies (e.g., Puzia et al. 2006; Schiavon 2007), and techniques for extracting mean abundances from high-resolution integrated spectra of extragalactic globular clusters have been developed by Colucci et al. (2009). A merger of the two approaches, matching high-resolution spectroscopic data to synthetic spectra that are a sum over multiple distinct populations, will allow detailed searches for abundance variations in extragalactic globular clusters to very large distances.

It is becoming an accepted paradigm that the majority of, if not all, "normal" Galactic globular clusters contain stars with a range of light-element abundances, although they are resolutely mono-metallic. This requires that clearly multi-metallic, multi-age clusters like $\omega$ Cen and M54 formed in different environments, and not as subsystems of the Milky Way. Rather, their extended star formation histories and ability to retain supernova feedback indicate that their early development occurred in a fairly high-mass environment. M54 lies quite close to the core of the Sagittarius dwarf galaxy, prompting some to claim that it formed as the nucleus of the galaxy (e.g., Layden & Sarajedini 2000), while others argue that M54 formed as a normal globular cluster but is being trapped in the galactic nucleus (e.g., Bellazzini et al. 2008). One group (Carretta et al. 2010a) has made the claim that M54 and $\omega$ Cen both originated as nuclear star clusters in dwarf galaxies, with $\omega$ Cen having been captured by the Milky Way earlier while M54 is still being removed from its galaxy of origin. The schematic model of multi-metallicity globular clusters having formed as nuclear star clusters in dwarf galaxies (e.g., Georgiev et al. 2009) is attractive: the dark-matter halo of the galaxy would permit the cluster to experience extended feedback and star formation, and present-day nuclear star clusters are similar to multi-metallicity globular clusters in several properties such as half-light radius, escape velocity and horizontal branch morphology.

Recent announcements of mild [Fe/H] and [Ca/Fe] variations in NGC 2419 (Cohen et al. 2010), along with the discovery of photometric complexity (which may be a result of age or metallicity variations) in several otherwise unexceptional globular clusters (e.g., Piotto 2009), raise the question of whether there is a class of globular clusters intermediate between "normal" mono-metallic, light element-variable globular clusters and the more massive multi-metallicity clusters. Theoretical studies of supernova feedback in extremely massive proto-globular clusters would help to clarify the feasibility of claiming that clusters with mild metallicity variations constitute the high-mass end of typical globular cluster self-enrichment. Numerical simulations of interactions between nucleated dwarf galaxies and the Milky Way would provide an estimate of how many nuclear star clusters may have been captured into Milky Way orbit, and whether clusters like NGC 2419 can be considered as examples of captured nuclear star clusters with a history of low-efficiency feedback.

### 6.0.1 Acknowledgements

SLM would like to thank the Scientific Organizing Committee of the Astronomische Gesellschaft 2010 Annual Meeting for the invitation to speak on this subject.

# References

Anderson, J., Piotto, G., King, I.R., Bedin, L.R., Guhathakurta, P.: 2009, ApJ 697, L58

Arp, H.C., Johnson, H.L.: 1955, ApJ 122, 171

Bedin, L.R., Piotto, G., Anderson, J., et al.: 2004, ApJ 605, L125

Bell, R.A., Dickens, R.J.: 1980, ApJ 242, 657

Bellazzini, M., Ibata, R.A., Chapman, S.C., et al.: 2008, AJ 136, 1147

Bellini, A., Bedin, L.R., Piotto, G., et al.: 2010, AJ 140, 631

Bertelli, G., Girardi, L., Marigo, P., Nasi, E.: 2008, A&A 484, 815

Bogdanov, S., van den Berg, M., Heinke, C.O., et al.: 2010, ApJ 709, 241

Boudreault, S., Caballero, J.A.: 2010, astro-ph/1011.0983

Boyer, M.L., Woodward, C.E., van Loon, J.T., et al.: 2006, AJ 132, 1415

Briley, M.M., Cohen, J.G., Stetson, P.B.: 2004, AJ 127, 1579

Butler, D., Carbon, D., Kraft, R.P.: 1975, BAAS 7, 239

Canterna, R.: 1976, AJ 81, 228

Carretta, E., Bragaglia, A., Gratton, R., D'Orazi, V., Lucatello, S.: 2009a, A&A 508, 695

Carretta, E., Bragaglia, A., Gratton, R.G., et al.: 2009b, A&A 505, 117

Carretta, E., Bragaglia, A., Gratton, R.G., et al.: 2010a, ApJ 714, L7

Carretta, E., Bragaglia, A., Gratton, R.G., et al.: 2010b, A&A 516, A55

Cassisi, S., Salaris, M., Pietrinferni, A., et al.: 2008, ApJ 672, L115

Cohen, J.G.: 1978, ApJ 223, 487

Cohen, J.G., Kirby, E.N., Simon, J.D., Geha, M.: 2010, ApJ 725, 288

Colucci, J.E., Bernstein, R.A., Cameron, S., McWilliam, A., Cohen, J.G.: 2009, ApJ 704, 385

Conroy, C., Spergel, D.N.: 2011, ApJ 726, 36

Cottrell, P.L., Da Costa, G.S.: 1981, ApJ 245, L79

D'Antona, F., Gratton, R., Chieffi, A.: 1983, Mem. Soc. Astron. Ital. 54, 173

De Marchi, G., Paresce, F., Portegies Zwart, S.: 2010, ApJ 718, 105

de Mink, S.E., Pols, O.R., Langer, N., Izzard, R.G.: 2009, A&A 507, L1

Decressin, T., Charbonnel, C., Meynet, G.: 2007a, A&A 475, 859

Decressin, T., Meynet, G., Charbonnel, C., Prantzos, N., Ekström, S.: 2007b, A&A 464, 1029

D'Ercole, A., D'Antona, F., Ventura, P., Vesperini, E., McMillan, S.L.W.: 2010, MNRAS 407, 854

D'Ercole, A., Vesperini, E., D'Antona, F., McMillan, S.L.W., Recchi, S.: 2008, MNRAS 391, 825

Dickens, R.J., Bell, R.A.: 1976, ApJ 207, 506

D'Orazi, V., Marino, A.F.: 2010, ApJ 716, L166

Dotter, A., Chaboyer, B., Ferguson, J.W., et al.: 2007, ApJ 666, 403

Dotter, A., Chaboyer, B., Jevremović, D., et al.: 2008, ApJS 178, 89

Evans, A., Stickel, M., van Loon, J.T., et al.: 2003, A&A 408, L9

Gennaro, M., Brandner, W., Stolte, A., Henning, T.: 2011, MNRAS 412, 2469

Georgiev, I.Y., Hilker, M., Puzia, T.H., Goudfrooij, P., Baumgardt, H.: 2009, MNRAS 396, 1075

Gratton, R.G., Sneden, C., Carretta, E., Bragaglia, A.: 2000, A&A 354, 169

Han, S., Kim, Y., Lee, Y., et al.: 2009a, in: T. Richtler, S. Larsen (eds.), *Globular Clusters – Guides to Galaxies*, p. 33

Han, S., Lee, Y., Joo, S., et al.: 2009b, ApJ 707, L190

Harbeck, D., Smith, G.H., Grebel, E.K.: 2003, AJ 125, 197

Hesser, J.E., Bell, R.A., Harris, G.L.H., Cannon, R.D.: 1982, AJ 87, 1470

Iben, Jr., I.: 1965, ApJ 142, 1447

Iben, Jr., I.: 1975, ApJ 196, 525

Kayser, A., Hilker, M., Grebel, E.K., Willemsen, P.G.: 2008, A&A 486, 437

Kraft, R.P., Ivans, I.I.: 2003, PASP 115, 143

Kroupa, P., Tout, C.A., Gilmore, G.: 1993, MNRAS 262, 545

Langer, G.E.: 1985, PASP 97, 382

Langer, G.E., Suntzeff, N.B., Kraft, R.P.: 1992, PASP 104, 523

Lardo, C., Bellazzini, M., Pancino, E., et al.: 2011, A&A 525, A114

Layden, A.C., Sarajedini, A.: 2000, AJ 119, 1760

Letarte, B., Hill, V., Tolstoy, E., et al.: 2010, A&A 523, A17

Lorimer, D.R.: 2010, ArXiv e-prints **which?**

Mallia, E.A.: 1975, MNRAS 170, 57P

Marino, A.F., Villanova, S., Piotto, G., et al.: 2008, A&A 490, 625

Marks, M., Kroupa, P.: 2010, MNRAS 406, 2000

Martell, S.L., Grebel, E.K.: 2010, A&A 519, A14

McClure, R.D.: 1973, in: C. Fehrenbach, B.E. Westerlund (eds.), *Spectral Classification and Multicolour Photometry*, IAU Symp. 50, p. 162

McClure, R.D.: 1979, Mem. Soc. Astron. Ital. 50, 15

McLaughlin, D.E., van der Marel, R.P.: 2005, ApJS 161, 304

McWilliam, A., Smecker-Hane, T.A.: 2005, in: T.G. Barnes III, F.N. Bash (eds.), *Cosmic Abundances as Records of Stellar Evolution and Nucleosynthesis*, ASPC 336, p. 221

Milone, A.P., Bedin, L.R., Piotto, G., et al.: 2008, ApJ 673, 241

Milone, A.P., Bedin, L.R., Piotto, G., Anderson, J.: 2009, A&A 497, 755

Milone, A.P., Piotto, G., King, I.R., et al.: 2010, ApJ 709, 1183

Norris, J.: 1978, in: A.G.D. Philip, D.S. Hayes (eds.), *The HR Diagram – The 100th Anniversary of Henry Norris Russell*, IAU Symp. 80, p. 195

Norris, J.: 1987, ApJ 313, L65

Norris, J., Cottrell, P.L.: 1979, ApJ 229, L69

Norris, J., Zinn, R.: 1977, ApJ 215, 74

Norris, J., Cottrell, P.L., Freeman, K.C., Da Costa, G.S.: 1981, ApJ 244, 205

Norris, J., Freeman, K.C., Da Costa, G.S.: 1984, ApJ 277, 615

Palouš, J., Wünsch, R., Tenorio-Tagle, G., Silich, S.: 2009, in: J. Andersen, J. Bland-Hawthorn, B. Nordström (eds.), *The Galaxy Disk in Cosmological Context*, IAU Symp. 254, p. 233

Pancino, E., Rejkuba, M., Zoccali, M., Carrera, R.: 2010, A&A 524, A44

Parmentier, G., Gilmore, G.: 2005, MNRAS 363, 326

Parmentier, G., Jehin, E., Magain, P., et al.: 1999, A&A 352, 138

Peterson, R.C.: 1980, ApJ 237, L87

Pilachowski, C.A., Sneden, C., Kraft, R.P.: 1996a, AJ 111, 1689

Pilachowski, C.A., Sneden, C., Kraft, R.P., Langer, G.E.: 1996b, AJ 112, 545

Piotto, G.: 2009, in: E.E. Mamajek, D.R. Soderblom, R.F.G. Wyse (eds.), *The Ages of Stars* , IAU Symp. 258, p. 233

Piotto, G., Bedin, L.R., Anderson, J., et al.: 2007, ApJ 661, L53

Preston, G.W.: 1961, ApJ 134, 651

Puzia, T.H., Kissler-Patig, M., Goudfrooij, P.: 2006, ApJ 648, 383

Sandage, A.: 1958, Ricerche Astronomiche 5, 41

Sandage, A., Wallerstein, G.: 1960, ApJ 131, 598

Sbordone, L., Bonifacio, P., Buonanno, R., et al.: 2007, A&A 465, 815

Schiavon, R.P.: 2007, ApJS 171, 146

Shen, Z., Bonifacio, P., Pasquini, L., Zaggia, S.: 2010, A&A 524, L2

Shetrone, M., Venn, K.A., Tolstoy, E., et al.: 2003, AJ 125, 684

Shetrone, M.D.: 1996, AJ 112, 1517

Siegel, M.H., Dotter, A., Majewski, S.R., et al.: 2007, ApJ 667, L57

Sills, A., Glebbeek, E.: 2010, MNRAS 407, 277

Smith, G.H.: 2008, PASP 120, 952

Smith, G.H.: 2010, PASP 122, 1171

Sneden, C., Kraft, R.P., Prosser, C.F., Langer, G.E.: 1992, AJ 104, 2121

Sollima, A., Ferraro, F.R., Bellazzini, M., et al.: 2007, ApJ 654, 915

Strömgren, B.: 1963, Quarterly Journal of the Royal Astronomical Society 4, 8

Suntzeff, N.B.: 1981, ApJS 47, 1

Sweigart, A.V., Mengel, J.G.: 1979, ApJ 229, 624

van Loon, J.T., Stanimirović, S., Evans, A., Muller, E.: 2006, MNRAS 365, 1277

Villanova, S., Piotto, G., King, I.R., et al.: 2007, ApJ 663, 296

Yanny, B., Rockosi, C., Newberg, H.J., et al.: 2009, AJ 137, 4377

Yong, D., Grundahl, F.: 2008, ApJ 672, L29

Yong, D., Karakas, A.I., Lambert, D.L., Chieffi, A., Limongi, M.: 2008, ApJ 689, 1031

Yong, D., Grundahl, F., D'Antona, F., et al.: 2009, ApJ 695, L62

Zinn, R.: 1973, A&A 25, 409

Zinn, R.: 1977, ApJ 218, 96

# Massive black holes and the evolution of galaxies

Marta Volonteri and Jillian Bellovary

Astronomy Department, University of Michigan
500 Church Street, Ann Arbor, MI, 48109, USA

martav@umich.edu

**Abstract**

*MBHs weighing million solar masses and above have been recognized as the engines that power quasars. Dynamical evidence also indicates that MBHs with masses of millions to billions of solar masses ordinarily dwell at the centers of nearby galaxies. The masses of today's MBHs define surprisingly clear correlations with the properties of their host galaxies (luminosity, mass, and stellar velocity dispersion), suggesting a connected mechanism for assembling MBHs and forming galaxies. The evidence therefore favors a joint galaxy and MBH evolution. The advances in observational cosmology and extragalactic astronomy have provided a large amount of data on quasars and their engines. We do know that MBHs are there, but we do not know how they got there. We also do not know if we are detecting all the existing MBHs. The outstanding questions are then how and when MBHs formed, what is the interplay between MBH and galaxy assembly, and how MBHs (via AGN feedback) affect the baryon content and the star formation history of their hosts.*

## 1 Introduction

MBHs weighing million solar masses and above have been recognized as the engines that power quasars. Dynamical evidence also indicates that MBHs with masses in the range $\sim 10^6 - 10^9\ M_\odot$ ordinarily dwell at the centers of nearby galaxies (Ferrarese & Ford 2005; Richstone et al. 1998). The masses of today's MBHs define surprisingly clear correlations with the properties of their host galaxies (luminosity, mass, and stellar velocity dispersion), suggesting a connected mechanism for assembling MBHs and forming galaxies. The evidence therefore favors a joint galaxy and MBH evolution.

- Where and at what redshift do MBHs form? Several scenarios for MBH formation have been proposed (e.g., Begelman et al. 2006; Bromm & Loeb 2003; Devecchi & Volonteri 2009; Lodato & Natarajan 2006; Loeb & Rasio 1994; Madau & Rees 2001), typically focusing on runaway accumulation of gas during the assembly of galaxies. *The details of the resulting early MBH population, and, crucially, the observable signatures that ALMA might be able to*

*Reviews in Modern Astronomy 23: Zoomimg in: The Cosmos at High Resolution.* First Edition.
Edited by Regina von Berlepsch.

*pick up have not been coherently determined yet.* The high gas density in and around high-redshift galaxies makes them natural nests for MBH formation. Most of the recent literature on MBH formation focuses on gas-dynamical processes under metal-free conditions, as metal ennriched gas has more efficient cooling, which favors fragmentation and star formation over MBH formation. In most theories MBH formation terminates when the interstellar medium becomes enriched by supernova winds. These models need to be accurately implemented in hydrodynamical simulations.

- How do MBHs grow? How does the host galaxy environment affect baryon accretion on MBHs? It is often assumed that MBHs gain their mass in merger-driven accretion episodes (Di Matteo et al. 2008, 2005). Hence, the MBH mass-growth follows closely the host merger history. In this case, we would expect that, for instance, the massive end of the $M_{\rm BH} - \sigma$ relation is established early, and lower mass MBHs migrate onto it. However, cold flows (cold gas that flows rapidly to the center of galaxies from filamentary structures around haloes) play a major role in the buildup of galaxy disks at high redshift (Brooks et al. 2009; Governato et al. 2009). Is the same process at work for MBHs? In those previous studies gas accreted through cold flows arrives to the galaxy center on a time-scale a few Gyr shorter than gas that is first shocked to the virial temperature of the host halo and then cools on to the disc, leading to the creation of a large reservoir of cold gas. The competition between star formation and MBH feeding for gas consumption must be studied in a full cosmological environment, especially at high redshift.

MBHs have unique properties that make them perfect "beacons" at early cosmic times. MBHs that are active, that is, shining as quasars, are the most luminous sources in the Universe, making them lighthouses in the early stages of galaxy assembly, when the starlight from galaxies is too dim to be detected by our telescopes. Also, black holes are not only seen in the electromagnetic spectrum, but they are also sirens for gravitational wave detectors, bringing about a novel way of observing our Universe. In the following I delineate some possible observational tests of the earliest evolution of black holes in galaxies that can be performed with current and future instruments.

# 2 Massive black hole formation

Detection of luminous quasars at $z \approx 6$ (e.g, Fan 2001) implies that the first black holes must have formed less than a billion years after the Big Bang. Indeed, the luminosities of these quasars, well in excess of $10^{47}$ erg s$^{-1}$, imply black holes with masses $\sim 10^9\, M_\odot$ already in place when the Universe is only 1 Gyr old. If we assume that black holes powering these quasars accrete mass at the Eddington rate, their mass increases with time as

$$M(t) = M(0)\, \exp\left(\frac{1-\epsilon}{\epsilon}\frac{t}{t_{\rm Edd}}\right), \tag{1}$$

where $t_{\rm Edd} = 0.45\,{\rm Gyr}$ and $\epsilon$ is the radiative efficiency. For a 'standard' radiative efficiency $\epsilon \approx 0.1$, and a seed mass $M(0) = 10^2 - 10^5\ M_\odot$, it takes at least 0.5 Gyr of uninterrupted accretion to grow to $\simeq 10^9\ M_\odot$.

In the framework of CDM models the collapse of structures proceeds *bottom-up* on larger and larger scales, giving rise to a hierarchy of smaller structures that are incorporated into larger ones at later times (White & Rees 1978). It is within the first dark matter halos, only as massive as a dwarf galaxy today, that the first stars and the first black holes form.

One of the most popular scenarios for black hole formation associates their seeds with the remnants of the first generation of stars, formed out of zero metallicity gas. Simulations of the collapse of primordial gas clouds (Abel et al. 2000; Bromm et al. 1999, 2002; Gao et al. 2007; Yoshida et al. 2006) suggest that the first generation of stars contained many 'very massive stars' (VMSs) with $m_\star > 100\ M_\odot$(Carr et al. 1984). If the first stars retain their high mass until death, they will collapse after a short ($\approx$ Myrs) life-time. Stars with mass over 260 $M_\odot$on the main sequence become a black hole containing at least half of the initial stellar mass, is born inside the star (Fryer et al. 2001). It has been suggested by Madau & Rees (2001) that a numerous population of black holes may have been the endproduct of the first episode of pre-galactic star formation; since they form in density high peaks, relic black holes with mass $> 150\ M_\odot$would be predicted to cluster in the cores of more massive halos formed by subsequent mergers. Although this path to black hole formation seems very natural, large uncertainties exist on the final mass of PopIII stars. We do not know if PopIII stars are indeed very massive, and in particular if they are above the threshold ($\simeq 260\ M_\odot$) for black hole formation.

Another family of models for MBH formation relies on the collapse of supermassive objects formed directly out of dense gas (Begelman et al. 2006; Bromm & Loeb 2003; Eisenstein & Loeb 1995; Haehnelt & Rees 1993; Koushiappas et al. 2004; Lodato & Natarajan 2006; Loeb & Rasio 1994). It has been suggested that efficient gas collapse probably occurs only in massive halos with virial temperatures $T_{\rm vir} > 10^4$K under metal-free conditions where the formation of $H_2$ is inhibited (Bromm & Loeb 2003), or for gas enriched below the critical metallicity threshold for fragmentation (Santoro & Shull 2006). Highly turbulent systems are also likely to experience a limited amount of fragmentation, suggesting that efficient gas collapse could proceed also in metal-enriched galaxies at later cosmic epochs (Begelman & Shlosman 2009). In such halos where fragmentation is suppressed, and cooling proceeds gradually, the gaseous component can cool and contract until rotational support halts the collapse. Additional mechanisms inducing transport of angular momentum are needed to further condense the gas until conditions fostering MBH formation are achieved.

An appealing route to efficient angular momentum shedding is by global dynamical instabilities, such as the "bars-within-bars" mechanism, that relies on global gravitational instability and dynamical infall (Begelman et al. 2006; Shlosman et al. 1989). Self-gravitating gas clouds become bar-unstable when the level of rotational support surpasses a certain threshold. A bar can transport angular momentum outward on a dynamical timescale via gravitational and hydrodynamical torques, allowing the radius to shrink. Provided that the gas is able to cool, this shrinkage leads to even greater instability, on shorter timescales, and the process cascades. This mechanism is a very attractive candidate for collecting gas in the centers of halos, because it works on a dynamical time and can operate over many decades of radius. Global bar-driven instabilities have now been observed in high-resolution numerical simulations of gas–rich galaxies (Levine et al. 2008; Mayer et al. 2010; Regan & Haehnelt 2009; Wise et al. 2008). These simulations find strong inflows that occur before most of the gas fragments and forms stars also at solar metallicities. In particular, Mayer et al. (2010) find that extremely efficient inflows are triggered by gas-rich major mergers.

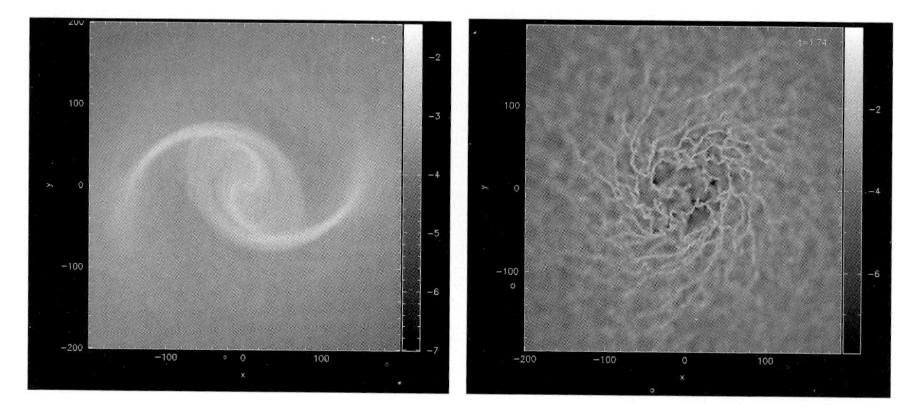

**Figure 1:** Color-coded density maps for two exponential gaseous discs, each embedded in a dark matter halo, soon after the discs reach the instability regime. Upper panel: in this case, fragmentation is suppressed, spiral waves develop in the disc, funnelling gas in the central few parsecs. Lower panel: strong cooling allows fragmentation in the disc to set in.

It has been also proposed that gas accumulation in the central regions of protogalaxies can be described by local, rather than global, instabilities. During the assembly of a galaxy disc, the disc can become self-gravitating. As soon as the disc becomes massive enough to be marginally stable, it will develop structures that will redistribute angular momentum and mass through the disc, preventing the surface density from becoming too large and the disc from becoming too unstable. To evaluate the stability of the disc, the Toomre stability parameter formalism can be used. The Toomre parameter is defined as $Q = \frac{c_s \kappa}{\pi G \Sigma}$, where $\Sigma$ is the surface mass density, $c_s$ is the sound speed, $\kappa = \sqrt{2} V_h / R$ is the epicyclic frequency, and $V$ is the circular velocity of the disc. When $Q$ approaches a critical value, $Q_c$, of order unity the disc is subject to gravitational instabilities. If the destabilization of the system is not too violent, instabilities lead to mass infall instead of fragmentation into bound clumps and global star formation in the entire disk (Lodato & Natarajan 2006). The efficiency of the mass assembly process ceases at large halo masses, where the mass-accretion rate from the halo is above the critical threshold for fragmentation and the disc undergoes global star formation instead (Figure 1, lower panel).

After it has efficiently accumulated in the center, the gas made available in the central compact region can form a central massive object. The typical masses of gas collected within the central few parsecs are of order $10^4 - 10^6\ M_\odot$. We expect that gas forms a supermassive star (mass above $\simeq 10^4\ M_\odot$), which would eventually collapse and form a black hole. Over time, the black hole can also grow at the expense of the gaseous envelope, until finally the growing luminosity succeeds in unbinding the envelope and the seed hole is unveiled.(Begelman 2009; Begelman et al. 2008).

An MBH seed could also form as a result of dynamical interactions in dense stellar systems (Begelman & Rees 1978; Ebisuzaki et al. 2001; Freitag et al. 2006a,b; Gürkan et al. 2006, 2004; Miller & Hamilton 2002; Portegies Zwart et al. 2004; Portegies Zwart & McMillan 2002). In dense systems where mass segregation can occur on short timescales, the most massive stars sink to the center of a star cluster. Star-star collisions can take place in a runaway fashion that ultimately leads to the growth of a VMS (Portegies Zwart et al. 1999) over a short timescale. Metal-enriched VMSs are expected to end their lives as black holes with mass $\sim$ few $M_\odot$ due to mass loss in winds Gaburov et al. (2009), while at sub-solar ($\simeq 10^{-3}$) metallicity the mass of the final remnant is large.

Devecchi & Volonteri (2009) investigate the formation of MBHs, remnants of VMS formed via stellar collisions in the very first stellar clusters at early cosmic times. In Toomre-unstable proto-galactic discs, mild instabilities lead to mass infall instead of fragmentation into bound clumps and global star formation in the entire disk. The gas inflow increases the central density, and within a certain, compact, region star formation ensues and a dense star cluster is formed. Most of these high redshift compact star clusters go into core collapse in $\lesssim 3$ Myr, and runaway collisions of stars form a VMS, leading to a MBH remnant. As the metallicity of the Universe increases, this process becomes inefficient and the process terminates.

# 3 Understanding the effect of environment on black hole growth

How does cosmic structure at galactic scales affect the growth of MBHs? The standard model for quasar activity and MBH growth assumes that these events are triggered by major galaxy mergers (Di Matteo et al. 2005). However, there is also evidence for AGN activity that is not merger-driven. Most Seyfert hosts are disk galaxies, which are unlikely to have undergone a recent major merger, hinting that secular processes or minor interactions may play an important role. *The advantage to studying MBH growth in cosmological simulations is that one can consistently include all the effects of major galaxy mergers as well as accretion of ambient cold gas from cold flows/filaments, or shock heated gas.* Each effect can be isolated by following the dynamical and thermodynamical history of the gas, as was done in Brooks et al. (2009) for galaxy disks, and we can determine how MBH growth is affected by each process. Bellovary et al. (2011) examine a set of galaxies with different masses and merger histories to determine under what circumstances their MBHs undergo the most efficient growth.

Figures 2 and 3 show results from two preliminary runs of a Milky Way and the high redshift progenitor of a massive elliptical (run to $z = 6$). Accretion that occurred via mergers is labeled as 'clumpy' accretion. Smoothly accreted gas particles may or may not undergo a shock as they reach the virial radius of the primary galaxy.

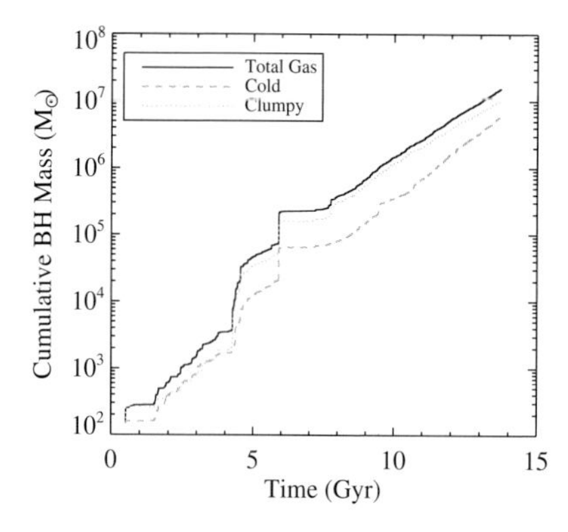


**Figure 2:** Cumulative mass accreted onto the central MBH in the galaxy *h239*. The main contribution occurs via gas brought in during mergers.

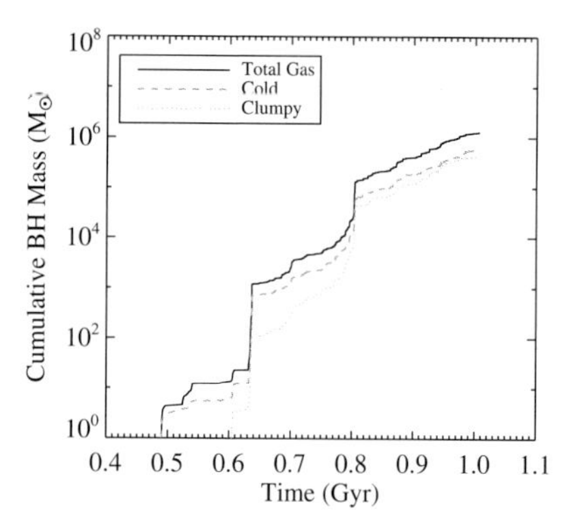


**Figure 3:** Cumulative mass accreted onto the central MBH in the galaxy *hz3*. The main contribution occurs via gas brought in by cold flows.

The presence of a shock depends mainly on the mass of the halo; generally, gas will shock if the galaxy halo is above $10^{10}\ M_{\odot}$. Cold gas, however, can flow into halos along the filaments inherent to the large-scale structure of the universe, and these filaments may be dense enough to penetrate and deliver gas directly to the galaxy. These filaments are critical in building up galaxy disks at high redshift (Brooks et al. 2007). Preliminary results show that the MBH in the Milky Way-size galaxy grows mainly through mergers, while the MBH in the massive high-redshift galaxy primarily accretes gas from cold flows, and it *reaches quasar luminosity at redshift $z = 6$ without undergoing a major merger, or even a minor one.* This early result might have important implications on the number of active quasars at high z.

# 4 Black holes and their high redshift hosts

The constraints on black hole masses at the highest redshifts currently probed, $z \simeq 6$, are few, and seem to provide seemingly conflicting results. (i) There seems to be little or no correlation with velocity dispersion, $\sigma$ (Wang et al. 2010) in the brightest radio-selected quasars, (ii) typically black holes are 'over-massive' at fixed galaxy mass/velocity dispersion compared to their $z = 0$ counterparts (e.g., Walter et al. 2004; at lower redshift see also Decarli et al. 2010; McLure & Dunlop 2004; Merloni 2010; Peng et al. 2006; Shields et al. 2006; Woo et al. 2008), on the other hand (iii) clustering and analysis of the mass/luminosity function suggest that either many massive galaxies do not have black holes, or these black holes are less massive than expected (Willott et al. 2010, W10). Here point (i) derives from observations of massive/rare systems, while point (iii) relies on deeper observations of slightly lower mass systems.

Expanding on point (ii), most authors propose that there is a *positive* evolution of the $M_{\rm BH}$−galaxy relationships, and quantify it as a change in *normalization*, in the sense that at fixed galaxy properties, black holes at high redshift are more massive

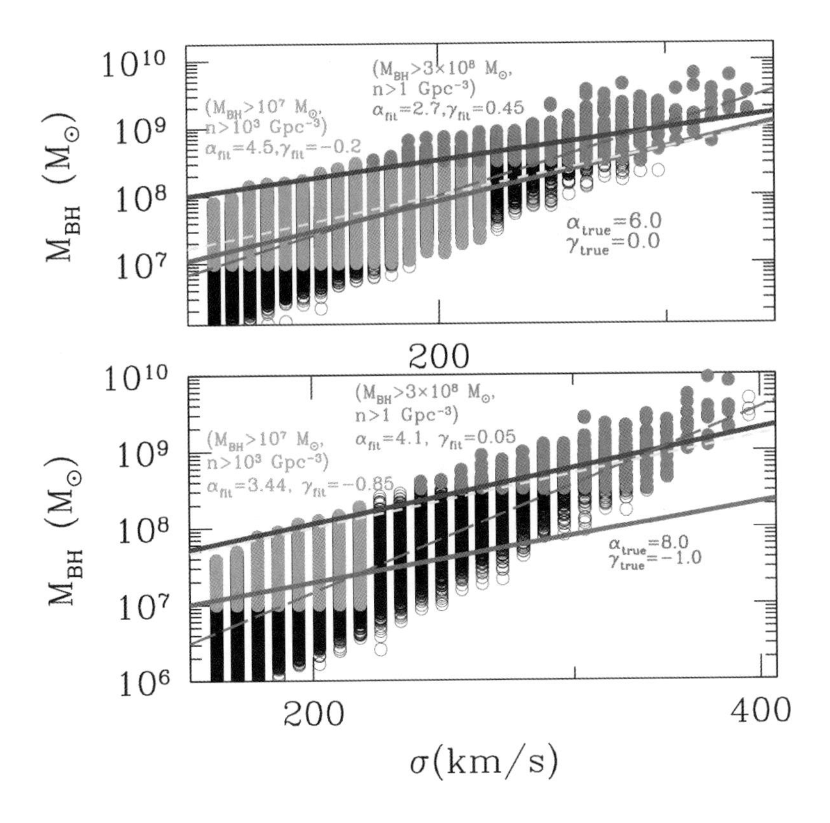


**Figure 4:** Top panel: $M_{\rm BH} - \sigma$ relation at $z = 6$, assuming $\alpha = 6$, $\gamma = 0$, and a scatter of 0.25 dex. Cyan dots: 'observable' population in a shallow survey. Blue line: linear fit to this 'observable' population, yielding $\alpha = 2.7$. Green dots: 'observable' population in a pencil-beam survey. Dark green line: linear fit to this 'observable' population, yielding $\alpha = 4.5$. Red line: fit to the whole population, yielding $\alpha = 6$. Yellow dashed line: $M_{\rm BH} - \sigma$ at $z = 0$ (Equation 1 with $\alpha = 4$ and $\gamma = 0$). Right panel: same for $\alpha = 8$, $\gamma = -1$.

than today. For instance, while Jahnke et al. (2009) find no difference between the $M_{\rm BH} - M_*$ relation at $z \simeq 1.4$ and the local $M_{\rm BH} - M_{\rm bulge}$, Merloni et al. (2010) propose that $M_{\rm BH} - M_*$ evolves with redshift as $(1+z)^{0.68}$, while Decarli et al. (2010) suggest $(1+z)^{0.28}$. Point (iii) above, however, is inconsistent with this suggestion, unless only about $1/100$ of galaxies with stellar mass $\simeq 10^{10} - 10^{11}\,M_\odot$ at $z = 6$ host a black hole (W10). These galaxies are nonetheless presumed to be the progenitors of today's massive ellipticals, which typically host central massive black holes.

The population of high redshift black holes we can study is however a small, rather biased, subsample . On the one hand, only the most massive black holes, powering the most luminous quasars can be picked up at such high redshift (flux-related bias; Shen et al. 2008; Vestergaard et al. 2008). On the other hand, they must be hosted in relatively common galaxies (survey size-related bias). These biases imply that only black holes in a narrow range of masses and host properties are selected.

We can show the effects of selection biases with a simple exercise. Let us assume an evolution of the $M_{\rm BH} - \sigma$ relationship of the form:

$$M_{\rm BH} = 10^8\,{\rm M_\odot}\left(\frac{\sigma}{200{\rm km\,s^{-1}}}\right)^{\alpha}(1+z)^{\gamma}, \qquad (2)$$

where $\alpha$ is a function of redshift. Let us now also assume that at fixed $\sigma$ the scatter in black hole mass is $\Delta$ =0.5 dex (see, e.g., Gultekin et al. 2009, Merloni et al. 2010 ) with a gaussian distribution in $\log\Delta$ (results are qualitatively unchanged for a uniform distribution in $\log\Delta$).

In Figure 4 we show a Monte Carlo simulation of the $M_{\rm BH} - \sigma$ relation at $z = 6$ one would find assuming $\alpha = 6$ and $\gamma = 0$, or $\alpha = 8$ and $\gamma = -1$. In section 3 we will show that these particular choices of $\alpha$ and $\gamma$ are motivated by our attempt to fit the black hole mass function of W10. For this exercise we run a number of realizations $N \propto 1/n$, where $n$ is the number density of halos of a given mass. We then select only systems that are likely to be observed ('observable' population), that is, black holes with a sizeable mass, implying that large luminosities can be achieved, $M_{\rm BH} > 3 \times 10^8\ M_\odot$ (see, e.g. Salviander et al. 2007; Lauer et al. 2007b, Vestergaard et al. 2008, Shen et al. 2008 for a discussion of this bias), hosted in halos with space density $n > 1$ Gpc$^{-3}$, so that they have a chance of being identified in existing large surveys. For instance the SDSS quasar catalogue selects sources with luminosities larger than $M_i = -22.0$ ($\simeq 10^{45}$ erg s$^{-1}$) over an area of 9380 deg$^2$, corresponding to a volume of almost 7 comoving Gpc$^3$ at z=6. Pencil beam surveys, such as GOODS, can probe fainter systems, but at the cost of a smaller area, e.g. the 2 Ms Chandra Deep Fields cover a combined volume of $\simeq 10^5$ comoving Mpc$^3$ at $z = 6$ and reach flux limits of $\simeq 10^{-17}$ and $\simeq 10^{-16}$ erg cm$^{-2}$ s$^{-1}$ in the 0.5-2.0 and 2-8 keV bands, respectively (the flux limit corresponds to a luminosity $\simeq 10^{43}$ and $\simeq 10^{44}$ erg s$^{-1}$at $z = 6$). As an example of a pencil beam survey, we select black holes with mass $M_{\rm BH} > 10^7\ M_\odot$ hosted in halos with density $n > 10^3$ Gpc$^{-3}$

To select 'observable' sources in number density, we link the velocity dispersion, $\sigma$, to the mass of the host dark matter halo by assuming that $\sigma = V_c/\sqrt{3}$, where $V_c$ is the virial circular velocity of the host halo[1]. Since the virial circular velocity is a measure of the total mass of the dark matter halo of the host, one can relate in simple ways the mass of the central black hole to the mass of its host halo, and estimate the number density from the Press & Schechter formalism (Sheth & Tormen 1999).

When we fit, in log-log space, the 'observable' population for the $M_{\rm BH} - \sigma$ relation of black holes in the 'shallow' survey, we find that the best fit has $\alpha_{\rm fit} \simeq 2.7 \pm 0.2$ and $\gamma_{\rm fit} \simeq 0.45 \pm 0.04$. The apparent normalization of the relationship therefore increases by 0.35 dex (all the blue points lie above the yellow line in the top panel of figure 4). So, while the underlying population is characterized only by a change in slope (with respect to the $z = 0$ relationship), what would be recovered

[1]We note that in an isothermal sphere $\sigma = V_c/\sqrt{2}$. This assumption is discussed further in Section 3 below. The results of this experiment are not strongly dependent on this specific assumption.

from the 'observable' population is a shallower slope and a positive evolution of the normalization only (in agreement with point (ii) above). We note, additionally, that the smaller the range in $M_{BH}$ that is probed, the more likely it is that the scatter $\Delta$ hides *any* correlation, likely explaining the lack of correlation (point (i) above) found by Wang et al. (2010)[2]. Finally, below the 'hinge' of Equation 2 black holes are under-massive at fixed host mass, with respect to their $z = 0$ counterparts. In fact the best fit has $\alpha_{\rm fit} \simeq 4.5$ and $\gamma_{\rm fit} \simeq -0.2$. This implies that low-mass black holes are 'pushed' into large galaxies, decreasing their number density (see point (iii) in the Introduction), and as we will show in Section 4 this helps reproducing the mass function of $z = 6$ black holes proposed by W10. However, the larger the scatter, the more the density of black holes at the *high-mass end* increases, as discussed in Lauer et al. (2007) and Gültekin et al. (2009).

We can repeat the same exercise for, e.g., $\alpha = 8$ and $\gamma = -1$, and although the underlying population has a much steeper slope and a *negative* evolution of the normalization of the $M_{\rm BH} - \sigma$ relation with redshift, the 'observable' population in shallow survey would display no evolution at all (blue vs yellow lines in Fig. 4). If we set $\alpha = 4$ and $\gamma = 0$ (no evolution), we find $\alpha_{\rm fit} \simeq 1$ in the 'shallow' survey (with almost all 'observable' black holes lying above the $\alpha = 4$ and $\gamma = 0$ line, suggesting 'overmassive' black holes, only because of the mass threshold), and $\alpha_{\rm fit} \simeq 2.5$ in the 'pencil beam' survey.

# 5 Blazars at early cosmic times

Ajello et al. (2009, hereafter A09) recently published the list of blazars detected in the all sky survey by the Burst Alert Telescope (BAT) onboard the *Swift* satellite, between March 2005 and March 2008. BAT is a coded mask designed to detect Gamma Ray Bursts (GRBs), has a large field of view ($120^\circ \times 90^\circ$, partially coded) and is sensitive in the [15–150 keV] energy range. This instrument was specifically designed to detect GRBs, but since GRBs are distributed isotropically in the sky, BAT, as a by–product, performed an all sky survey with a reasonably uniform sky coverage, at a limiting sensitivity of the order of 1 mCrab in the 15–55 keV range (equivalent to $1.27 \times 10^{-11}$ erg cm$^{-2}$ s$^{-1}$) in 1 Ms exposure (A09). Taking the period March 2005 – March 2008, and evaluating the image resulting from the superposition of all observations in this period, BAT detected 38 blazars (A09), of which 26 are Flat Spectrum Radio Quasars (FSRQs) ad 12 are BL Lac objects, once the Galactic plane ($|b < 15^\circ|$) is excluded from the analysis. A09 reported an average exposure of 4.3 Ms, and considered the [15–55 keV] energy range, to avoid background problems at higher energies. The well defined sky coverage and sources selection criteria makes the list of the found blazars a complete, flux limited, sample, that enabled A09 to calculate the luminosity function (LF) and the possible cosmic evolutions of FSRQs and BL Lacs, together with their contribution to the hard X–ray background. A09 also stressed the fact that the detected BAT blazars at high redshift are among the most

[2]Wang et al. did not attempt any fit to the $M_{\rm BH} - \sigma$ relation. They note that they find significant scatter, extending to over 3 orders of magnitude, and that most of the quasar black hole masses lie above the local relationship. See also Shields et al. (2006) for quasars at $z = 3$.

powerful blazars and could be associated with powerful accreting systems. Within the BAT sample, there are 10 blazars (all FSRQs) at redshift greater than 2, and 5 at redshift between 3 and 4.

(Ghisellini et al. 2010, hereafter G10) find that *all* the 10 BAT blazars at $z > 2$ are powered by a black hole heavier than $10^9\ M_\odot$. Since these objects are at high redshifts, our finding has important implications on the number density of heavy black holes, especially if we consider that for each blazar pointing at us, there must be hundreds of similar sources (having black holes of similar masses) pointing elsewhere.

G10 derive the expected volume density of high redshift blazars hosting a black hole of mass larger than $10^9\ M_\odot$ using the cosmological evolution model of A09 along with its high–$z$ cut–off (i.e. "minimal") version, assuming that all blazars with $L_X > 2 \times 10^{47}$ erg s$^{-1}$ have a $M > 10^9\ M_\odot$ black hole (G10). We cannot exclude that blazars with lower X–ray luminosity also host massive black holes, so the "observational" points, strictly speaking, are lower limits.

Lower limits to the density of high redshift blazars powered by black holes with $M > 10^9\ M_\odot$ are placed by the existence of at least 4 blazars at $4 < z < 5$ for which G10 have estimated a black hole mass larger than $10^9\ M_\odot$. These blazars are RXJ 1028.6–0844 ($z = 4.276$; Yuan et al. 2005); GB 1508+5714 ($z = 4.3$; Hook et al. 1995); PMN J0525–3343 ($z = 4.41$; Worsley et al. 2004a) and GB 1428+4217 ($z = 4.72$; Worsley et al. 2004b). The lower limit in the 5–6 redshift range corresponds to the existence of at least one blazar, Q0906+6930 at $z = 5.47$, with an estimated black hole mass of $2 \times 10^9\ M_\odot$ (Romani 2006).

All these points concerns sources pointing at us. The real number density of heavy black holes, $\Phi(z, M > 10^9 M_\odot)$, must account for the much larger population of misaligned sources. We have then multiplied the mass function of blazars and the other lower limits by $2\Gamma^2 = 450$, i.e. we have assumed an average $\Gamma$–factor of 15, appropriate for the BAT blazars analysed here.

Summarizing, the BAT blazar survey allowed to meaningfully construct the hard X–ray LF of blazars. G10 have also constructed the minimal evolution consistent with the existing data and the (few) existing lower limits. At the high luminosity end the LF can be translated into the mass function of black holes with more than one billion solar masses. In Fig. 5 we show $\Phi(z, M > 10^9 M_\odot)$ as derived from the cosmological evolution model of A09 as red squares, and that derived from the "minimal" LF as green squares.

The four blazars detected in the the $4 < z < 5$ redshift bin did not come from a complete all sky survey, and therefore should be taken as a *lower limit* to the real number of blazars, Correspondingly, we can derive a lower limit on the expected number density of radio–loud sources in the same redshift bin: over the all sky we should have more than $4 \times 2\Gamma^2 \sim 1,800\,(\Gamma/15)^2$ radio–loud sources in the same redshift bin (and at least $\sim$450 at $z > 5$).

In order to check this expectation, we consider the Quasar Catalog of the Sloan Digital Sky Survey (SDSS, Schneider et al. 2010) Data Release Seven (DR7) that includes information on radio detection in and the Faint Images of the Radio Sky at Twenty–cm survey (FIRST, Becker et al. 1995). The region of the sky covered by both surveys is $\sim$8770 square degrees. We adopt the public catalog with quasar prop-

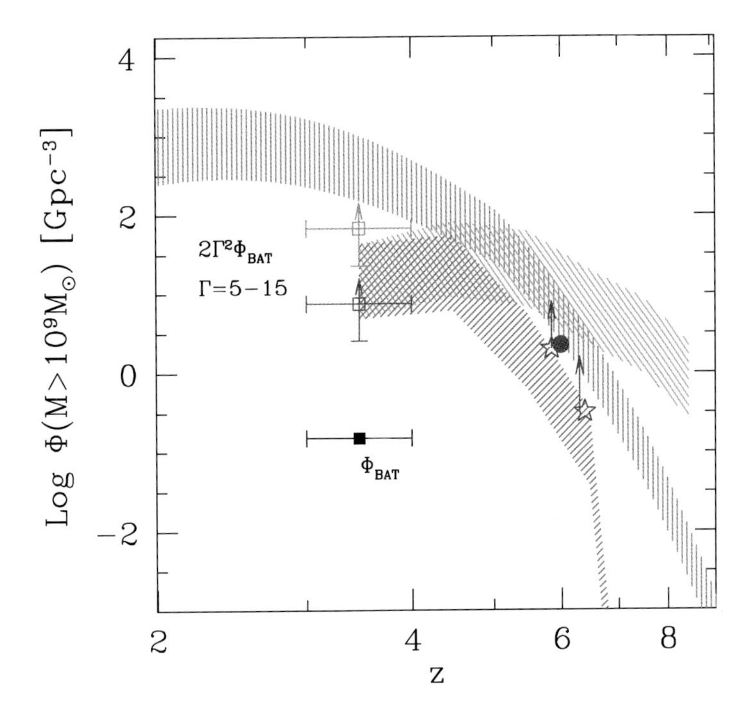


**Figure 5:** Number density of black holes with $M > 10^9 M_\odot$ as a function of redshift. The filled black square in the $3 < z < 4$ bin is taken directly from Fig. 10 of A09. The two empty squares account for the population of misaligned sources, multiplying by $2\Gamma^2$, with $\Gamma = 5$ and $\Gamma = 15$. Red hatched area: A09 cosmic evolution of the LF of blazars, assuming $\Gamma = 5$ (lower curve), or $\Gamma = 15$ (upper curve). Green hatched area: "minimal" cosmic evolution of the LF of blazars, with $\Gamma = 5$ (lower curve), or $\Gamma = 15$ (upper curve). Blue hatched area: radio-quiet quasars (LF and its evolution from Hopkins et al. (2007). Upper boundary: the average Eddington fraction of radio–quiet quasars is $f_{\rm Edd} = 0.3$. Lower boundary: average Eddington fraction of radio-quiet quasars is $f_{\rm Edd} = 1$). Blue stars: number density of $M > 10^9 M_\odot$ derived from the existence of the black holes analyzed in Kurk et al. (2007). Purple dot: number density of $M > 10^9 M_\odot$ derived from the mass function of black holes at $z = 6$ proposed by Willott et al. 2010.

erties described in Shen et al. (2010), which includes quasars bolometric luminosity (using bolometric corrections derived from the composite spectral energy distributions from Richards et al. 2006). The catalog also provides the radio flux density at rest-frame 6cm and the optical flux density at restframe 2500 angstrom that can be used to calculate 'radio-loudness'. Following Jiang et al. (2007) we define a source radio-loud if it has radio to optical flux ratio, $R$, larger than 10. For a handful of sources where optical quantities are not provided, we supplement the 'raw' catalog by calculating the bolometric luminosity from the absolute i-band magnitude, assuming a bolometric correction of 2.5. This bolometric correction is derived by matching the average bolometric luminosity provided by Shen et al. with the bolometric luminosity calculated from the absolute i-band magnitude for sources where the catalog lists both quantities. We then calculate the restframe optical flux from the luminosity, in order to derive an estimate of the radio-loudness. This 'extended' catalog will be our reference. We select all sources that are in the FIRST+SDSS footprint, have an optical bolometric luminosity $> 10^{47}$ erg s$^{-1}$ and radio to optical flux ratio larger than 10. We also require the quasars to be selected uniformly using the final quasar target selection algorithm described in Richards et al. (2002).

We quantify the redshift evolution of the ratio of radio–loud vs radio–quiet sources in Fig. 5 and Fig. 6. *We stress that up to $z = 4$, where we do see blazars, the cosmological evolution model, as derived by A09, is secure.* Beyond $z = 4$ it depends strongly on the assumed evolution. Since, however, the "minimal" evolution provides a lower limit to the number of radio-loud systems, we can be assured that *the radio–loud vs radio–quiet fraction remains at least close to constant, and near unity, up until $z \simeq 6$.* We find that the fraction of jetted sources increases from $z = 3.5$ to $z = 4.5$ by roughly an order of magnitude. Figure 5 also shows that for $M > 10^9\,M_\odot$ and $L > 10^{13} L_\odot$ ($L \gtrsim 10^{47}$ erg s$^{-1}$) the number density of radio–loud quasars approaches and possibly prevails over that of radio quiet–quasars, if we take face value the extrapolation of the cosmic evolution suggested by A09.

We further check our results via a comparison of the radio–loud fraction that we derive from the FIRST+SDSS sample we uniformly selected. This radio–loud fraction is shown in Fig. 6 (blue squares). We compare our estimate with the results for quasars of similar luminosities of Jiang et al. 2007, shown as gray triangles (their Fig. 3, lower left panel). The agreement between our selection in the SDSS+FIRST and Jiang's is excellent where the analyses overlap ($z < 4$). A striking result we find is that, while at $z < 2.5$ the "parent population" of SDSS+FIRST radio loud quasars traces almost perfectly the BAT blazars, assuming $\Gamma = 15$ and $f_{\rm Edd} = 1$, the two selections deviate at higher redshift.

We stress once again that at $z < 2.5$ the blazar population, with $\Gamma = 15$, joined with the radio-quiet population with $f_{\rm Edd} = 1$, is in excellent agreement with SDSS/FIRST data (both our analysis and Jiang et al. 2007 analysis). At $z = 3.5$ the number density of blazars is derived from observed sources (no redshift extrapolation), and our only assumption is the value of $\Gamma$. We are therefore confident that there *must* be either a transition in the astrophysical properties of the population or a selection bias. Below we discuss some possibilities to explain the found discrepancy.

The solution of the discrepancy might be in the SDSS/FIRST selection missing part of the parent population of blazars at $z > 3$. This may be the results of optical absorption, or else of collimation of the optical emission of the disk. In both cases, misaligned objects would be biased against and their apparent luminosity would be below the linit of $10^{47}$ erg s$^{-1}$ we have adopted. In this case we miss both radio-loud and radio–quiet objects. The radio–loud fraction could be right (if the bias apply equally to both kind of sources), but both classes are underrepresented by an order of magnitude, at least at $z > 3$.

Alternatively, the value $\Gamma \sim 15$ for the average bulk Lorentz factor of blazars is too large. A value of $\Gamma \sim 5$ would make the predicted numbers of misaligned radio sources to decrease by an order of magnitude, becoming then consistent with the SDSS+FIRST detected radio–loud sources. We have checked that fitting the SED of

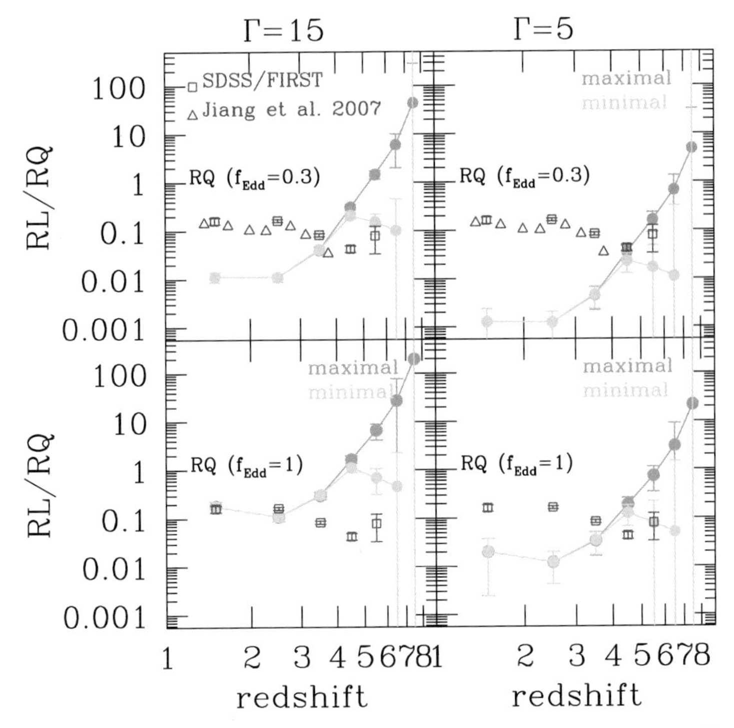


**Figure 6:** Ratio of the number density of $M > 10^9\,M_\odot$ and $L > 10^{13} L_\odot$ ($L \gtrsim 10^{47}$ erg s$^{-1}$) (radio-loud) blazars and (radio-quiet) quasars. Red: A09 cosmic evolution of the LF of blazars. Green: "minimal" cosmic evolution of the LF of blazars. The radio-quiet population is estimated from the LF of radio-quiet quasars, and its redshift evolution, from Hopkins et al. Top panel: average Eddington fraction of radio-quiet quasars is $f_{\rm Edd} = 0.3$. Bottom panel: average Eddington fraction of radio-quiet quasars is $f_{\rm Edd} = 1$. Left panels: $\Gamma = 15$. Right panels: $\Gamma = 5$.

our blazars with $\Gamma = 5$ gives reasonable results, but this value of $\Gamma$ cannot account for all the measured apparent superluminal velocity and, furthermore, it implies that the jet is away from the conditions of minimum jet power requirements (see Ghisellini & Tavecchio 2010). We note, however, that the discrepancy appears as redshift increases, requiring an evolution of $\Gamma$ with cosmic time ($\Gamma$ would increase with decreasing redshift).

The next step is to derive the mass function of black holes powering radio–quiet sources comparable to those powering high-redshift blazars. One can derive an upper limit by turning the bolometric LF of radio-quiet sources (Hopkins et al. 2007) into a mass function. One can simply assume that quasars radiate at an average fraction $f_{\rm Edd}$ of the Eddington limit so that:

$$\frac{M}{10^9\,M_\odot} = 3 \times 10^{-14} \frac{1}{f_{\rm Edd}} \frac{L}{L_\odot}. \qquad (3)$$

To derive $\Phi(z, M > 10^9 M_\odot)$ one then integrates the LF above the luminosity threshold, $L_{\rm min}$, corresponding to $M = 10^9 M_\odot$. We here assume a unity duty cycle, $x_{\rm dc} = 1$, corresponding to the fraction of black holes that are active (related to the ratio of the lifetime of quasars to the Hubble time). If the duty cycle is significantly less than one, then for each quasar there should be $1/x_{\rm dc}$ inactive black holes, thus increasing $\Phi(z, M > 10^9 M_\odot)$, and exacerbating the issues discussed below.

Estimates of $\Phi(z, M > 10^9 M_\odot)$ for radio-quiet quasars are shown in Fig. 5 for $f_{\rm Edd} = 0.3$ ($L_{\rm min} = 10^{13} L_\odot$) and $f_{\rm Edd} = 1$ ($L_{\rm min} = 3.4 \times 10^{13} L_\odot$). Note that the lower the average $f_{\rm Edd}$, the lower the luminosity of a quasar that hosts a $M > 10^9 M_\odot$ black hole is. Therefore, decreasing $f_{\rm Edd}$ allows one to integrate the LF down to lower luminosities, thus increasing the overall $\Phi(z, M > 10^9 M_\odot)$. However, if we were to assume, say, $f_{\rm Edd} = 0.1$, then the range of luminosities where we compare radio-loud and radio quiet quasars would differ, as blazars are selected to have $L_X > 2 \times 10^{47}$ erg s$^{-1}$.

Finally, notice that if the SDSS misses quasars because of obscuration biasing optical selection (see section 2 and Treister et al. 2011), than this mass function is in reality a lower limit, as more active black holes might exist.

We quantify the redshift evolution of the ratio of radio–loud vs radio–quiet sources in Fig. 5 and Fig. 6. *We stress that up to $z = 4$, where we do see blazars, the cosmological evolution model, as derived by A09, is secure.* Beyond $z = 4$ it depends strongly on the assumed evolution. Since, however, the "minimal" evolution provides a lower limit to the number of radio-loud systems, we can be assured that *the radio–loud vs radio–quiet fraction remains at least close to constant, and near unity, up until $z \simeq 6$.* We find that the fraction of jetted sources increases from $z = 3.5$ to $z = 4.5$ by roughly an order of magnitude. Figure 5 also shows that for $M > 10^9\ M_\odot$ and $L > 10^{13} L_\odot$ ($L \gtrsim 10^{47}$ erg s$^{-1}$) the number density of radio-loud quasars approaches and possibly prevails over that of radio quiet-quasars, if we take face value the extrapolation of the cosmic evolution suggested by A09.

We further check our results via a comparison of the radio-loud fraction that we derive from the FIRST+SDSS sample we uniformly selected. This radio–loud fraction is shown in Fig. 6 with blue squares. We compare our estimate with the results for quasars of similar luminosities of Jiang et al. 2007, shown as gray triangles (their Fig. 3, lower left panel). The agreement between our selection in the SDSS+FIRST and Jiang's is excellent where the analyses overlap ($z < 4$). A striking result we find is that, while at $z < 2.5$ the "parent population" of SDSS+FIRST radio loud quasars traces almost perfectly the BAT blazars, assuming $\Gamma = 15$ and $f_{\rm Edd} = 1$, the two selections deviate at higher redshift.

In section 2 we discussed possible incompleteness of the SDSS+FIRST selection, based on absolute numbers. This second analysis, which relies on a uniformly selected sample, does not require completeness as we are now dealing with fractions. We still find the a dearth of radio-loud sources at high redshift. On top of the four possible solutions described in section 2, we notice that the discrepancy can be alleviated if the average Eddington fraction of radio-quiet quasars with $L \gtrsim 10^{47}$ erg s$^{-1}$ decreases from $f_{\rm Edd} = 1$ to $f_{\rm Edd} = 0.3$ around $z = 3$.

We stress once again that at $z < 2.5$ the blazar population, with $\Gamma = 15$, joined with the radio-quiet population with $f_{\rm Edd} = 1$, is in excellent agreement with SDSS/FIRST data (both our analysis and Jiang et al. 2007 analysis). At $z = 3.5$ the number density of blazars is derived from observed sources (no redshift extrapolation), and our only assumption is the value of $\Gamma$. We are therefore confident that there *must* be either a transition in the astrophysical properties of the population or a selection bias.

The existence of these blazars possibly implies that normal 'feedback' might not be at play at the highest redshifts. A possible explanation is that high-accretion rate events, distinctively possible during the violent early cosmic times, trigger the formation of collimated outflows (e.g. blazars) that do not cause feedback directly on the host, which is pierced through. These jets will instead deposit their kinetic energy at large distances, leaving the host unscathed. This is likely if at large accretion rates photon trapping decreases the disk luminosity, while concurrently the presence of a jet helps dissipating angular momentum, thus promoting efficient accretion. This model may explain why high-redshift MBHs can accrete at very high rates without triggering self-regulation mechanisms.

# 6 Conclusions

The fast paced development of observational and theoretical tools is providing us with glimpses in the into the evolution of galaxies and black holes in the first billion years of the Universe.

In the near future the synergy of *JWST* and *ALMA* can zoom in on quasars and their hosts respectively informing us of their relationship and how the $M_{\rm BH} - \sigma$ relation is established, or how the accretion properties depend on the black hole or halo mass. In the near-IR, *JWST* will have the technical capabilities to detect quasars at $z \gtrsim 6$ down to a mass limit as low as $10^5 - 10^6\ M_\odot$, owing to its large field of view and high sensitivity. At the expected sensitivity of *JWST*, $\simeq 1$ nJy, almost $7 \times 10^3$ deg$^{-2}$ sources at $z > 6$ should be detected (Salvaterra et al. 2007). At the same time, the exquisite angular resolution and sensitivity of *ALMA* can be used in order to explore black hole growth up to high redshift even in galaxies with high obscuration and active star formation. To date the best studies of the hosts of $z \simeq 6$ quasar have been performed at cm-wavelength (Walter et al. 2004; Wang et al. 2010). The best studied case is J1148+5251 at z = 6.42. The host has been detected in thermal dust, non-thermal radio continuum, and CO line emission (Bertoldi et al. 2003; Carilli et al. 2004; Walter et al. 2004). *ALMA* will be able to detect the thermal emission from a source like J1148+5251 in a few seconds at sub-kpc resolution (Carilli et al. 2008).

Detection of gravitational waves from seeds merging at the redshift of formation (Sesana et al. 2007) is also probably one of the best ways to discriminate among formation mechanisms. *LISA* in principle is sensible to gravitational waves from binary MBHs with masses in the range $10^3 - 10^6\ M_\odot$ basically at any redshift of interest. A large fraction of coalescences will be directly observable by *LISA*, and on the basis of the detection rate, constraints can be put on the MBH formation process.

In the meantime, we need to develop dedicated cosmological simulations of black hole formation and early growth that can aid the interpretation of these data. The suggestion that the accretion rate of massive black holes depends on their environment, (through the host halo and its cosmic bias) must be tested with cosmological simulations that implement physically-motivated accretion and feedback prescriptions. We also need to derive predictions for the occupation fraction of black holes in galaxies based on black hole formation models (Bellovary et al. in prep.).

# 7 Acknowledgments

MV acknowledges support from SAO Award TM9-0006X and NASA award ATP NNX10AC84G.

# References

Abel T., Bryan G. L., Norman M. L., 2000, ApJ, 540, 39

Becker R. H., White R. L., Helfand D. J., 1995, ApJ , 450, 559

Begelman M. C., 2009, ArXiv e-prints

Begelman M. C., Rees M. J., 1978, MNRAS , 185, 847

Begelman M. C., Rossi E. M., Armitage P. J., 2008, MNRAS , 387, 1649

Begelman M. C., Shlosman I., 2009, ApJ , 702, L5

Begelman M. C., Volonteri M., Rees M. J., 2006, MNRAS, 370, 289

Bertoldi F., Cox P., Neri R., Carilli C. L., Walter F., Omont A., Beelen A., Henkel C., Fan X., Strauss M. A., Menten K. M., 2003, A&A, 409, L47

Bromm V., Coppi P. S., Larson R. B., 1999, ApJL, 527, L5

Bromm V., Coppi P. S., Larson R. B., 2002, ApJ, 564, 23

Bromm V., Loeb A., 2003, ApJ, 596, 34

Brooks A. M., Governato F., Booth C. M., Willman B., Gardner J. P., Wadsley J., Stinson G., Quinn T., 2007, ApJ , 655, L17

Brooks A. M., Governato F., Quinn T., Brook C. B., Wadsley J., 2009, ApJ , 694, 396

Carilli C. L., Walter F., Bertoldi F., Menten K. M., Fan X., Lewis G. F., Strauss M. A., Cox P., Beelen A., Omont A., Mohan N., 2004, AJ , 128, 997

Carilli C. L., Walter F., Wang R., Wootten A., Menten K., Bertoldi F., Schinnerer E., Cox P., Beelen A., Omont A., 2008, Ap&SS , 313, 307

Carr B. J., Bond J. R., Arnett W. D., 1984, ApJ, 277, 445

Decarli R., Falomo R., Treves A., Labita M., Kotilainen J. K., Scarpa R., 2010, MNRAS , 402, 2453

Devecchi B., Volonteri M., 2009, ApJ , 694, 302

Di Matteo T., Colberg J., Springel V., Hernquist L., Sijacki D., 2008, ApJ, 676, 33

Di Matteo T., Springel V., Hernquist L., 2005, Nature, 433, 604

Ebisuzaki T., Makino J., Tsuru T. G., Funato Y., Portegies Zwart S., Hut P., McMillan S., Matsushita S., Matsumoto H., Kawabe R., 2001, ApJL, 562, L19

Eisenstein D. J., Loeb A., 1995, ApJ , 443, 11

Fan X. e. a., 2001, AJ , 121, 54

Ferrarese L., Ford H., 2005, Space Science Reviews, 116, 523

Freitag M., Rasio F. A., Baumgardt H., 2006, MNRAS, 368, 121

Freitag M., Gürkan M. A., Rasio F. A., 2006, MNRAS, 368, 141

Fryer C. L., Woosley S. E., Heger A., 2001, ApJ, 550, 372

Gaburov E., Lombardi J., Portegies Zwart S., 2009, ArXiv e-prints

Gao L., Yoshida N., Abel T., Frenk C. S., Jenkins A., Springel V., 2007, MNRAS , 378, 449

Ghisellini G., Della Ceca R., Volonteri M., Ghirlanda G., Tavecchio F., Foschini L., Tagliaferri G., Haardt F., Pareschi G., Grindlay J., 2010, MNRAS , 405, 387

Ghisellini G., Tavecchio F., 2010, MNRAS , 409, L79

Governato F., Brook C. B., Brooks A. M., Mayer L., Willman B., Jonsson P., Stilp A. M., Pope L., Christensen C., Wadsley J., Quinn T., 2009, MNRAS , 398, 312

Gültekin K., Richstone D. O., Gebhardt K., Lauer T. R., Tremaine S., Aller M. C., Bender R., Dressler A., Faber S. M., Filippenko A. V., Green R., Ho L. C., Kormendy J., Magorrian J., Pinkney J., Siopis C., 2009, ApJ , 698, 198

Gürkan M. A., Fregeau J. M., Rasio F. A., 2006, ApJ , 640, L39

Gürkan M. A., Freitag M., Rasio F. A., 2004, ApJ , 604, 632

Haehnelt M. G., Rees M. J., 1993, MNRAS , 263, 168

Hopkins P. F., Richards G. T., Hernquist L., 2007, ApJ , 654, 731

Jahnke K., et al., 2009, ApJ , 706, L215

Jiang L., Fan X., Ivezić Ž., Richards G. T., Schneider D. P., Strauss M. A., Kelly B. C., 2007, ApJ , 656, 680

Koushiappas S. M., Bullock J. S., Dekel A., 2004, MNRAS , 354, 292

Lauer T. R., Faber S. M., Richstone D., Gebhardt K., Tremaine S., Postman M., Dressler A., Aller M. C., Filippenko A. V., Green R., Ho L. C., Kormendy J., Magorrian J., Pinkney J., 2007, ApJ , 662, 808

Levine R., Gnedin N. Y., Hamilton A. J. S., Kravtsov A. V., 2008, ApJ , 678, 154

Lodato G., Natarajan P., 2006, MNRAS , 371, 1813

Loeb A., Rasio F. A., 1994, ApJ , 432, 52

Madau P., Rees M. J., 2001, ApJ , 551, L27

Mayer L., Kazantzidis S., Escala A., Callegari S., 2010, Nature, 466, 1082

McLure R. J., Dunlop J. S., 2004, MNRAS , 352, 1390

Merloni A. e. a., 2010, ApJ , 708, 137

Miller M. C., Hamilton D. P., 2002, MNRAS , 330, 232

Peng C. Y., Impey C. D., Rix H.-W., Kochanek C. S., Keeton C. R., Falco E. E., Lehár J., McLeod B. A., 2006, ApJ , 649, 616

Portegies Zwart S. F., Baumgardt H., Hut P., Makino J., McMillan S. L. W., 2004, Nature, 428, 724

Portegies Zwart S. F., Makino J., McMillan S. L. W., Hut P., 1999, A&A, 348, 117

Portegies Zwart S. F., McMillan S. L. W., 2002, ApJ, 576, 899

Regan J. A., Haehnelt M. G., 2009, MNRAS , 396, 343

Richards G. T., et al., 2002, AJ , 123, 2945

Richards G. T., Lacy M., Storrie-Lombardi L. J., Hall P. B., Gallagher S. C., Hines D. C., Fan X., Papovich C., Vanden Berk D. E., Trammell G. B., Schneider D. P., Vestergaard M., York D. G., Jester S., Anderson S. F., Budavári T., Szalay A. S., 2006, ApJS , 166, 470

Richstone D., et al., 1998, Nature, 395, A14+

Salvaterra R., Haardt F., Volonteri M., 2007, MNRAS, 374, 761

Santoro F., Shull J. M., 2006, ApJ, 643, 26

Schneider D. P., et al., 2010, AJ , 139, 2360

Sesana A., Volonteri M., Haardt F., 2007, MNRAS , 377, 1711

Shen Y., Greene J. E., Strauss M. A., Richards G. T., Schneider D. P., 2008, ApJ , 680, 169

Shen Y., Hall P. B., Richards G. T., Schneider D. P., Strauss M. A., Snedden S., Bizyaev D., Brewington H., Malanushenko V., Malanushenko E., Oravetz D., Pan K., Simmons A., 2010, ArXiv e-prints

Shields G. A., Menezes K. L., Massart C. A., Vanden Bout P., 2006, ApJ , 641, 683

Shlosman I., Frank J., Begelman M. C., 1989, Nature, 338, 45

Vestergaard M., Fan X., Tremonti C. A., Osmer P. S., Richards G. T., 2008, ApJ , 674, L1

Walter F., Carilli C., Bertoldi F., Menten K., Cox P., Lo K. Y., Fan X., Strauss M. A., 2004, ApJ , 615, L17

Wang R., Carilli C. L., Neri R., Riechers D. A., Wagg J., Walter F., Bertoldi F., Menten K. M., Omont A., Cox P., Fan X., 2010, ApJ , 714, 699

White S. D. M., Rees M. J., 1978, MNRAS , 183, 341

Willott C. J., Albert L., Arzoumanian D., Bergeron J., Crampton D., Delorme P., Hutchings J. B., Omont A., Reylé C., Schade D., 2010, AJ , 140, 546

Wise J. H., Turk M. J., Abel T., 2008, ApJ , 682, 745

Woo J., Treu T., Malkan M. A., Blandford R. D., 2008, ApJ , 681, 925

Yoshida N., Omukai K., Hernquist L., Abel T., 2006, ApJ , 652, 6

# High-energy astrophysics

Martin Pohl

Institut für Physik und Astronomie,
Universität Potsdam,
14476 Potsdam-Golm, Germany

and

DESY, 15738 Zeuthen, Germany

pohlmadq@gmail.com

**Abstract**

*High-energy astrophysics, or astroparticle physics as it is sometimes called, is a vibrant and active research field that has many connections to other branches of physics and astronomy. I review 3 topics of current interest in high-energy astrophysics: the origin of cosmic rays at GeV–TeV energies, the origin of higher-energy particles near* $10^{17}$ *eV, and the hunt of dark-matter signatures in cosmic ray data.*

## 1 Introduction

Nearly 100 years have passed since Viktor Hess conducted the balloon flights that led to the discovery of cosmic rays, an achievement that earned him a Noble Prize some 20 years later. Hess' discovery of energetic particles of extraterrestrial origin also marks the birth of a new field in physics, connecting particle physics with astrophysics, and hence often called astroparticle physics. An equally apt name is high-energy astrophysics, for it is the high-energy world that is under study.

Measurement techniques have evolved considerably in the past century, and so we now have at our disposal both direct and indirect information on energetic charged particles, or cosmic rays for short, the latter arising from observing their decay or interaction products such as gamma rays. We have realized that cosmic rays can be found almost anywhere in the Universe, but they appear to be most abundant in system harboring substantial bulk flows, such as Supernova remnants (SNR), pulsar-wind nebulae (PWN), or active galactic nuclei (AGN).

The basic questions that one asks in high-energy astrophysics are

- Where are cosmic rays produced? This question has two aspects: first, what systems are predominantly responsible for providing the particles that we see locally, i.e. with ground-based of space-based particle detectors. Second, where exactly in SNR or AGN are the particles produced, mainly at shocks or in a more distributed fashion?

*Reviews in Modern Astronomy 23: Zoomimg in: The Cosmos at High Resolution.* First Edition.
Edited by Regina von Berlepsch.

- How are the particles produced? This question is not completely independent of the first, but does ask for the detailed micro physical processes that endow a few particles with a tremendous amount of energy.

- How do they influence their environment? Cosmic rays carry about as much energy density as radiation, the magnetic field, and the thermal gas. They emit and absorb photons, they can damp and excite electromagnetic turbulence, thus shaping the magnetic environment, and they can heat, ionize, and exert pressure on gas. What role that plays in the evolution of, e.g., galaxies is the subject of active research.

- Do they carry signatures of new physics? This area has seen lots of attention recently, almost matching the early days of particle physics, when, prior to the establishment of powerful accelerators, fundamental particles such as the neutron, the positron, and the pion were detected in cosmic-ray showers. Today, the intention is to find a new type of weakly interacting particle that could constitute the elusive dark matter.

In this paper I shall review 3 topics of current interest in high-energy astrophysics: the origin of cosmic rays at GeV–TeV energies, the origin of higher-energy particles near $10^{1}7$ eV, and the hunt of dark-matter signatures in cosmic ray data.

## 2 The origin of cosmic rays in the GeV–TeV energy band

Shell-type Supernova remnants (SNR) have long been thought to be the Galactic accelerators of CRs (Ginzburg & Syrovatsky 1961), but we are still missing direct observations that would prove the role of SNRs in producing the bulk of galactic cosmic rays.

High-energy gamma rays are a unique probe of cosmic rays. Observations in the gamma-ray band allow us to measure the properties of energetic particles anywhere in the Universe, such as their number, composition, and spectrum. Among the many types of Galactic gamma-ray sources, observations of high-energy emission from SNRs are particularly beneficial because only with them do we have an opportunity to perform spatially resolved studies in systems with known geometry, and the plasma physics deduced from these observations will help us to understand other systems where rapid particle acceleration is believed to occur and where observations as detailed as those of SNRs are not possible.

The maturity of high-energy gamma-ray astrophysics is best illustrated by the ability of current atmospheric Cherenkov detectors such as HESS, MAGIC, and VERITAS to resolve sources and to map the brightness distribution in TeV-band gamma rays. The interpretation of these TeV observations is complicated because two competing radiation processes, pion-decay photons from ion-ion interactions and Inverse-Compton (IC) emission from TeV electrons scattering off the cosmic microwave background and the ambient galactic radiation, can produce similar fluxes in the GeV-TeV energy range.

SNR modeling may be carried out in several ways. The most direct approach is to perform a complete simulation of the plasma by following individual particles in magnetic fields and a gas-flow pattern derived from hydrodynamic simulation. Particle acceleration may be added by coupling a semi-analytical kinetic model of acceleration with plasma simulations. This approach, however, requires a huge computational effort, in fact hundreds of processors are needed to run for many hours to perform a single 2D/3D calculation. Another approach for SNR modeling including particle acceleration is to solve the convection-diffusion equation for the CR momentum distribution together with the hydrodynamic equations. Although it is a more efficient, it is still computationally difficult. Until now only plane-parallel and 1-D spherically symmetric calculations have been published. Finally, two other approaches have found limited application. Monte Carlo simulations based on scattering models are applicable to steady-state shocks, and two-fluid descriptions have a strong dependence on the closure parameters, injection rate, an so forth.

The cosmic-ray transport equation is a diffusion-convection equation in both space and momentum:

$$\frac{\partial N}{\partial t} = \nabla(D_r \nabla N - \vec{v} N) - \frac{\partial}{\partial p}\left((N \dot{p}) - \frac{\nabla \vec{v}}{3} N p\right) + Q \qquad (1)$$

where $N$ is the differential number density of cosmic rays, $D_r$ is the spatial diffusion coefficient, $\vec{v}$ is the gas velocity, $\dot{p}$ are the energy losses, $D_p$ is the momentum-space diffusion coefficient, and $Q$ is the source term representing the injection of the thermal particles into the acceleration process.

The main difficulty for the numerical solution of this equation is that the diffusion coefficient $D_r$ may be strongly dependent on the particle momentum ($D_r = D_r(p) \sim p$) and hence spans over a wide range of values ($\sim$ 10 orders of magnitude). Formally, the largest space-step must be less then the smallest diffusion length defined as $D_r(p_{min})/V_s$, where $V_s$ is the shock speed. Therefore, to "resolve the acceleration" of the lowest energy particles which have a small diffusion length and which are actually injected at the shock, the grid must be very fine, and it is advantageous to transform to a new spatial variable that provide very good resolution at the shock. In the transport equation (1), the actual shock acceleration is treated as adiabatic compression, but for a realistic flow profile $\div\vec{V} \neq 0$ for large part of the SNR volume. We therefore combine the cosmic-ray continuity equation with self-similar solutions of SNR evolution or 1-D hydrodynamical simulations thereof, where needed.

Particle acceleration can arise both at the forward and at the reverse shock. A number of effects can modify the spectrum of accelerated particles compared with the simple textbook case (for a review see Reynolds 2008). Bulk-momentum transfer to the upstream plasma, magnetic-field amplification, and the escape of particles at the highest energies can make the acceleration more efficient and lead to relatively hard spectra at high energy. Conversely, neutral particles, heating, and turbulent motion in the upstream plasma, as well as magnetic-field amplification again, can lead to softer spectra. The competition of these and other processes may provide spectral structure that makes difficult even a supposedly simple task such as extracting the maximum energy of accelerated particles.

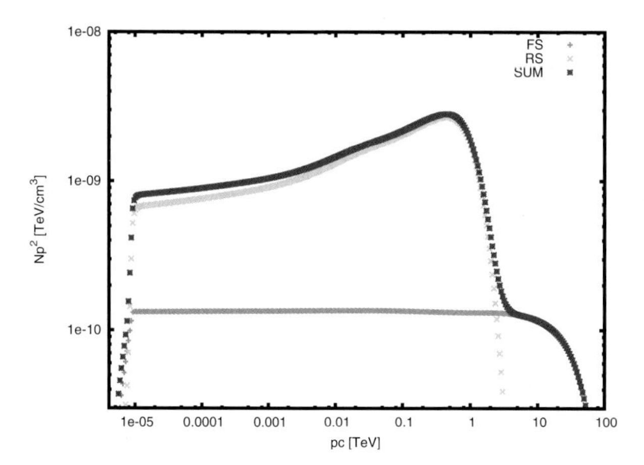


**Figure 1:** Proton spectra in the remnant of a generic type-II Supernova. The red line refers to particles accelerated at the forward shock, whereas the green line is for particles from the reverse shock; the blue line is the sum.

Figure 1 shows an example, the parameters being those of a remnant of a generic type-II Supernova a few hundred years after the explosion, a situation not unlike that of Cas A. Particle acceleration at the reverse shock involves many particles, but is slower than that at the forward shock, and therefore dominates in the GeV band. The propagation of the reverse shock backward from the contact discontinuity is responsible for a continuous hardening of the particle spectrum that could be mistaken as evidence of a cosmic-ray modified shock. In fact the spectra result from shoch acceleration in the test-particle limit.

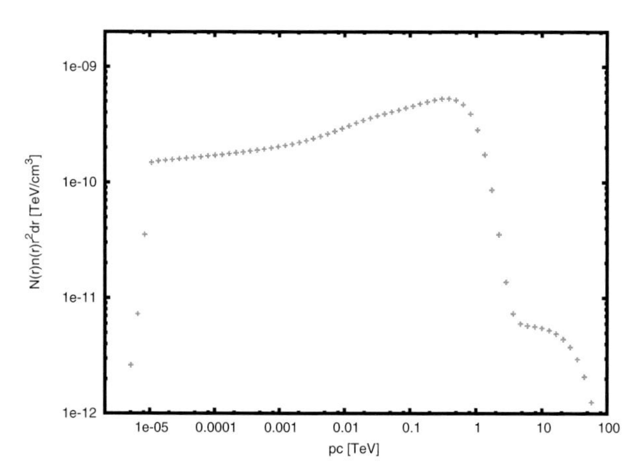


**Figure 2:** Mock gamma-ray spectrum from a generic type-II SNR calculated as a density-weighted volume integral of the particle spectra shown in Figure 1, thus mimicking hadronic gamma-ray emission.

At high energies we only see particles from the forward shock, albeit at a lower flux. The mock gamma-ray spectrum shown in Figure 2 therefore shows a complicated structure, including a step-like feature at high energies. It is particularly interesting to see that spectral curvature, often taken as evidence of acceleration at cosmic-ray-modified shocks, can arise naturally. The interpretation of observed gamma-ray spectra is thus more complicated than previously thought.

# 3 The origin of cosmic rays above $10^{17}$ eV

An open problem in cosmic-ray astrophysics is at what energy we observe the transition from a galactic to an extragalactic origin of particles. Related is the question which sources in the Galaxy contribute Ultra-high-energy cosmic rays (UHECR) at the highest energies. It is already known that long GRB likely have insufficient power to account for UHECR at GZK energies (Eichler et al. 2010). One can also show that at energies below the ankle the contribution from galactic GRB should on average be stronger by about 3 orders of magnitude than that of extragalactic GRB, thus further constraining models involving a GRB origin of GZK-band particles (Eichler & Pohl 2011).

Here we study the time-dependent diffusive transport of UHECR in the Galaxy using the method of Monte-Carlo to account for the unknown location and explosion time of GRB or other sources with similar population statistic. This approach permits us to accurately account for intermittency effects in the local UHECR spectrum and thus goes beyond the scope of earlier publications (Calvez et al. 2010; Levinson & Eichler 1993; Wick et al. 2004).

We assume the propagation in the Galaxy of cosmic rays at energies $10^{15}$ eV to $10^{18}$ eV can be accurately described as isotropic diffusion. This requires, a) that the particle Larmor radii are considerably smaller than the largest scale on which the galactic magnetic field is turbulent, and b) that the particle mean free path, $\lambda_{\rm mfp}$, is much smaller than a few kpc, the typical distance between the solar system and a GRB in the Galaxy. The Larmor radius of a $Z = 1$ particle in a 10 $\mu$G field reaches $\sim 100$ pc at $10^{18}$ eV, and therefore the first approximation should hold for UHECRs of any composition below $\approx 10^{18}$ eV. The second approximation requires that $\lambda_{\rm mfp}$ be within a factor of $\sim 10$ of the Larmor radius.

To evaluate the level of systematic uncertainties in our model description, we explore various geometric forms of the propagation volume of UHECR in the Galaxy. We find that a disk-like geometry, which appears more likely to be accurate than the assumption of spherical symmetry, renders the observational constraints on anisotropy and composition more difficult to meet.

Instead of a computationally expensive full solution of the diffusion problem in disk geometry (Büsching et al. 2005), we use a steady-state solution to rewrite the propagation equation in terms of the mid-plane cosmic-ray density, $N_0$, as well as turn the diffusive flux at the halo boundaries ($z = \pm H$) into a simple catastrophic loss term,

$$\tau_{\rm esc} = \frac{H^2}{2\,D} \simeq (1.2 \cdot 10^6\ {\rm yr}) \left(\frac{H}{5\ {\rm kpc}}\right)^2 \left(\frac{\lambda_{\rm mfp}}{0.1\ {\rm kpc}}\right)^{-1} . \qquad (2)$$

In the energy band of interest, escape is the dominant loss process of cosmic rays in the Galaxy. Ignoring variations in the diffusion coefficient within the Galactic plane, the problem only depends on the in-plane distance between source (GRB) and observer, $\rho$, and can be recast as 2-D diffusion equation for the mid-plane cosmic-ray density around a point source,

$$\frac{\partial N_0}{\partial t} + \frac{N_0}{\tau_{\rm esc}} - \frac{1}{\rho}\frac{\partial}{\partial \rho}\left(\rho\, D\,\frac{\partial N_0}{\partial \rho}\right) = Q(E)\,\delta(t)\,\frac{\delta(\rho)}{2\pi\,\rho\,H}\,, \tag{3}$$

whose solution is

$$N_0(\rho,t,E) = \exp\left(-\frac{t}{\tau_{\rm esc}}\right)\frac{\Theta(t)}{4\pi\,D\,t}\,\frac{Q(E)}{H}\,\exp\left(-\frac{\rho^2}{4\,D\,t}\right)\,. \tag{4}$$

The anisotropy in the case of a single GRB is

$$\delta \simeq \lambda_{\rm mfp}\,\frac{1}{N_0}\left|\vec{\nabla} N_0\right| \simeq \frac{3\,\rho}{2\,c\,t} \tag{5}$$

The halo size, $H$, is not well known. We use $H = 5$ kpc, which is at the high end of the range of likely values. We thus probably underestimate the flux suppression arising from a finite halo size; consequently our results on spectral structure and anisotropy are conservative.

Heavier nuclei have a smaller rigidity at the same total energy, $R \propto E/Z$. The mean free path of an ultra-high-energy particle should only depend on the rigidity, and in the absence of energy losses a nucleus of charge $Z$ and energy $E_Z$ should behave like a proton of energy $E = E_Z/Z$. Thus equation 4 also describes the distribution of heavy nuclei in the Galaxy, provided the appropriate scaling is applied to the energy and the source rate.

Generally, GRBs in the Galaxy are expected every million years or so, the exact rate depending on the beaming fraction and the detailed scaling of long GRB with star formation and metallicity (For a detailed review see Gehrels et al. 2009). Therefore, only a small number of GRB can contribute to the particle flux at the solar circle, and their relative contribution depends on the location and explosion time of the GRB. Variations in the local particle flux must be expected, and neither the particle spectrum from an individual GRB nor the spectrum calculated for a homogeneous source distribution are good proxies. To fully account for discreteness of GRBs in space and time, we can use the method of Monte-Carlo to randomly place GRBs in the Galaxy with given spatial probability distribution in galactocentric radius

$$P(r_{GC}) = \frac{2\,r_{GC}}{r_0^2}\,\exp\left(-\frac{r_{GC}^2}{r_0^2}\right) \tag{6}$$

with scale $r_0 = 5$ kpc, and with given GRB rate. More details are found in an upcoming publication (Pohl & Eichler, in prep.).

We have calculated spectra for $10^4$ random sets of GRBs, going back 6 Gyr in time, but ignoring energy losses through inelastic collisions. Overall, the method is the same as that used to model the transport of cosmic-ray electrons in the Galaxy

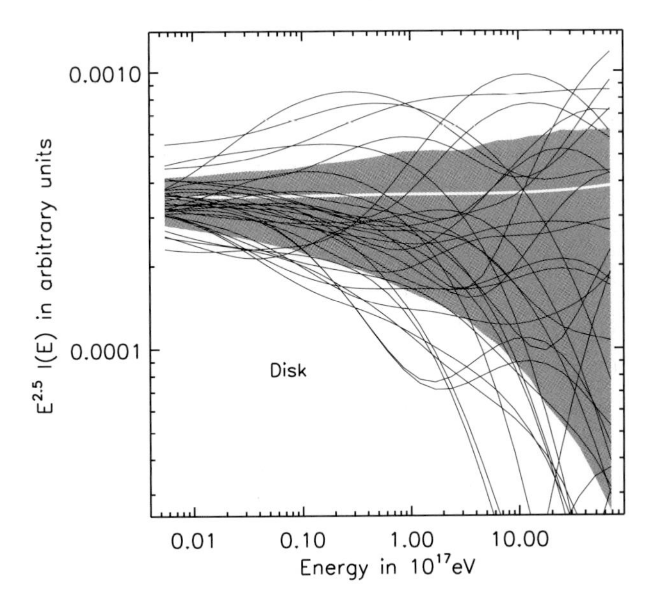


**Figure 3:** Proton spectra at the solar circle expected for a diffusion coefficient scaling with $\sqrt{E}$ in disk geometry. The red band indicates the central 68% containment region for the particle flux at the given energy.The GRB rate is set to 5 $\mathrm{Myr}^{-1}$. We have plotted 31 individual, randomly selected spectra. The average spectrum is given by the thick white line.

(Pohl et al. 2003; Pohl & Esposito 1998). Results are shown in Figure 3, where we have assumed an injection index $s = 2$.

It is the energy dependence of the diffusion coefficient that determines the particle spectrum. Structure in the observed spectrum could thus arise from changes in the energy dependence, e.g. from shallow at lower energies to Bohmian at higher energies, without requiring any structure in the source spectrum (see also Calvez et al. 2010). Intermittency is strong for a GRB rate below 1 per Myr, in particular for the more realistic disk geometry. In essence, the local UHECR spectrum from galactic GRBs is unpredictable if the scattering mean free path exceeds about 100 pc, which for the parameters used here is the case above $10^{17}$ eV for protons, and above $3 \cdot 10^{18}$ eV for iron. Model fits of single-source spectra can thus be very misleading (cf. Wick et al. 2004). The actually expected spectra display bumps unrelated to both source and propagation physics, some of which may indeed be observed (Arteaga-Velázquez et al. 2010). The absence of very large bumps in the observed UHECR spectra suggests that either the mean free path for scattering is smaller than assumed here, or the rate of cosmic-ray producing GRBs in the Galaxy exceeds 1 per Myr, at which the amplitude of such bumps becomes smaller. Generally, careful accounting of the statistical fluctuations is mandatory for proper estimating the local UHECR spectrum from GRBs (cf. Calvez et al. 2010).

We now try to construct a model that reproduces the spectrum of cosmic rays between $10^{15}$ eV and $10^{19}$ eV together with the anisotropy limits and the composition.

We use data of the Kascade-Grande collaboration (Apel et al. 2009), HiRes (Abbasi et al. 2005; High Resolution Fly's Eye Collaboration et al. 2009), and the Auger collaboration (Abraham et al. 2008; Pierre AUGER Collaboration et al. 2010). At $10^{18}$ eV the anisotropy is low, $\delta \leq 0.01$ (the 99% upper limit is 0.02), and the composition is light, but not necessarily dominated by protons (Abraham et al. 2010).

Figure 4 shows the spectra for a possible model configuration, where for simplicity we display only spectra for protons, helium, and carbon as proxies for light and heavy nuclei, respectively. The GRB rate is set to $P(t) = 1\ \mathrm{Myr}^{-1}$ and the source spectral index is $s = 2.1$. The mean free path transitions from a shallow energy dependence to Bohmian scaling ($\propto E$) as the particle energy increases,

$$\lambda_{\rm mfp} = \lambda_0\, E^{0.3} \left[1 + \frac{E}{60\,\mathrm{PeV}}\right]^{0.7} \tag{7}$$

where $\lambda_0 = \eta\, r_L$ is chosen so a proton has a mean free path that is a certain multiple $\eta$ (unity in Figure 4) of 11 pc at $10^{17}$ eV, its Larmor radius in a 10-$\mu$G magnetic field. The spectrum below $10^{17}$ eV is far below the data to accommodate other galactic

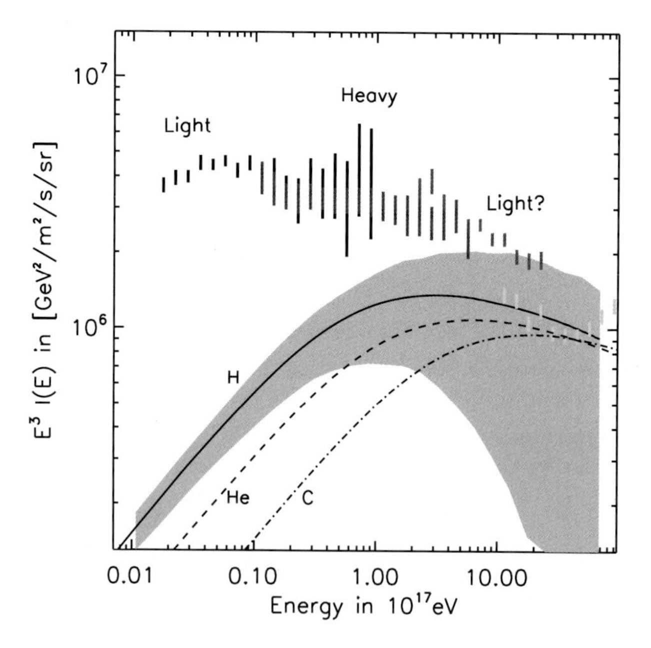


**Figure 4:** Example of model spectra for cosmic-ray protons, helium, and carbon nuclei, including the 68% variation range in the case of protons. The solid line denotes the average spectrum of hydrogen, the dashed line displays the same for helium, and the dash-dotted line is for carbon. The mean free path follows equation 7 and is Bohmian above $10^{17}$ eV. Also shown are the spectra measured with KASCADE-Grande, HiRes, and Auger, together with labels indicating the composition. The offset between spectra from different experiments is likely due to errors in the absolute energy scale.

sources of cosmic rays, such as SNR or PWN. The fluctuation amplitude at energies above $10^{17}$ eV is large for the GRB rate used here, one per Myr.

The average spectrum depends weakly on the absolute value of the mean free path at high energies. We can calculate the cosmic-ray source power required to sustain the observed flux of UHRCRs at $10^{18}$ eV, which does depend on the mean free path. For an injection spectrum $\propto E^{-2.1}$ extending from the GeV band to the highest energies, fitting the observed flux of UHRCRs at $10^{18}$ eV requires the source power

$$P_{\mathrm{CR}} = P(t) \int_{1\ \mathrm{GeV}} dE\, E\, Q_0\, E^{-s} \simeq \frac{\lambda_{\mathrm{mfp}}}{r_L}\ (10^{37}\ \mathrm{erg/s}) \qquad (8)$$

The source power in the energy interval $[10^{17}, 10^{18}]$ eV alone is

$$P_{\mathrm{EeV}} \simeq \frac{\lambda_{\mathrm{mfp}}}{r_L}\ (3.5 \cdot 10^{35}\ \mathrm{erg/s}) \qquad (9)$$

The anisotropy at $10^{18}$ eV is not easy to keep below the upper limit established with Auger data. Figure 5 shows the anisotropy for protons, helium nuclei, and carbon for a mean free path follows equation 7 and $\lambda_0 = \eta\, r_L$ chosen so that at high

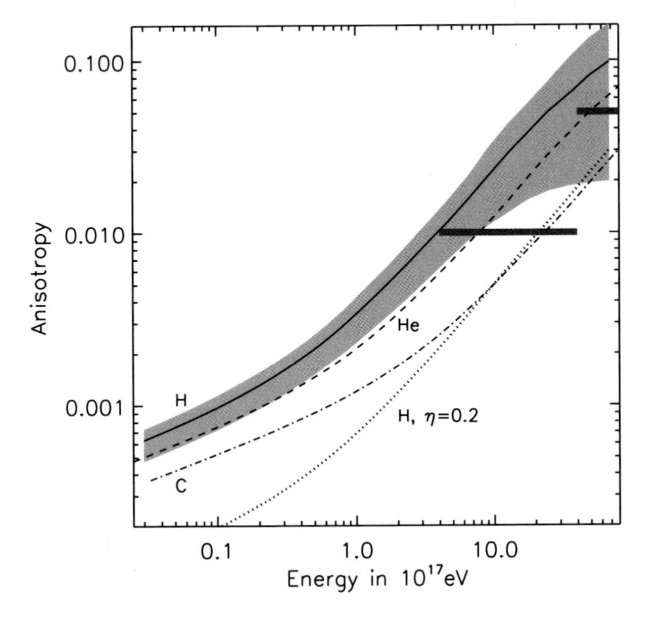


**Figure 5:** The expected anisotropy for cosmic-ray protons, helium, and carbon nuclei, including the 68% variation range in the case of protons. The solid line denotes the average spectrum of hydrogen, the dashed line displays the same for helium, and the dash-dotted line is for carbon. The mean free path follows equation 7 and is equal to the particle Larmor radius above $10^{17}$ eV (for protons). For comparison, we also show the hydrogen anisotropy for a reduced mean free path, $\lambda_{\mathrm{mfp}} = 0.2\, r_L$ at high energies. The horizontal bars indicate observational upper limits.

energy $\eta = 1$ and, for hydrogen only, $\eta = 0.2$, respectively. The observational upper limits are indicated by horizontal bars and reflect averages over a finite energy range that is weighted with the UHECR flux and experiment-specific efficiency function. To be noted from the figures is that the Auger limits, $\delta \leq 0.01$ at $10^{18}$ eV, is marginally violated by helium nuclei in the case of Bohm diffusion, and the mean free path would have to be at least a factor 5 smaller than the particle Larmor radius, if the dominant particle species were protons.

On average, galactic long GRB only need to contribute about $10^{37}$ erg/s in accelerated particles to fully account for the observed particle flux at $10^{18}$ eV, assuming a Bohmian mean free path. UHECR from galactic long GRB can meet the observational limits on anisotropy, if the mean free path for scattering is sufficiently small. Contributing the observed sub-ankle particles (at $10^{18}$ eV) requires Bohmian diffusion if the UHECR are as heavy as carbon. A light composition such as protons or helium requires sub-Bohmian diffusion, which is a highly unlikely situation for isotropic diffusion. We have not investigated the effects of a galactic guiding field that may modify the probability of escape from the Galactic disk.

Auger data suggest that at $10^{18}$ eV the composition is indeed light, thus posing a problem for the notion that galactic GRB (or any other source class with similar population statistic) produce the observed UHECR up to the ankle. This measurement if not undisputed, though, for the KASCADE-Grande collaboration has just published their analysis results which seem ot favor a relatively heavy composition nearly up to $10^{18}$ eV (Arteaga-Velázquez et al. 2010). The UHECR composition is a very critical constraint, but its measurement is subject to considerable systematic uncertainties arising its dependence on models for the development of air showers. It is imperative that measures be taken to better understand the air-shower physics near $10^{18}$ eV. Much of the UHECR anisotropy arises from the expected location of long GRB in the inner Galaxy. As there is no power problem with galactic GRB, it may be worthwhile to consider short GRB. They provide supposedly less power as a population, but they may have a very extended spatial distribution in the Galaxy, thus strongly reducing the anisotropy (Berger 2010).

# 4 The hunt for dark-matter signals

The annihilation or decay of dark matter provides a means for the indirect detection of these elusive particles that, in contrast to direct detection in elastic-scattering experiments or at the LHC, can also permit the measurement of the sky distribution of dark matter. Of particular interest are reaction products that are not abundantly produced by other, more conventional astrophysical processes. Antiparticles are thus promising messengers of potential dark-matter signals, but care must be exercised to properly account for the propagation from their sources to our detectors (e.g. Pohl 2009).

The positron excess in Galactic cosmic ray positrons between 1 and 100 GeV, despite a history of conflicting results, appears to be confirmed by PAMELA (Adriani et al. 2009a). There are, of course, several possible astrophysical explanations. Nearby sources (e.g. supernova or pulsars) of positrons, for example, which suffer fewer losses than typical Galactic positrons because they are younger (e.g. Eichler & Maor 2005; Profumo 2008), contribute harder spectra. Nevertheless, dark matter annihilation (e.g. Hooper et al. 2009; Tylka and Eichler 1987; Tylka 1989) or decay (Arvanitaki et al. 2009; Eichler 1989) have long been suggested as a possible source.

Annihilation, however, encounters a number of difficulties, as it requires a substantial boosting either from clumping (e.g. Kuhlen & Malyshev 2009) or from a Sommerfeld enhancement (Arkani-Hamed et al. 2009), and would lead to intense photon emission from the Galactic-Center region (e.g. Eichler & Maor 2005; Zhang et al. 2009) or extragalactic background radiation from the superposition of haloes (Profumo & Jeltema 2009), that appears to exceed current observational limits.

Here we investigate dark-matter decay as the source of the excess positrons observed with PAMELA (Pohl & Eichler 2010). Our purpose is not a comprehensive theory of the dark-matter decay, in particular not the viability of leptophilic decay, which is one of the challenges (Arkani-Hamed et al. 2009), given that standard hadronization scenarios are already excluded by the lack of an excess in the cosmic-ray antiproton data (Adriani et al. 2009b). The decay source function scales with the dark-matter density, not the density squared as does annihilation, and therefore the limits on decaying dark matter from observations of high-energy $\gamma$ rays from, e.g., dwarf galaxies are weak (Essig et al. 2009), but diffuse emission may provide more stringent constraints (Chen et al. 2009; Ishiwata et al. 2009). We calculate the diffuse galactic $\gamma$-ray emission of the excess leptons, presuming they arise from dark-matter decay. The $\gamma$-ray intensity thus derived is more model-independent and also a lower limit to the true emission level, because we do not count $\gamma$ rays that are directly produced in the dark-matter decay, possibly via other unstable particles.

The propagation length of positrons and electrons at energies $\geq 50$ GeV is $L_p \leq 600$ pc (Grasso et al. 2009; Pohl et al. 2003), considerably less than the scale-length of the dark-matter halo in the Galaxy, implying that the spatial redistribution of those electrons by diffusive and convective transport can be neglected. The differential number density of positrons therefore has, in good approximation, the same spatial profile as the dark-matter density. The PAMELA collaboration has measured a positron fraction that is rapidly increasing above 10 GeV. The highest-energy data point is about 10 times that expected from secondary production in cosmic-ray interactions. The total electron spectrum has not been determined with PAMELA yet, but we may use the excellent Fermi-LAT data that between 20 GeV and 1 TeV are well represented by a single power law (Abdo et al. 2009).

For simplicity we assume the injection of monoenergetic electrons,

$$Q_{e^+/e^-} = Q_0\,\delta\,(E - E_{\max}) \tag{10}$$

where $E_{\max} \leq 500$ GeV to avoid a spectral feature in the total electron spectrum, which would be in conflict with the power-law fit to the Fermi data. We shall see that the $\gamma$-ray limits calculated below provide much tighter constraints on $E_{\max}$. Assuming an equal number of excess electrons and positrons, we therefore estimate the total differential density of electrons/positrons that may come from dark-matter decay as

$$N(E, \mathbf{x}) \propto \rho_{\rm DM}(\mathbf{x})\,E^{-2}\,\Theta\,(E_{\max} - E) \tag{11}$$

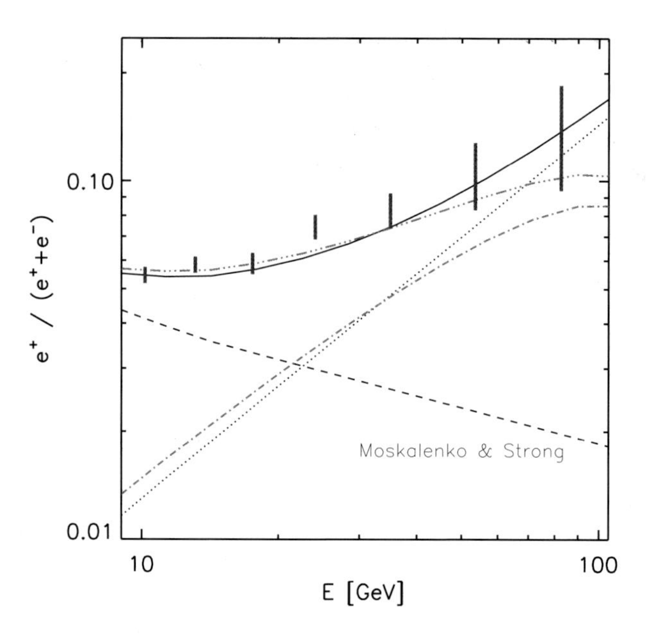


**Figure 6:** The measured positron fraction between 10 GeV and 82 GeV shown in comparison with a calculation for pure secondary production of positrons (Moskalenko & Strong 1998), the dark-matter contribution according to Eq. 11 (dotted line), and the total calculated positron fraction (solid line). For comparison, we also show the positron fraction for a flat injection spectrum up to 200 GeV (dot-dashed line) and the total positron fraction for that case (triple-dot-dashed line).

In Figure 6 we demonstrate that the sum of this modeled dark-matter decay component and the expected contribution of secondary positron production provides a good fit to the PAMELA data. For comparison, the figure also displays the positron fraction for the case of a flat injection spectrum, which may result from the decay of an intermediate particle with high kinetic energy.

$$N_{\mathrm{alt.}}(E,\mathbf{x}) \propto \rho_{\mathrm{DM}}(\mathbf{x}) \left(1 - \frac{E}{E_{\max}}\right) E^{-2}\,\Theta\left(E_{\max} - E\right) \tag{12}$$

Here $Q_0$ is chosen 20% larger than in the case of monoenergetic injection to improve the fit of the positron fraction. To be noted from the figure is that $E_{\max} \geq 200$ GeV is needed for flat injection to well reproduce the positron fraction measured with PAMELA.

The electrons and positrons in the halo will scatter soft photons into the $\gamma$-ray band where they can be observed with, e.g., the Fermi-LAT detector. The target photon field includes the microwave background, galactic infrared emission, and galactic optical emission, only the first of which is isotropic. The latter two will be backscattered toward the Galaxy, thus somewhat increasing the scattering rate compared with the isotropic case. The radiation field has been recently modeled out to $z = 5$ kpc

(Porter & Strong 2005). Surprisingly, the energy density of optical light is higher in the halo than in the midplane on account of the thin-disk distribution of absorbers.

Whereas no significant intensity above 1 GeV is expected to come from beyond a distance of 5 kpc, the limit to which we have integrated the galactic emission, this is not true for the upscattering of microwave-background photons into the 100-MeV band. In fact, in the outer halo inverse-Compton scattering of the microwave background accounts for a larger share of the electron energy losses than near the Galactic plane, because the infrared and optical photon fields quickly lose intensity beyond 5 kpc above the plane of the Galaxy. Also, the magnetic-field strength is expected to fall off, although we do not know at what point it drops to 3 $\mu$G, below which the synchrotron energy losses are subdominant. We will estimate the intensity in the 100-MeV band from galactic dark-matter decay in the outer halo in the next section, together with the extragalactic component.

The majority of dark matter is in fact located sufficiently far away from galaxies ($\geq$ 20 kpc) that positrons and electrons from its decay would primarily interact with the microwave background. Since all electrons suffer the same fate and the electron source rate scales linearly with the dark-matter density, we can ignore any density structure and use spatially averaged quantities, i.e. a conservative fraction 90% of $\Omega_m$ times the critical density. The resulting expected $\gamma$-ray intensity is plotted in Figure 7, together with that produced in the outer parts of the Milky-Way halo at height $z \geq 10$ kpc where IC scattering off the CMB is also dominant.

At 200 MeV $\gamma$-ray energy, the predicted intensity is close to that observed with Fermi (Abdo et al. 2010). The uncertainty in the measured intensity is typically 15% and predominantly systematic in origin. The predicted $\gamma$-ray intensity would exceed the observational limits if the total magnetic-field in the solar vicinity were stronger than 10 $\mu$G, the value we have assumed. It would also exceed the observational limits if the characteristic energy of the injected pairs were higher than 250 GeV. For comparison, Figure 7 also shows the expected intensity for $E_{\rm max} = 300$ GeV to be twice that observed at 200 MeV $\gamma$-ray energy.

While the predicted GeV-band intensity from the inverse-Compton scattering of infrared and optical photons is below a preliminary estimate of the extragalactic $\gamma$-ray background based on Fermi data (Abdo et al. 2010), the *extragalactic* background from the decay of intergalactic dark matter would produce a bump at 100–300 MeV that is close to the observed extragalactic background at these energies, even one estimated from the older EGRET data. Dark-matter decay therefore does not seem to be a viable explanation of the positron excess, unless the characteristic energy of the pairs produced in dark-matter decay is only in a narrow window between the lower limit $\sim$ 100 GeV imposed by high-energy limit of the Pamela measurement and the high energy limit of $\sim$ 200 GeV imposed by the extragalactic 100 to 200 MeV $\gamma$-ray background.

We stress that the predicted 100-200 MeV $\gamma$-ray intensity from dark-matter decay is nearly model-independent, because it depends only on the total dark-matter density in regions outside those of strong galactic magnetic fields and starlight, i.e. in regions where inverse-Compton scattering of microwave background photons is a calorimeter for intergalactic electrons and positrons on account of its dominance among the energy-loss processes. The only assumptions we have made are time-

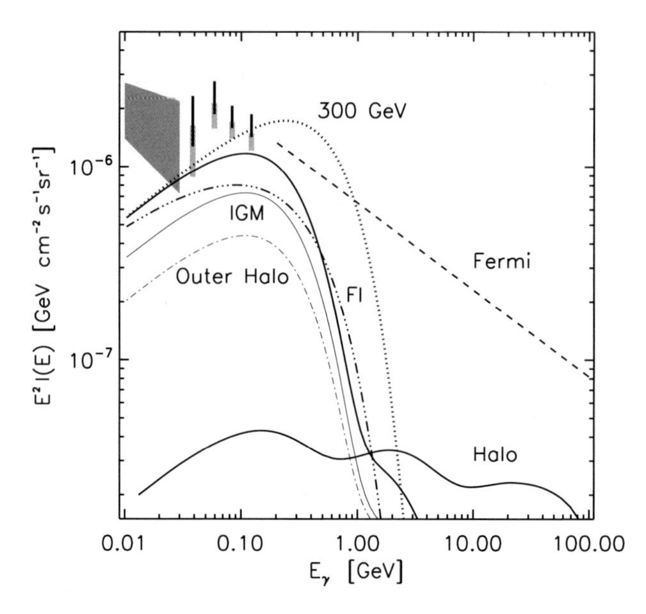


**Figure 7:** Comparison of the predicted $\gamma$-ray intensity observed from the galactic poles with the preliminary estimate of the extragalactic isotropic background observed with Fermi, indicated by the dashed line (Abdo et al. 2010), and older data from COMPTEL EGRET. The thick line labeled Halo shows galactic emission and the other lines are for $\gamma$-ray emission from intergalactic dark matter, all for $E_{\rm max} = 200$ GeV. In detail, the thin solid line is extragalactic emission, the dot-dashed line is emission from the outer Milky-Way halo beyond $z = 10$ kpc, the thick solid line the sum of the two, and the dotted line is the same for $E_{\rm max} = 300$ GeV. The triple-dot-dashed line labeled FI is for flat injection up to 300 GeV.

independence of the dark-matter lifetime, which follows from conventional elementary particle physics, and the best estimates for the average magnetic field of the Galaxy in the solar neighborhood.

### Acknowledgements

Much of the work described here has been performed in collaboration with David Eichler, Vikram Dwarkadas, and Igor Telezhinsky. Any errors are mine entirely.

## References

Abbasi, R., et al. 2005, Physics Letters B, 619, 271

Abdo, A. A., et al. 2010, Physical Review Letters, 104, 101101

Abdo, A.A. et al. (The Fermi collaboration) 2009, PRL 102, 181101

Abraham, J., et al. 2010, Physical Review Letters, 104, 091101

Abraham, J., et al. 2008, Physical Review Letters, 101, 061101

Adriani, O., Barbarino, G.C., Bazilevskaya, G.A., et al. 2009a, Nature 458, 607

Adriani, O., Barbarino, G.C., Bazilevskaya, G.A., et al. 2009b, PRL 102, 051101

Apel, W. D., et al. 2009, Astroparticle Physics, 31, 86

Arkani-Hamed, N., Finkbeiner, D.P., Slatyer, T.R., Weiner, N. 2009, PRD 79, 015014

Arteaga-Velázquez, J. C., et al. 2010, arXiv:1009.4716

Arvanitaki, A. Dimopoulos, S., Dubovsky, S., Graham, P.; Harnik, R., Rajendran, S. 2009, Phys. Rev. D80.055011

Berger, E. 2010, ApJ , 722, 1946

Büsching, I., Kopp, A., Pohl, M., Schlickeiser, R., Perrot, C., & Grenier, I. 2005, ApJ , 619, 314

Calvez, A., Kusenko, A., & Nagataki, S. 2010, Physical Review Letters, 105, 091101

Chen, C.R., Mandal, S.K., Takahashi, F. 2009, (arXiv:0910.2639v2)

Eichler, D., Guetta,D. & Pohl, M. 2010, ApJ 722, 543

Eichler, D., & Pohl, M. 2011, ApJ submitted

Eichler, D. 1989, PRL 63, 2440

Eichler, D., and Maor, I. 2005, (arXiv:astro-ph/0501096)

Essig, R., Sehgal, N., Strigari, L.E. 2009, PRD 80, 223506

Gehrels, N., Ramirez-Ruiz, E., & Fox, D. B. 2009, ARA&A , 47, 567

Ginzburg, V. L., & Syrovatsky, S. I. 1961, Progress of Theoretical Physics Supplement, 20, 1

Grasso, D., Profumo, S., Strong, A.W., et al. 2009, Astrop. Phys. 32, 140

The High Resolution Fly'S Eye Collaboration, et al. 2009, Astroparticle Physics, 32, 53

Hooper, D., Stebbins, A., Zurek, K.N. 2009, PRD 79, 103513

Ishiwata, K. Matsumoto, S., Moroi, T. 2009, Phys. Lett. B 679-1, 1

Kuhlen, M., & Malyshev, D. 2009, PRD 79, 123517

Levinson, A. & Eichler, D. 1993, ApJ 418, L386

Moskalenko, I.V., Strong, A.W. 1998, ApJ 493, 694

Pierre AUGER Collaboration, et al. 2010, Physics Letters B, 685, 239

Pohl, M., & Eichler, D. 2010, ApJ , 712, L53

Pohl, M. 2009, PRD, 79, 041301

Pohl, M., Perrot, C., Grenier, I., Digel, S. 2003, A& A 409, 581

Pohl, M., & Esposito, J. A. 1998, ApJ , 507, 327

Porter, T.A. & Strong, A.W. 2005, Proc. of the 29th ICRC, Pune, (arXiv:astro-ph/0507119)

Profumo, S., Jeltema, T.E. 2009, JCAP 07, 020

Profumo, S. 2008, arXiv:0812.4457v2

Reynolds, S. P. 2008, Ann. Rev. Astron. Astroph., 46, 89
Tylka, A.J and Eichler, D. (2007) U. Md. Technical Report,
Tylka, A.J. 1989, PRL 63, 40
Wick, S. D., Dermer, C. D., & Atoyan, A. 2004, Astroparticle Physics, 21, 125
Zhang, J. et al. 2009, PRD 80, 023007

# Star Formation at High Resolution: Zooming into the Carina Nebula, the nearest laboratory of massive star feedback

Thomas Preibisch

Universitäts-Sternwarte München
Ludwig-Maximilians-Universität
Scheinerstr. 1, 81679 München, Germany

preibisch@usm.uni-muenchen.de

**Abstract**

*Most stars form in rich clusters and therefore in close proximity to massive stars, that strongly affect their environment by ionizing radiation, winds, and supernova explosions. Due to the relatively large distances of massive star forming regions, detailed studies of these feedback effects require very high sensitivity and angular resolution. The latest generations of large telescopes have now made studies of the full (i.e. high-* and *low-mass) stellar populations of distant ($D > 2$ kpc) star forming regions feasible, and can provide high enough angular resolution to reveal the small-scale structure of their clouds. In this paper, I present first results of a recent deep multi-wavelength study of the Great Nebula in Carina. The Carina Nebula contains some of the most massive and luminous stars in our Galaxy and is an ideal site to study in detail the physics of violent massive star formation and the resulting feedback effects, i.e. cloud dispersal and triggering of star formation. With a distance of 2.3 kpc, it constitutes our best bridge between nearby regions like Orion and the much more massive, but also more distant extragalactic starburst systems like 30 Doradus. Our new X-ray and infrared data reveal, for the first time, the low-mass stellar population in the Carina Nebula, and allow us to study the ages, mass function, and disk properties of the young stars. With sub-mm observations we also probed the morphology of the cold dusty molecular clouds throughout the complex and obtained new insight into the interaction between massive stars and clouds. These observational data will be compared to detailed numerical radiation-hydrodynamic simulations of the effects of stellar feedback on molecular cloud dynamics and turbulence. This will show how ionizing radiation and stellar winds disperse the clouds and trigger the formation of a new generation of stars.*

## 1 Introduction

Stars are the fundamental building blocks of the universe and created all the heavy elements in the cosmos. The process of star formation is thus of central importance in

*Reviews in Modern Astronomy 23: Zoomimg in: The Cosmos at High Resolution.* First Edition.
Edited by Regina von Berlepsch.

astrophysics and determines the evolution of cosmic matter. The formation of planetary systems, another highly important topic of current interest, is directly linked to the star formation process.

Until recently, almost all detailed observations of star formation were restricted to very nearby ($D \lesssim 400$ pc) and thus easily observable regions like Taurus. This has accumulated enormous amounts of very important information, but the view of star formation conveyed by these studies is biased to the particularly quiescent physical conditions in these nearby regions, most of which contain only low- and intermediate mass stars, but no high-mass ($M \gtrsim 20\,M_\odot$) stars. Even in the Orion Nebula Cluster (the most nearby region of recent high-mass star formation) the most massive stellar member, $\theta^1$ C Ori, has a mass of just $\approx 37\,M_\odot$ (Kraus et al. 2009), which is considerably less than the $\approx 100 - 150\,M_\odot$ of the most massive known stars in our Galaxy (Schnurr et al. 2008). Today, it is clear that these nearby *quiescent regions of low-mass star formation are* ***not*** *representative*, because most stars in our Galaxy form in a very different environment: in giant molecular clouds and rich stellar clusters, and therefore *in close proximity to massive stars*. This also applies to the origin of our solar system, for which recent investigations found convincing evidence that it formed in a large cluster, consisting of (at least) several thousand stars, and that the original solar nebula was directly affected by nearby massive stars (e.g., Adams 2010). As a consequence, the role of environment is now an essential topic in studies of star and planet formation.

The presence of high-mass stars can lead to physical conditions that are vastly different from those in regions where only low-mass stars form. The very luminous O-type stars profoundly influence their environment by their strong ionizing radiation, powerful stellar winds, and, finally, by supernova explosions. This feedback can disperse the surrounding natal molecular clouds, and thus terminate the star formation process (= negative feedback). However, ionization fronts and expanding superbubbles can also compress nearby clouds and thereby trigger the formation of new generations of stars (= positive feedback). While this general picture is now well established, the details of the feedback processes, i.e. cloud dispersal on the one hand, and triggering of star formation on the other hand, are still only poorly understood. One major observational difficulty results from the wide range of different temperatures and spatial scales that are involved in these processes. Studies of the interaction of the hot OB star winds ($T \sim 10^6$ K) and ionization fronts ($T \sim 10^4$ K) with the surrounding cold molecular clouds ($T \sim 10$ K) require observations over a wide range of wavelengths. A multi-wavelength approach is fundamentally required, where each wavelength regime contributes to the understanding of different aspects of the physical processes. Another requirement concerns the spatial scales that have to be covered by the observations. In order to get a comprehensive picture, one has to resolve scales from less than one tenth of a parsec (the typical length scale of individual cloud cores) up the several tens or even hundreds of parsec (the full spatial extent of massive star formation complexes). This often requires to obtain large mosaics of images with high angular resolution. The last and most fundamental problem is that nearly all massive star-forming clusters with high levels of feedback are quite far away and therefore difficult to study. At distances of $D > 5$ kpc, the detection and characterization of the *full* stellar populations is very difficult (if not impossible);

often, only the bright high- and intermediate-mass stars can be studied, leaving the low-mass stars (which constitute the vast majority of the stellar population) unexplored or even undetected. Furthermore, the small scale structure of the clouds (on scales below one parsec) in such distant regions cannot be resolved.

With respect to these observational difficulties, the Great Nebula in Carina (NGC 3372; see Smith & Brooks 2008, for an overview and full references) provides a unique target for studies of massive star feedback. The Carina Nebula Complex (CNC, hereafter) is located at a moderate and very well known distance of 2.3 kpc. With 65 known O-type stars, it represents the nearest southern region with a large massive stellar population. Among these are several of the most massive ($M \gtrsim 100\,M_\odot$) and luminous stars known in our Galaxy, e.g. the famous Luminous Blue Variable $\eta$ Car, the O2 If* star HD 93129Aa, several O3 main sequence stars and Wolf-Rayet stars. The CNC is *the most* ***nearby*** *region that samples the* ***top of the stellar mass function***. The presence of stars with $M \gtrsim 100\,M_\odot$ implies that the level of feedback in the CNC is already close to that in more extreme extragalactic starburst regions, while at the same time its comparatively moderate distance guarantees that we still can study details of the cluster and cloud structure at good enough spatial resolution and detect and characterize the low-mass stellar populations. Due to this unique combination of properties, the CNC represents the best galactic analogue of giant extragalactic H II and starburst regions.

Most of the very massive stars in the CNC reside in several loose clusters, including Tr 14, 15, and 16, which have ages ranging from $<1$ to $\sim 8$ Myr. The combined hydrogen ionizing luminosity of the massive stars in the Carina Nebula is about 150 times higher than in the Orion Nebula. In the central region, around $\eta$ Car and the cluster Tr 16, the molecular clouds have already been largely dispersed by the stellar feedback. Southeast of $\eta$ Car, in the so-called "South Pillars" region, the clouds are eroded and shaped by the radiation and winds from $\eta$ Car and Tr 16, giving rise to numerous giant dust pillars, which feature very prominently in the mid-infrared images made with the *Spitzer* Space Observatory (Smith et al. 2010b). On larger scales, the combined action of the ionizing radiation and winds of the numerous OB stars drive an expanding roughly bipolar superbubble with a size of $\sim 50$ pc.

In the past, the Carina Nebula was usually considered to be just an evolved HSchwarzschildII region, devoid of active star formation. However, new sensitive observations have changed this view drastically during recent years. The region contains more than $10^5\,M_\odot$ of gas and dust (see Preibisch et al. 2011a; Smith & Brooks 2008; Yonekura et al. 2005), and deep infrared observations showed clear evidence of ongoing star formation in these clouds. Several very young stellar objects (e.g., Mottram et al. 2007) and a spectacular young embedded cluster (the "Treasure Chest Cluster"; see Smith et al. 2005) have been found in the molecular clouds. A deep *HST* H$\alpha$ imaging survey revealed dozens of jet-driving young stellar objects (Smith et al. 2010a), and *Spitzer* surveys located numerous embedded protostars throughout the Carina complex (Povich et al. 2011; Smith et al. 2010b). The formation of this substantial population of very young ($\leq 1$ Myr), partly embedded stars was probably triggered by the advancing ionization fronts that originate from the (several Myr old) high-mass stars in the CNC.

The CNC thus shows clear evidence for negative feedback (= cloud destruction; primarily in the central part, very close to the massive stars) as well as for positive feedback (= triggered star formation; primarily at the periphery of the complex). This makes it an ideal target in which to conduct a detailed study of the feedback through UV radiation and stellar winds from very massive stars during the formation of an OB association.

While the un-obscured population of high-mass stars ($M \geq 20\,M_\odot$) in the CNC is well known and characterized, the (much fainter) low-mass ($M \leq 2\,M_\odot$) stellar population remained largely unexplored until now. This is mainly related to the difficulties of distinguishing low-mass CNC members from the numerous galactic field stars in the area. However, a good knowledge of the low-mass stellar content is essential for any determination of the global properties of the complex, which are the key towards understanding the star formation process in the CNC. Important aspects, for which the low-mass stellar population plays a crucial role, include the following:

**(1) The initial mass function (IMF) of the complex.**
The IMF is dominated by low-mass stars. According to representations of the standard galactic field IMF (see Kroupa 2002), every O-type star is associated by several hundred low-mass stars. While the bright high-mass stars dominate the total luminosity of the stellar population, most of the total mass is in the low-mass stars, which are therefore very important for the dynamical evolution of the embedded star clusters as they emerge from their natal cloud. One of the most fundamental open questions of star-formation theory is whether the IMF is the same everywhere or whether there are systematic IMF variations in different environments (see Bastian et al. 2010). It was often claimed that some (very) massive star forming regions have a *truncated IMF*, i.e. contain much smaller numbers of low-mass stars than expected from the field IMF (see, e.g., Leitherer 1998). However, most of the more recent and sensitive studies of massive star forming regions (see, e.g., Espinoza et al. 2009; Liu et al. 2009) found high numbers of low-mass stars, as expected from the "normal" field star IMF, suggesting that the previously reported apparent low-mass star deficit was related to observational problems. A direct determination of the low-mass IMF in the CNC would be an extremely valuable contribution to this long-standing question.

**(2) The spatial distribution of the low-mass stars,** which contain most of the stellar mass in the complex, yields important information about the structure, dynamics, and evolution of the region. Revealing the low-mass stellar population is crucial for all studies of possible mass segregation and the question whether most of the stars form in a clustered mode or in a dispersed mode. Detailed studies of the stellar content and the spatial structure of many nearby star forming regions are now available (e.g., Gutermuth et al. 2009), and provide the basis for a comparison to the more massive CNC.

**(3) The relation between the high- and low-mass stars.**
For many years, star formation was supposed to be a bimodal process (e.g. Shu & Lizano 1988) according to which high- and low-mass stars should form in separate processes and in different sites. It also has often been claimed that in many clusters the low-mass stars would be systematically older than the high-mass stars. However, most recent studies did *not* confirm such claims, and showed that high- and low-mass

stars form generally together (see, e.g., Briceno et al. 2007; Preibisch & Zinnecker 1999). If the ages of the low-mass stellar populations in the CNC can be estimated, a comparison to the (independently determined) ages of the high-mass stars will provide important insight into this question.

**(4) Evolution of low-mass stars and their protoplanetary disks in a harsh environment**. The intense UV radiation from massive stars may remove considerable amounts of the circumstellar material from nearby young stellar objects. This may limit the final masses of the low-mass stars (see Whitworth & Zinnecker 2004) and should also affect the formation of planets (see, e.g., Throop & Bally 2005). The CNC provides an excellent target to investigate these effects on the formation and evolution of low-mass stars (and their forming planetary systems) in a harsh environment during ages between $\lesssim 1$ Myr and several Myr.

The obvious first step of any study of the low-mass stellar population is, of course, the identification of the individual low-mass stars in the CNC. Although the statement sounds trivial, this identification is very difficult for several reasons. Firstly, due to its location very close to the galactic plane and near the tangent point of a spiral arm, any optical and infrared observations of the CNC suffer from extremely strong field star contamination and confusion problems. Secondly, due to their intrinsic faintness, an individual spectroscopic identification of the young low-mass stars (e.g. by their Lithium lines or gravity-sensitive lines; see, e.g., Preibisch et al. 2002; Slesnick et al. 2008) is unfeasible. Thirdly, these problems are amplified by the strongly variable and highly position dependent pattern of cloud extinction across the CNC, and the bright and complex nebular emission from interstellar gas ionized by the CNC OB stars.

Until recently, essentially all known young low-mass stars in the CNC (as well as in most other massive star forming regions) were identified by infrared excess emission, which is a tracer of circumstellar material. However, there are (at least) two problems with excess-selected samples: First, it is well known that infrared excess emission in young stars disappears on timescales of just a few Myr (e.g., Briceno et al. 2007); at an age of $\sim 3$ Myr only $\sim 50\%$ of the young stars still show near-infrared excesses, and by $\sim 5$ Myr this is reduced to $\sim 15\%$. Since the expected ages of most young stars in the CNC are a few Myr, *any excess-selected sample will be highly incomplete*. On the other hand, excess-selected samples can be *strongly contaminated* by background sources, since evolved Be stars, carbon stars, planetary nebulae, star-forming galaxies, and even AGN can show near-infrared excesses very similar to those of young stars (e.g., Oliveira et al. 2009; Rebull et al. 2010). This background contamination is particularly strong in Carina due to its position on the galactic plane and the moderate level of cloud extinction. These factors strongly limit the usefulness of a near-infrared excess selected sample of young stars in the case of the CNC.

Sensitive X-ray observations provide a very good solution of this problem, since one can detect the young stars by their strong X-ray emission (e.g., Feigelson et al. 2007) and efficiently discriminate them from the numerous older field stars in the survey area. X-rays are equally sensitive to young stars which have already dispersed their circumstellar disks, thus avoiding the bias introduced when selecting samples based only on infrared excess. Many X-ray studies of star forming regions have demonstrated the success of this method (see, e.g., Broos et al. 2007; Forbrich & Preibisch 2007; Preibisch & Zinnecker 2002; Preibisch, Zinnecker & Herbig 1996; Wang et al. 2010). Also, the relations between the X-ray properties and basic stellar properties in young stellar populations are now very well established from very deep X-ray observations such as the *Chandra* Orion Ultradeep Project (COUP) (see Getman et al. 2005; Preibisch et al. 2005).

# 2 The *Chandra* Carina Complex Project

Although the CNC has been observed with basically all X-ray observatories of the last few decades, only *Chandra* has good enough angular resolution ($< 1''$) to allow a proper identification of the individual X-ray sources in such a heavily crowded region. The *Chandra* Carina Complex Project has recently mapped the CNC with a mosaic of 22 individual ACIS-I pointings, each with an exposure time of $\sim 60$ ksec ($\sim$ 17 hours). The total observing time of all the *Chandra* data used in this project sums up to 1.34 Megaseconds (15.5 days); the total observed area covers more than 1.4 square-degrees. A complete overview of the project can be found in Townsley et al. (2011), which is the introduction to a set of 16 papers in a Special Issue of the *Astrophysical Journal Supplements* devoted to the *Chandra* Carina Complex Project. Figure 1 shows a comparison of a small part of the X-ray image to infrared and optical images of the region.

The on-axis completeness limit is $L_X \approx 10^{29.9}$ erg/s in the $0.5 - 8$ keV band for lightly absorbed sources, but the sensitivity is several times worse near the edges of the individual fields. After extensive source detection efforts, a final merged list of 14 368 individual X-ray sources was compiled (Broos et al. 2011a). As in any X-ray observation, there is a small degree of contamination by foreground stars as well as by background stars and extragalactic sources (see Getman et al. 2011). Since these different classes of contaminants have different typical X-ray, optical, and infrared properties, a statistical approach was used to compute the probability that, based on its individual source properties, a given X-ray source is a member of the CNC or one of three different contaminant classes. This classification showed that 10 714 X-ray sources are very likely young stars in the CNC (Broos et al. 2011b).

The spatial distribution and clustering properties of the X-ray sources was studied by Feigelson et al. (2011). They identified 20 principal clusters of X-ray stars (most of which correspond to known optical clusters in the CNC), 31 small groups of X-ray stars outside these major clusters, and a widely dispersed, but highly populous, distribution of more than 5000 X-ray stars. This implies that about half of the total stellar population in the CNC is located in one of numerous clusters (with sizes ranging from a dozen to several thousand members), while the other half constitutes a widely distributed, non-clustered population.

Extrapolation of the number of X-ray detected Carina members suggests a total population of about 74 000 stars with a total stellar mass of $\sim 40\,000\,M_\odot$ in the CNC. This implies that *the CNC is one of the most massive known star forming*

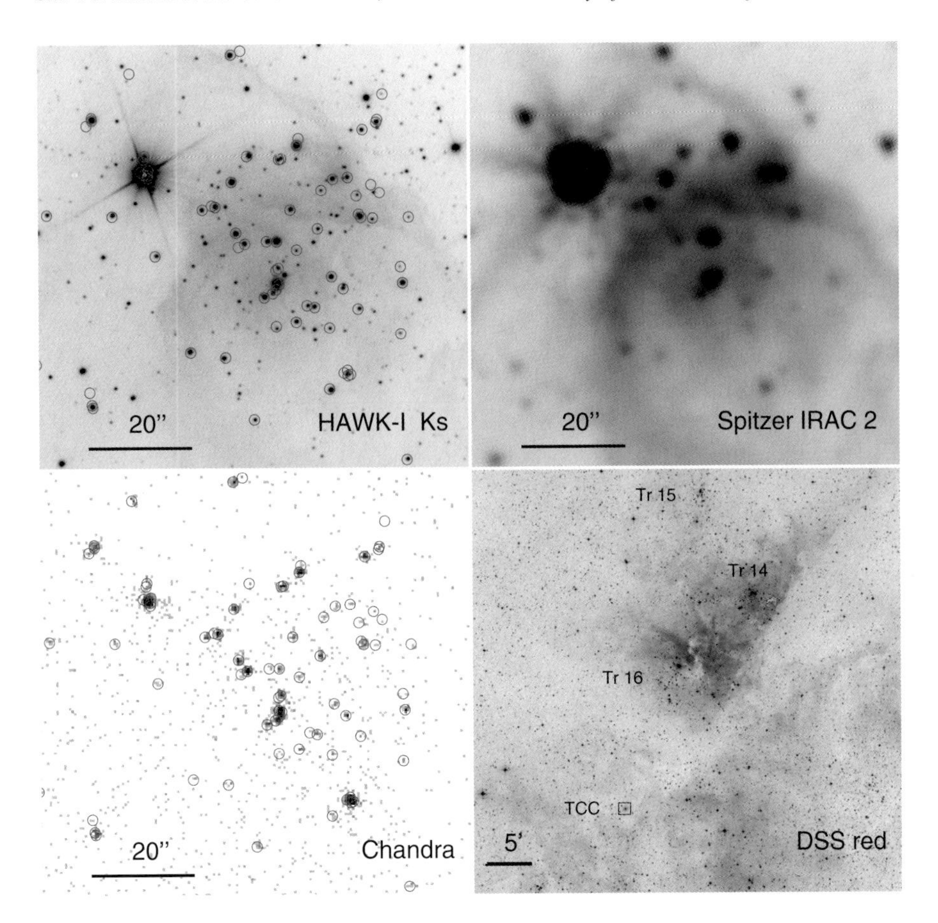


**Figure 1:** As an example for our data we show here a small part of the infrared and X-ray images of the CNC, centered on the Treasure Chest Cluster. The left column shows our HAWK-I $K_s$-band image (top) and the *Chandra* image (bottom); X-ray source positions are marked by circles. The upper right panel shows the *Spitzer* IRAC2 image of the same field. The lower right image shows a wide-field view of the CNC constructed from the red optical DSS image; the location of the Treasure Chest Cluster is marked by a box.

*complexes in our Galaxy*, not far from the mass range of extragalactic Super-Star Clusters such as R 136 in the LMC.

While these *Chandra* data provide, for the first time, an unbiased (although luminosity-limited) sample of the young stars in the region, the X-ray data alone do not yield much information about the properties of the individual stars. For these purposes, *deep optical or near-infrared data are fundamentally required* in order to determine key stellar parameters including mass, age, and circumstellar disk properties. The 2MASS data are clearly not sensitive enough for this purpose: Only 6194 (43.1%) of the 14 368 X-ray sources have counterparts in the 2MASS catalog, and just 4502 of these have valid $J$-, $H$-, and $K_s$-band photometry. The fact that 68.7% of the X-ray sources have no or incomplete NIR photometry from 2MASS clearly illustrates the need for a much deeper near-infrared survey of the CNC.

## 3 HAWK-I near-infrared observations of the Carina Nebula Complex

We have therefore used HAWK-I, the new near-infrared imager at the ESO 8 m Very Large Telescope, to perform a deep wide-field survey of the CNC. Our survey consists of a mosaic of 24 contiguous HAWK-I fields covering a total area of about 1280 square-arcminutes and includes the central part of the Nebula with $\eta$ Car and Tr 16, the clusters Tr 14 and Tr 15, as well as large parts of the South Pillars region. As an example, we show in Fig. 1 HAWK-I, Spitzer, and Chandra images of the Treasure Chest cluster. All HAWK-I data were processed and calibrated by the Cambridge Astronomical Survey Unit. Since objects as faint as $J \sim 23$, $H \sim 22$, and $K_s \sim 21$ are detected with S/N $\geq 3$, our survey represents *the largest and deepest near-infrared survey of the CNC obtained so far*. The HAWK-I images are deep enough to detect *all* stars with masses down to $0.1\,M_\odot$ and an age of 3 Myr through extinctions of $A_V = 15$ mag. The very good seeing conditions during our observations (FWHM $\lesssim 0.5''$) and the $0.106''$ pixel scale of HAWK-I resulted in a superb image quality, revealed numerous interesting cloud structures in unprecedented detail, and lead to the detection of a circumstellar disk around an embedded young stellar object (Preibisch et al. 2011c).

Our final HAWK-I photometric catalog lists 600 336 individual objects, about 20 times more than the number of 2MASS sources in the same area. Most (502 714) catalog objects are simultaneously detected in the $J$-, $H$-, and the $K_s$-band. The area of the HAWK-I mosaic covers 27% of the area of the *Chandra* survey and includes 52% (7472) of all 14 368 *Chandra* sources. Only 2534 (33.9%) of these 7472 *Chandra* sources had matches with 2MASS counterparts with valid $J$-, $H$-, and $K_s$-band photometry. Our new HAWK-I catalog strongly increases the number of known infrared counterparts to the *Chandra* sources in the HAWK-I field to 6636 (88.8% of all X-ray sources).

A complete discussion of the HAWK-I data and the results for the infrared counterparts of the X-ray sources detected in the *Chandra* Carina Survey is given in Preibisch et al. (2011b) and Preibisch et al. (2011d). The number of X-ray detected low-mass stars is at least as high as expected from the number of known high-mass stars and scaling by the field star IMF. *There is thus clearly* ***no deficit*** *of low-mass stars in the CNC.* This result directly confirms the notion of Miller & Scalo (1978) that galactic star formation is dominated by OB associations. We also find that the shape of the K-band luminosity function of the X-ray selected Carina members agrees very well with that derived for the Orion Nebula Cluster; this suggests that (at least down to the X-ray detection limit around $0.5\,M_\odot$) the *shape of the IMF in Carina is consistent with that in Orion (and thus the field IMF).*

Our analysis of the HAWK-I data reveals considerable variations in the near-infrared excess fractions of the different parts of the CNC. While the excess fractions are 7% and 10% for Tr 16 and Tr 14, it is only 2% for Tr 15 (which is several Myr older than Tr 16 and Tr 14). For the X-ray selected members of the very young ($\leq$ 1 Myr) Treasure Chest cluster, a much higher near-infrared excess fraction of 32% is found. There is thus a clear temporal anti-correlation between cluster age and excess fraction, qualitatively similar to what is known from other galactic clusters. However, the absolute values of the near-infrared excess fractions for the clusters in the CNC are clearly lower that those typical for nearby, less massive clusters of similar age (e.g., Briceno et al. 2007). This suggests that *the process of circumstellar disk dispersal proceeds on a faster timescale in the CNC than in the more quiescent regions*, and is most likely the consequence of the very high level of massive star feedback in the CNC.

We also analyzed color-magnitude diagrams to estimate the typical ages of the stars. For the clusters Tr 14, 15, and 16, our age estimates are consistent with the assumption that the low-mass stars in the individual clusters are *coeval* with their high-mass members, i.e. *high- and low-mass stars have formed at the same time.* The widely distributed, non-clustered, population of X-ray emitting stars seems to show a broader distribution of ages up to $\sim$ 8 Myr, with most of the objects being $\lesssim$ 4 Myr old. This agrees with the idea that the formation of most of these distributed young stars was triggered by the advancing ionization fronts created by the massive stars in Tr 16.

# 4 LABOCA sub-mm mapping of the Carina Nebula Complex

A comprehensive investigation clearly also requires information on the cool dust and gas in the (molecular) clouds, and the deeply embedded protostars within these clouds. During the very earliest stages of star formation, these dense gas clumps and cores remain very cold ($10 - 30$ K), and therefore escape detection at near- and mid-infrared wavelengths, even with instruments as sensitive as *Spitzer*. The (sub-) millimeter emission from cool dust allows for an almost un-hindered, unique view onto the processes in the dense clouds. In order to meaningfully complement the extraordinary quality of the recent X-ray, optical, and infrared observations, (sub-)mm observations with high spatial resolution, high sensitivity, and large spatial coverage (at least 1 square-degree) were clearly required.

We have therefore performed sub-mm observations of the CNC with the APEX telescope (Güsten et al. 2006), using the instrument LABOCA (see Siringo et al. 2009) that operates in the atmospheric window at 870 $\mu$m (345 GHz). The angular resolution of LABOCA is $18.6''$ and corresponds to a linear dimension of 0.2 pc at the distance of the Carina Nebula; this is considerably better than all existing wide-field (sub)-mm- and radio-maps of the region. Our LABOCA map (see Fig. 2) provides the first detailed and very deep wide-field survey of the sub-mm emission in the CNC. The results of this observation are described in Preibisch et al. (2011a). We find that the cold dust in the complex is distributed in a wide variety of structures, from the very massive ($\sim$ 15 000 $M_{\odot}$) and dense cloud complex to the west of Tr 14, over several clumps of a few hundred solar masses, to numerous small clumps

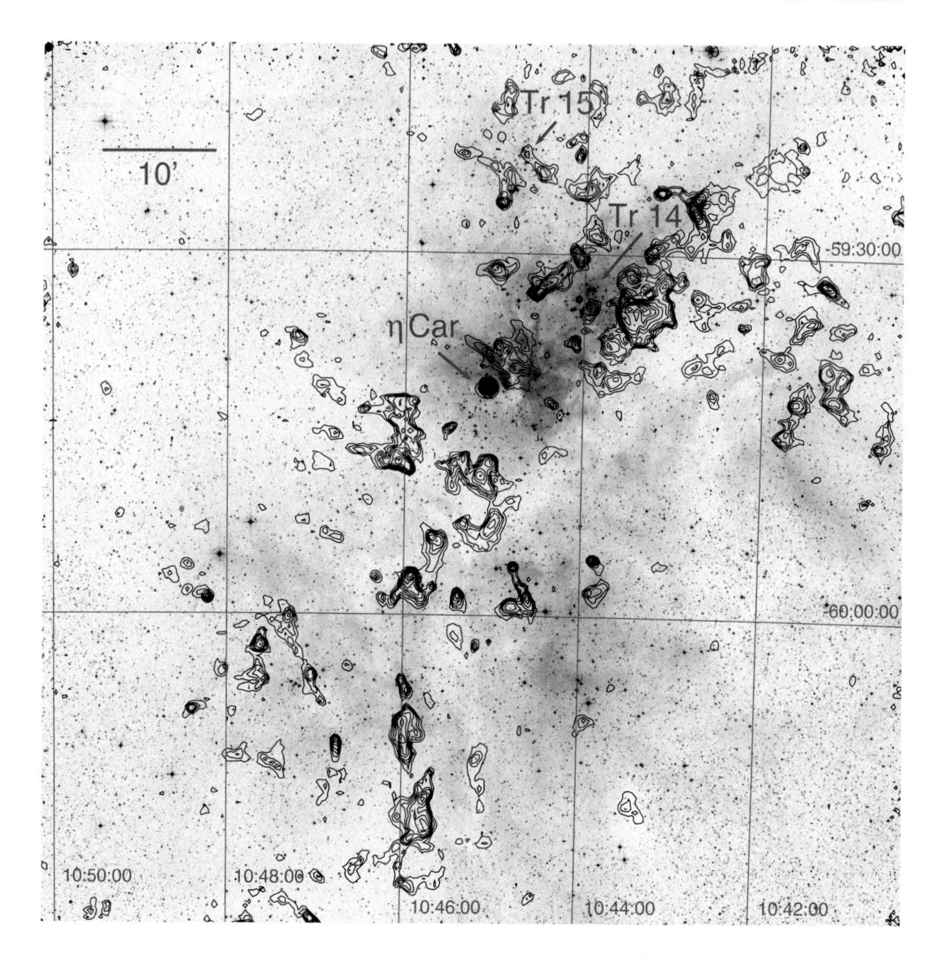


**Figure 2:** Negative grayscale representation of the red optical Digitized Sky Survey image with contours of the LABOCA map overplotted.

containing only a few solar masses of gas and dust. Many of the clouds show clear indications that their structure is shaped by the very strong ionizing radiation and possibly the stellar winds of the massive stars.

The total mass of the dense clouds to which LABOCA is sensitive is $\sim 60\,000\,M_\odot$. This value agrees fairly well with the mass estimates for the well localized molecular gas traced by $^{13}$CO (Yonekura et al. 2005). The CNC also contains a considerable amount of widely distributed atomic gas, which is neither recovered in our LABOCA map, nor can be seen in the CO data. Our radiative transfer modeling of the global, optical to mm-wavelength, spectral energy distribution of the CNC (see Preibisch et al. 2011a) suggests that the total mass of such an distributed atomic gas component does probably not exceed the total molecular gas mass in the region ($140\,000\,M_\odot$) traced by CO (Yonekura et al. 2005). Thus, the overall (atomic + molecular) gas mass in the field of our LABOCA map is probably $\lesssim 300\,000\,M_\odot$.

Our analysis of the LABOCA data shows that only a small fraction ($\sim 10\%$) of the gas in the CNC is currently in dense and massive enough clouds to be available for further star formation. Most clouds have masses of less than a few $1000\,M_\odot$ and will thus most likely *not* form any very massive stars ($M_* > 50\,M_\odot$), as present in large numbers in the older stellar generation in the CNC. This suggests a clear quantitative difference between the currently ongoing process of mostly triggered star-formation and the process that formed the very massive stars in the clusters Tr 14 and 16 a few Myr ago.

If we compare our estimate of the total cloud mass ($\lesssim 300\,000\,M_\odot$) to the total stellar mass of $\sim 28\,000\,M_\odot$ estimated from the X-ray and near-infrared data (Preibisch et al. 2011d), we find that (so far) only $\sim 10\%$ of the total cloud mass in the complex have been transformed into stars. This fraction is similar to the typical values of the global star-formation efficiency determined for other OB associations (Briceno et al. 2007).

# 5 Future Herschel observations of the Carina Nebula Complex

We have been awarded *Herschel* observing time to map the full spatial extent of the CNC (5.5 square-degrees) with the instruments SPIRE and PACS. These *Herschel* maps will yield fluxes at the critical far-IR wavelengths of 75, 170, 250, 350, and $500\,\mu$m and will allow us to reliably determine cloud temperatures, column densities, and, finally, the cloud masses. This will yield a complete inventory of all individual clouds in the complex, down to cloud masses of $1\,M_\odot$, and allow us to detect the youngest and most deeply embedded protostars (down to $0.1\,M_\odot$). The *Herschel* data will also reveal the important small-scale structure of the irradiated clouds. We will be able to map large-scale temperature gradients and changes in the dust properties that are expected as a consequence of the strong feedback by the massive stars, and establish and compare the clump mass functions in different parts of the complex. By comparison with similar *Herschel* data for other star forming regions, we can address the question of how the particularly high levels of massive star feedback influence the evolution of the clouds and the star formation process.

# 6 Conclusions and Outlook

The CNC is clearly a unique laboratory for studies of cloud destruction and triggered star formation by the radiative and wind feedback from very massive stars. Our comprehensive multi-wavelength project has already provided a set of deep, wide-field, and high angular resolution data in the X-ray, near-infrared, and sub-mm regime. This observational database will be further extended by the upcoming *Herschel* observations as well as the inclusion of other new and archival data sets.

A very important aspect of the project is that we will combine the observational data with detailed numerical simulations of how molecular clouds evolve under the influence of strong massive star feedback (see Gritschneder et al. 2010). We will use the known positions, UV fluxes and wind parameters of the O stars to reproduce as closely as possible the shape and physical properties of the gas pillars and the distribution of clumps and protostars in the region with high-resolution numerical simulations. This comparison will allow us to obtain new and detailed insights into fundamental processes such as the disruption of molecular clouds by massive stars, the origin of the observed complex pillar-like structures at the interfaces between molecular clouds and HII regions, the effect of stellar feedback on molecular cloud dynamics and turbulence, and how ionizing radiation and stellar winds trigger the formation of a second generation of stars.

To conclude this article, we provide an outlook to the future of the CNC, to highlight the relevance of the specific conditions for the currently forming stars. Within less than $\sim 1$ Myr, $\eta$ Car will explode as a supernovae. This event will be followed by series of at least some 70 further supernova explosions from the massive stars in the complex. Each of these explosions will send shockwaves through the surrounding clouds. While such shockwaves are very destructive for clouds very close to the supernova, they decay quickly into much slower and weaker shocks after traveling distances of $\gtrsim 1$ pc. Today, most of the molecular clouds in the CNC are already located at the periphery of the complex, typically a few pc away from the massive stars; these clouds will then be compressed, but probably not destroyed by the crossing "evolved" shockwaves. At locations where suitable conditions are met, vigorous star formation activity can then be expected (see, e.g., Preibisch & Zinnecker 2007, for the example of the Scorpius-Centaurus Association). These supernova shockwaves will also inject short-lived radionuclides such as $^{60}$Fe into collapsing protostellar clouds (Boss & Keiser 2010) as well as in forming planetary systems around young stellar objects. The analysis of elemental abundances in meteorites provided direct evidence that such short-lived radionuclides were incorporated into the solar nebula material during the formation of our solar system (Bizzarro et al. 2007). In this context, the strongly irradiated clouds in the CNC, where stars are currently forming and will soon be exposed to supernova ejecta, may actually provide a very good template in which to study the initial conditions for the formation of our solar system.

## Acknowledgements

I would like to thank Hans Zinnecker for many years of advice and numerous stimulating discussion, and Karl Menten for his support with the LABOCA observations. This work is supported by the Deutsche Forschungsgemeinschaft in the project PR 569/9-1. Additional support came from funds from the Munich Cluster of Excellence: "Origin and Structure of the Universe". The analysis of the X-ray data was supported by Chandra X-ray Observatory grant GO8-9131X (PI: L. Townsley) and by the ACIS Instrument Team contract SV4-74018 (PI: G. Garmire), issued by the *Chandra* X-ray Center, which is operated by the Smithsonian Astrophysical Observatory for and on behalf of NASA under contract NAS8-03060. The near-infrared observations for this project were collected with the HAWK-I instrument on the VLT at Paranal Observatory, Chile, under ESO program 60.A-9284(K). The Atacama Pathfinder Experiment (APEX) is a collaboration between the Max-Planck-

Institut für Radioastronomie, the European Southern Observatory, and the Onsala Space Observatory.

## References

Adams, F. C. 2010, ARA&A , 48, 47

Bastian, N., Covey, K. R., & Meyer, M. R. 2010, ARA&A , 48, 339

Bizzarro, M., Ulfbeck, D., Trinquier, A., Thrane, K., Connelly, J. N., & Meyer, B. S. 2007, Science, 316, 1178

Boss, A. P., & Keiser, S. A. 2010, ApJ , 717, L1

Briceno, C., Preibisch, Th., Sherry, W., Mamajek, E., Mathieu, R., Walter, F., Zinnecker, H. 2007, in: Protostars & Planets V, eds. B. Reipurth, D. Jewitt, & K. Keil, University of Arizona Press, Tucson, p. 345

Broos, P. S., Feigelson, E. D., Townsley, L. K., Getman, K. V., Wang, J., Garmire, G. P., Jiang, Z., & Tsuboi, Y. 2007, ApJS , 169, 353

Broos, P. S., Townsley, L., Feigelson, E.D. et al. 2011a, ApJS , 194, 2

Broos, P. S., Getman, K.V., Povich, M.S., et al. 2011b, ApJS , 194, 4

Espinoza, P., Selman, F. J., & Melnick, J. 2009, A&A, 501, 563

Feigelson, E. D., Townsley, L., Güdel, M., Stassun, K. 2007, Protostars & Planets V, eds. B. Reipurth, D. Jewitt, and K. Keil, Univ. Arizona Press, p. 313

Feigelson, E. D., Getman, K.V., Townsley, L.K., et al. 2011, ApJS , 194, 9

Forbrich, J., & Preibisch, T. 2007, A&A, 475, 959

Getman, K. V., Flaccomio, E., Broos, P.S., et al. 2005, ApJS , 160, 319

Getman, K. V., Broos, P.S., Feigelson, E.D., et al. 2011, ApJS , 194, 3

Gritschneder, M., Burkert, A., Naab, T., & Walch, S. 2010, ApJ , 723, 971

Güsten, R., Nyman, L. Å., Schilke, P., Menten, K., Cesarsky, C., & Booth, R. 2006, A&A, 454, L13

Gutermuth, R. A., Megeath, S. T., Myers, P. C., Allen, L. E., Pipher, J. L., & Fazio, G. G. 2009, ApJS , 184, 18

Kraus, S., Weigelt, G., Balega, Y.Y., Docobo, J.A., Hofmann, K.-H., Preibisch, Th., et al. 2009, A&A, 497, 195

Kroupa, P. 2002, Science, Vol. 295, No. 5552, p. 82

Leitherer, C., 1998, in ASP Conf Ser. Vol. 142, The Stellar Initial Mass Function, eds. G. Gilmore, D. Howell, 61

Liu, Q., de Grijs, R., Deng, L. C., Hu, Y., Baraffe, I., & Beaulieu, S. F. 2009, MNRAS , 396, 1665

Miller, G. E. & Scalo, J.M. 1978, PASP 90, 506

Mottram, J. C., Hoare, M.G., Lumsden, S.L., et al. 2007, A&A, 476, 1019

Oliveira, I., Merin, B., Pontoppidan, K.M., et al. 2009, ApJ , 691, 672

Povich, M. S., Smith, N., Majewski, S.R., et al. 2011, ApJS , 194, 14

Preibisch, Th. & Zinnecker, H. 1999, AJ , 121, 1040

Preibisch, Th., & Zinnecker H. 2002, AJ , 123, 1613

Preibisch, Th. & Zinnecker, H. 2007, IAU Symposium 237, 270

Preibisch, Th., Zinnecker, H., Herbig, G.H. 1996, A&A, 310, 456

Preibisch, Th., Brown, A.G.A., Bridges, T., Guenther, E., & Zinnecker, H. 2002, AJ, 124, 404

Preibisch, Th., & Feigelson, E. D. 2005, ApJS , 160, 390

Preibisch, Th., Kim, Y.-C., Favata, F., et al. 2005, ApJS , 160, 401

Preibisch, Th., Schuller, F., Ohlendorf, H., Pekruhl, S., Menten, K.M., Zinnecker, H. 2011a, A&A, 525, A92

Preibisch, Th., Hodgkin, S., Irwin, M., Lewis, J.R., et al. 2011b, ApJS, 194, 10

Preibisch, Th., Ratzka, T., Gehring, T., et al. 2011c, A&A, 530, A40

Preibisch, Th., Ratzka, T., Kuderna, B., et al. 2011d, A&A, 530, A34

Rebull, L. M., Padgett, D.L., McCabe, C.-E., et al. 2010, ApJS , 186, 259

Schnurr, O., Casoli, J., Chené, A.-N., Moffat, A. F. J., & St-Louis, N. 2008, MNRAS , 389, L38

Shu, F.H., Lizano, S., 1988, in Stellar Matter, Moran J.M., Ho P.T.P. (eds.), Gordon & Breach, 65

Siringo, G., Kreysa, E., Kovacs, A., et al. 2009, A&A, 497, 945

Slesnick, C. L., Hillenbrand, L. A., & Carpenter, J. M. 2008, ApJ , 688, 377

Smith, N., Stassun, K.G., & Bally, J. 2005, AJ, 129, 888

Smith, N., & Brooks, K. J. 2008, Handbook of Star Forming Regions, Volume II: The Southern Sky, ASP Monograph Publications, Vol. 5. Edited by Bo Reipurth, p.138

Smith, N., Bally, J., & Walborn, N.R. 2010a, MNRAS , 405, 1153

Smith, N., Powich, M.S., Whitney, B.A., et al. 2010b, MNRAS , 789

Throop, H.B., & Bally, J. 2005, ApJ, 623, L149

Townsley, L. K., Broos, P., Corcoran, M.F. et al. 2011, ApJS , 194, 1

Wang, J., Feigelson, E. D., Townsley, L. K., Broos, P. S., Román-Zúñiga, C. G., Lada, E., & Garmire, G. 2010, ApJ , 716, 474

Whitworth, A.P., & Zinnecker, H. 2004, A&A, 427, 299

Yonekura, Y., Asayama, S., Kimura, K., et al. 2005, ApJ 634, 476

# Characteristic structures in circumstellar disks – Potential indicators of embedded planets

Sebastian Wolf

Universität zu Kiel
Institut für Theoretische Physik und Astrophysik
Leibnizstraße 15, 24098 Kiel, Germany
wolf@astrophysik.uni-kiel.de

**Abstract**

*During the next decade large ground-based interferometry arrays (such as the Atacama Large Millimeter Array) and space observatories (such as the James Webb Space Telescope) will become available which are expected to provide the spatial resolution, sensitivity, and dynamic range needed to constrain the existence and major parameters of giant protoplanets as well as the physical conditions and processes during their formation in circumstellar disks. In this article, a brief summary of the approaches to trace planets through their imprint on the structure of circumstellar disks is given.*

## 1 Introduction

Planetary systems are expected to form in gas-rich circumstellar disks around protostars which result from the conservation of angular momentum during mass infall from the ambient molecular cloud. The very successful planet searches performed during the last two decades already allow one to conclude that a large variety of planets and of structures of planetary systems exists (e.g., Udry et al. 2007). In order to understand this observed diversity, one has to investigate the formation and early dynamical evolution of planetary systems. High-angular resolution, multi-wavelength images of circumstellar disks have provided valuable insides into the large-scale structure (e.g., Watson et al. 2007; Sauter et al. 2010; Wolf et al. 2003, 2008), dynamics (e.g., Dutrey et al. 2007), dust properties (e.g., Schegerer et al. 2006, Natta et al. 2007, Ricci et al. 2011), and chemical composition (e.g., Bergin et al. 2007) of circumstellar disks. The investigation of small-scale structures ($\approx$ 1-100 mas corresponding to $\approx$ 0.1 AU-10 AU in nearby star-forming regions, such as in the Taurus-Auriga molecular cloud) by the means of long-baseline interferometry in the infrared wavelength range even allows one to study the potential planet-forming region in circumstellar disks (e.g., Schegerer et al. 2008, 2009). However, the latter technique is still limited to the brightest targets and does not allow one yet to reconstruct complex

*Reviews in Modern Astronomy 23: Zoomimg in: The Cosmos at High Resolution.* First Edition.
Edited by Regina von Berlepsch.

structures due to a sparse coverage of the uv plane, achieved with existing interferometers.

During the next decade, new observatories and instruments will become available which will overcome the above limitations, allowing one to potentially trace protoplanets at least indirectly through their interaction with the circumstellar disk. These observations will bridge the still existing gap between the observation of grain growth as the proposed earliest stage of planet formation (e.g., Dominik et al. 2007) and "mature" planetary systems. This article provides a summary of a review[1] about observational signatures of the planet-disk interaction both in young gas-rich disks, as well as in more evolved, gas-poor debris disks. For a more detailed review about this subject see Wolf et al. (2007).

# 2 (Proto-)Planets in young, gas-rich disks

Finding and imaging planets which are still embedded in a young, gas-rich circumstellar disk is by far more difficult than the already challenging task to image planets around main-sequence stars (e.g., Kalas et al. 2008, Lagrange et al. 2009, Marois et al. 2008/2010). This is because the dust continuum radiation dominates the entire spectral range (ultraviolet to near-infrared: scattering of the stellar radiation, mid-infrared to millimeter: thermal reemission). Thus, in the preparation of observations of planets in disks even more constraints have to be taken into account, describing the properties of the disk, such as the optical parameters of the dust (chemical composition, size distribution, grain structure) and its spatial temperature and density distribution. In the case of young (primordial), optically thick circumstellar disks, the disk inclination and possible accretion on the planet have to be considered as well. So far, the only debated, but only indirectly detected planet in a young, gas-rich disk is the potential Hot Jupiter-type planet in the circumstellar disk of TW Hydrae (Setiawan et al. 2008, Huelamo et al. 2008).

Despite these obvious problems to investigate planets during the final stages of their formation and early evolution through direct imaging, the process of planet-disk interaction offers another approach. During recent years, numerical simulations and analytical studies of planet-disk interactions have shown that planets may cause characteristic large-scale signatures in the density distribution of circumstellar disks. In young circumstellar disks with a structure dominated by gas dynamics, the most important of these signatures are gaps and spiral density waves (e.g., Bryden et al. 1999, Bate et al. 2003, Nelson & Papaloizou 2003, Winters et al. 2003, Papaloizou et al. 2007, Marzari & Nelson 2009). These features may act as tracers for embedded young planets and thus provide constraints on the processes and timescales of planet formation (e.g., Wolf & D'Angelo 2005, Jang-Condell 2008/2009).

Recently obtained high spatial resolution images in scattered light and thermal reemission indicate the presence of overdensities, local gaps, and large-scale spiral waves in these disks (e.g., AB Aurigae: Fukagawa et al. 2004, Lin et al. 2006, Oppenheimer et al. 2008; Andrews et al. 2009, Bonavita et al. 2010). So far, these

[1]The review was presented at the Annual Meeting of the Astronomical Society in Bonn (Germany, 2010).

features have been found in outer disk regions, almost an order of magnitude larger than the potential planet forming region. However, with the advent of new observatories and instruments, such as the Atacama Large Millimeter Array (ALMA) and the $2^{\mathrm{nd}}$ generation instrumentation at the Very Large Telescope Interferometer (VLTI; e.g., Lopez et al. 2008, Malbet et al. 2008, Eisenhauer et al. 2008), the direct observational study of the planet-disk interaction will become feasible within the next 5–10 years.

In addition to pure intensity measurements, the polarization state of the stellar radiation scattered on the disk upper layers provides independent constraints on the disk structure. For example, differential polarimetry has been recently used to constrain the structure of the disk around TW Hydrae as close as $0.1''$ to the central star (Apai et al. 2004) and to evaluate the nature of apparent structures due to planet-disk interaction seen in intensity maps of AB Aurigae (Perrin et al. 2009).

# 3 Planets in debris disks

Infrared to millimeter surveys indicate that the primordial disks around young stellar objects disperse on timescales of about 5-10 Myr (e.g., Haisch et al. 2001, Jayawardhana et al. 2006, Hillenbrand 2008). Collisions of remaining asteroid - like bodies as well as the activity of comets then produce a second-generation dust disk which explains the infrared excess observed around many main-sequence stars (see, e.g., Krivov 2010 for a recent review of observations, essential physics, and theoretical models for debris disks).

## 3.1 Signatures of planets in spatially unresolved debris disks

Wolf & Hillenbrand (2003) and Moro-Martín et al. (2005) investigated the influence of various disk parameters and planet-disk configurations on the net spectral energy distribution (SED), taking scattering, absorption, and reemission by the circumstellar dust into account. That study was aimed at discussing the feasibility to distinguish between different disks and planet-disk configurations based on their infrared to submillimeter spectra. Most importantly, the mid-infrared continuum allows one to conclude if an inner hole / depletion of small dust grains exists. These cavities may be created by gravitational scattering with an inner planet: Dust grains drifting inwards due to the Poynting-Robertson effect are likely to be scattered into larger orbits resulting in a lower dust number density within the planet's orbit. Thus, beside high-resolution imaging of debris disks, multi-wavelength photometry and (low-resolution) spectroscopy aimed at deriving the mid-infrared SED of debris disks are valuable tools to deduce the existence of an inner gap. This is because the deficiency of hot dust in the stellar vicinity causes a decrease of the mid-infrared flux compared to an undisturbed disk. The analysis of the mid-infrared SED of the debris disks recently discovered with the Spitzer Space Telescope shows that the occurrence of inner regions with strong dust depletion is a frequent phenomenon in these systems (e.g., Meyer et al. 2004, Kim et al. 2005, Hines et al. 2006, Silverstone et al. 2006, Su et al. 2009).

However, there exist degeneracies that can complicate the interpretation of the SED in terms of determining the location of embedded planets. For example, the SED of a dust disk dominated by weakly absorbing grains (e.g., Fe-poor silicates) has its minimum at wavelengths longer than those of a disk dominated by strongly absorbing grains (e.g., carbonaceous and Fe-rich silicate). Because the SED minimum also shifts to longer wavelengths when the gap radius increases, there might be a degeneracy between the chemical composition of the dust and the semimajor axis of the planet clearing the gap. The degeneracy in the SED analysis illustrates the importance of obtaining high-resolution images, allowing to spatially resolve the debris disk structure, and/or spectroscopic observations constraining the chemical composition of the dust.

## 3.2 Spatial structure of debris disks with embedded planets

High-angular resolution images of debris disks in scattered light in the optical and near-infrared wavelength range and in thermal reemission at mid-infrared to millimeter wavelengths show complex structures, such as rings, gaps, arcs, warps, offset asymmetries and clumps of dust (e.g., Greaves et al. 1998, Schneider et al. 2006, Churcher et al. 2010; for a recent summary see Moro-Martín 2008). In evolved, optically thin debris disks some of these features are likely to be the result of gravitational perturbations by one or more massive planets on the dust disk.

When exploiting the predicted large-scale resonances in order to trace planets in debris disks, it is important to note that the appearance of the described structures depends on the wavelength regime of the observations. In contrast to far-infrared / millimeter observations, tracing the dust density and temperature structure, the relative brightness distribution of individual clumps in optical to near-infrared scattered light images may sensitively depend on the disk inclination. This effect is due to the asymmetry of the scattering function of the dust grains, which strongly depends on the individual grain size parameter.

# References


Andrews, S. M., Wilner, D. J., Hughes, A. M., Qi, C., Dullemond, C. P.: 2009, ApJ 700, 1502

Apai, D., Pascucci, I., Brandner, W., Henning, T., Lenzen, R., Potter, D. E., Lagrange, A.-M., Rousset, G.: 2004, A&A 415, 671

Bate, M. R., Lubow, S. H., Ogilvie, G. I., Miller, K. A.: 2003, MNRAS 341, 213

Bergin, E. A., Aikawa, Y., Blake, G. A., van Dishoeck, E. F.: 2007, Protostars and Planets V, B. Reipurth, D. Jewitt, and K. Keil (eds.), 751

Bonavita, M., Chauvin, G., Boccaletti, A., et al.: 2010, A&A 522, A2

Bryden, G., Chen, X., Lin, D. N. C., Nelson, R. P., Papaloizou, J. C. B.: 1999, ApJ 514, 344

Churcher, L., Wyatt, M., Smith, R.: 2011, MNRAS 410, 2

Dominik, C., Blum, J., Cuzzi, J. N., Wurm, G.: 2007, Protostars and Planets V, B. Reipurth, D. Jewitt, and K. Keil (eds.), 783

Dutrey, A., Guilloteau, S., Ho, P.: 2007, Protostars and Planets V, B. Reipurth, D. Jewitt, and K. Keil (eds.), 495

Eisenhauer, F., Perrin, G., Straubmeier, C., et al.: 2008, Proceedings of the International Astronomical Union, IAU Symposium 248, 100

Fukagawa, M., Hayashi, M., Tamura, M., et al.: 2004, ApJ 605, L53

Greaves, J. S., Holland, W. S., Moriarty-Schieven, G., et al.: 1998, ApJ 506, L133

Haisch, Jr., K. E., Lada, E. A., Lada, C. J.: 2001, ApJ 553, L153

Hines, D. C., Backman, D. E., Bouwman, J., et al.: 2006, ApJ 638, 1070

Hillenbrand, L.A.: 2008, Physica Scripta 130, 014024

Huélamo, N., Figueira, P., Bonfils, X., et al.: 2008, A&A 489, L9

Jang-Condell, H.: 2008, ApJ 679, 797

Jang-Condell, H.: 2009, ApJ 700, 820

Jayawardhana, R., Coffey, J., Scholz, A., Brandeker, A., van Kerkwijk, M. H.: 2006, ApJ 648, 1206

Kalas, P., Graham, J. R., Chiang, E., et al.: 2008, Science 322, 1345

Kim, J. S., Hines, D. C., Backman, D. E., et al.: 2005, ApJ 632, 659

Krivov, A. V.: 2010, Research in Astronomy and Astrophysics 10, 383

Lagrange, A.-M., Gratadour, D., Chauvin, G., et al.: 2009, A&A 493, L21

Lin, S.-Y., Ohashi, N., Lim, J., Ho, P. T. P., Fukagawa, M., Tamura, M.: 2006, ApJ 645, L1297

Lopez, B., Antonelli, P., Wolf, S., et al.: 2008, Optical and Infrared Interferometry. Schöller, M.; Danchi, W. C.; Delplancke, F. (eds.), Proceedings of the SPIE, Volume 7013, 70132

Malbet, F., Buscher, D., Weigelt, G., et al.: 2008, Optical and Infrared Interferometry. Schöller, M.; Danchi, W. C.; Delplancke, F. (eds.), Proceedings of the SPIE, Volume 7013, 701329

Marois, C., Macintosh, B., Barman, T., et al.: 2008, Science 322, 1348

Marois, C., Zuckerman, B., Konopacky, Q. M., Macintosh, B., Barman, T.: 2010, Nat. 468, 1080

Marzari, F., Nelson, A. F.: 2009, ApJ 705, 1575

Meyer, M. R., Hillenbrand, L. A., Backman, D. E., et al.: 2004, ApJ Suppl. 154, 422

Moro-Martín, A.: 2008, IAU Proc. 249, 347

Moro-Martín, A., Wolf, S., Malhotra, R.: 2005, ApJ 621, 1079

Natta, A., Testi, L., Calvet, N., Henning, Th., Waters, R., Wilner, D.: 2007, Protostars and Planets V, B. Reipurth, D. Jewitt, and K. Keil (eds.), 767

Nelson, R. P., Papaloizou, J. C. B.: 2003, MNRAS 339, 993

Oppenheimer, B. R., Brenner, D., Hinkley, S., et al.: 2008, ApJ 679, 1574

Papaloizou, J. C. B., Nelson, R. P., Kley, W., Masset, F. S., Artymowicz, P.: 2007, Protostars and Planets V, 655

Perrin, M. D., Schneider, G., Duchene, G., Pinte, C., Grady, C. A., Wisniewski, J. P., Hines, D. C.: 2009, ApJ 707, L132

Ricci, L., Mann, R. K., Testi, L., Williams, J. P., Isella, A., Robberto, M., Natta, A., Brooks, K. J.: 2011, A&A 525, A81

Sauter, J., Wolf, S., Launhardt, R., et al.: 2009, A&A 505, 1167

Schegerer, A. A., Wolf, S., Hummel, C. A., Quanz, S. P., Richichi, A.: 2009, A&A 502, 367

Schegerer, A. A., Wolf, S., Ratzka, T., Leinert, C.: 2008, A&A 478, 779

Schegerer, A., Wolf, S., Voshchinnikov, N. V., Przygodda, F., Kessler-Silacci, J. E.: 2006, A&A 456, 535

Schneider, G., Silverstone, M. D., Hines, D. C., et al.: 2006, ApJ 650, 414

Setiawan, J., Henning, T., Launhardt, R., Müller, A., Weise, P., Kürster, M.: 2008, Nature 451, 38

Silverstone, M. D., Meyer, M. R., Mamajek, E. E., et al.: 2006, ApJ 639, 1138

Su, K. Y. L., Rieke, G. H., Stapelfeldt, et al.: 2009, ApJ 705, 314

Udry, S., Santos, N. C.: 2007, ARA&A 45, 397

Winters, W. F., Balbus, S. A., Hawley, J. F.: 2003, ApJ 589, 543

Watson, A. M., Stapelfeldt, K. R., Wood, K., Menard, F.: 2007, Protostars and Planets V, B. Reipurth, D. Jewitt, and K. Keil (eds.), 523

Wolf, S., D'Angelo, G.: 2005, ApJ 619, 1114

Wolf, S., Hillenbrand, L. A.: 2003, ApJ 596, 603

Wolf, S., Moro-Martín, A., D'Angelo, G.: 2007, Plan. & Space Science 55, 569

Wolf, S., Padgett, D. L., Stapelfeldt, K. R.: 2003, ApJ 588, 373

Wolf, S., Schegerer, A., Beuther, H., Padgett, D. L., Stapelfeldt, K. R.: 2008, ApJ 674, L101

# Index of Contributors

# General Table of Contents

## Volume 1 (1988): Cosmic Chemistry



## Volume 2 (1989)

## Volume 3 (1990): Accretion and Winds

## Volume 4 (1991)

## Volume 5 (1992): Variabilities in Stars and Galaxies



## Volume 6 (1993): Stellar Evolution and Interstellar Matter

## Volume 7 (1994)



## Volume 8 (1995): Cosmic Magnetic Fields

## Volume 9 (1996): Positions, Motions, and Cosmic Evolution

## Volume 10 (1997): Gravitation

Volume 11 (1998): Stars and Galaxies



Volume 12 (1999):
Astronomical Instruments and Methods at the Turn of the 21st Century

## Volume 13 (2000): New Astrophysical Horizons

## Volume 14 (2001): Dynamic Stability and Instabilities in the Universe



## Volume 15 (2002): JENAM 2001 – Five Days of Creation: Astronomy with Large Telescopes from Ground and Space

## Volume 16 (2003): The Cosmic Circuit of Matter



## Volume 17 (2004): The Sun and Planetary Systems – Paradigms for the Universe

## Volume 18 (2005): From Cosmological Structures to the Milky Way



## Volume 19 (2006): The Many Facets of the Universe – Revelations by New Instruments.

Volume 20 (2008): Cosmic Matter.

## Volume 21 (2009): Formation and Evolution of Cosmic Structures.



## Volume 22 (2010): Deciphering the Universe through Spectroscopy.

## Volume 23 (2011): Zooming in: The Cosmos at High Resolution

# General Index of Contributors